Mechanical Engineering License Exam File

Fourth Edition

Richard K. Pefley
Santa Clara University

ENGINEERING PRESS, INC.   SAN JOSE, CALIFORNIA 95103-0001

The first three editions were titled
MECHANICAL ENGINEERING LICENSE REVIEW

Donald G. Newnan, Ph.D.
EXAM FILE Series Editor

Printed in the United States of America

2 3 4 5

Library of Congress Cataloging-In-Publication Data

Main entry under title:

Mechanical engineering license exam file.

Rev. ed. of: Mechanical engineering license review / edited by Richard K. Pefley, Donald G. Newnan. 3rd ed. c1980.

1. Mechanical engineering—Problems, exercises, etc.
2. Mechanical engineering—Examinations, questions, etc.
I. Pefley, Richard K. II. Mechanical engineering license review.
TJ159.M43 1986 621'.076 85-16230
ISBN 0-910554-52-8

Engineering Press, Inc. P.O. Box 1 San Jose, California 95103-0001

Contents

Preface

Becoming A Professional Engineer

1. Education

The obvious appropriate education is a B.S. degree in mechanical engineering from an accredited college or university. But probably this seldom is an absolute requirement. Alternative (but less acceptable) education is a B.S. degree in something other than mechanical engineering, or from a non-accredited institution, or four years of education but no degree.

2. Engineering Fundamentals (Engineer-In-Training) Exam

Most people are required to take and pass this 8-hour multiple-choice examination. Different states* call it by different names (Fundamentals of Engineering, E.I.T., or Intern Engineer exam) but the exam is the same in all states at the present time. It is prepared and graded by the National Council of Engineering Examiners (NCEE). Review materials for this exam are found in other books like Newnan: *Engineer-In-Training Exam File,* and are not included in this book.

3. Experience

Typically one must have four years of acceptable experience before being permitted to take the Professional Engineer exam. Some states may require less. It may, of course, take more than four years to acquire four years of acceptable experience.

4. Professional Engineering Exam

This second national exam is called *Principles and Practice of Engineering* by NCEE, but probably everyone else calls it the *Professional Engineer* or *P.E. exam.* The NCEE reports that all states (plus Guam, District of Columbia, and Puerto Rico) use the exam they prepare. There is a special *Principles and Practice of Engineering* exam for each of the major branches of engineering.

*The names and addresses of the various State Boards of Registration are in the Appendix.

Mechanical Engineering Professional Exam

The exam consists of two 4-hour sessions. You are required to work four of the ten problems given in each session. The combined morning and afternoon portions of the exam presently have the following distribution of problems:

Mechanical Design	8
Management	1
Energy Systems	6
Control Systems	1
Thermal and Fluid Processes	3
Engineering Economics	1

The NCEE establishes the dates on which its examinations are given throughout the country. The mechanical engineering exam is administered semiannually in April and October of each year. The announced dates are:

1987	April 9-11	October 29-31
1988	April 14-16	October 27-29
1989	April 13-15	October 26-28

The three day period allows the various states and other jurisdications some flexibility in selecting their exact exam dates. Although there are both Spring and Fall dates each year, it does not necessarily follow that a particular state gives the exam twice a year.

In a given year about 3400 mechanical engineers take the exam. In the last reported statistics, 70% of all the people who took the exam in California passed. (In the same period 65% of the people who took the E.I.T. exam passed.)

This Book

The task of preparing for the day-long mechanical engineer examinination is a formidable one. The logical approach is to undertake a systematic review of the categories within the mechanical engineering exam, and to do it at about the level of understanding required to successfully pass the exam. This book is designed to help with that task. The problems have been selected to represent an appropriate topic and level of difficulty. Many are actual exam problems, but are not from the NCEE exams, as they do not release their problems for publication.

The names of the contributors to this book appear on the first page of each chapter. Thanks to people who have studied the prior editions of this book, a number of errors have been identified and removed. If you note any errors in this edition, a letter, written to the Engineering Press address, would be greatly appreciated.

Richard K. Pefley

1

Mathematical Fundamentals

PETER A. SZEGO

A survey of the entirety of applied mathematics is beyond the scope of this review. Instead we focus attention on a number of common stumbling blocks. Although all of these are of an elementary character, experience has shown that they are at the root of a high proportion of errors in engineering mathematics.

FORMULA SUBSTITUTION

As simple a matter as substituting in formulas sometimes causes difficulties. Consider for example the quadratic equation, along with its solution, as expressed by the formulas:

$$ax^2 + bx + c = 0, \qquad x = \frac{-b \pm \sqrt{b^2 - 4ac}}{2a}$$

A numerical example, say $2x^2 - 3x + 5 = 0$, is handled in a straightforward manner. We either make the substitutions mentally or (to be quite certain) we make up a little equivalence table, thus:

$$a = 2,\ b = -3,\ c = 5 \quad \therefore x = \frac{-(-3) \pm \sqrt{(-3)^2 - 4(2)(5)}}{2(2)}$$

$$= \frac{3 \pm \sqrt{9-40}}{4} = \frac{3 \pm i\sqrt{31}}{4}$$

Sometimes, however, we must deal with a quadratic equation written as a formula (i.e., with symbol coefficients) and, moreover, some of the symbols may coincide with those in the "standard" form. For example, suppose we wish to solve for z in $2z^2 + az + c = 0$. To avoid confusion we need to distinguish the "a" and "c" of our problem from the like symbols in our reference formula. One way to do this is to use some special mark on the symbols of the reference formula, thus:

$$\bar{a}x^{-2} + \bar{b}\bar{x} + \bar{c} = 0, \qquad \bar{x} = \frac{-\bar{b} \pm \sqrt{\bar{b}^2 - 4\bar{a}\bar{c}}}{2\bar{a}}$$

Once more, we set up an equivalence table:

$$\bar{a} = 2,\ \bar{b} = a,\ \bar{c} = c,\ \bar{x} = z$$

so that

$$z = \frac{-a \pm \sqrt{a^2 - 4(2)c}}{2(2)}$$

The discussion of the quadratic equation is only by way of example. The problem of symbol substitution arises in all areas of mathematics. Texts and reference works typically use a unique set of symbols for certain categories of problems. For example, in calculus one usually sees "x" used as the independent variable and "y" used as the dependent variable. If in a particular problem some other symbols appear, we must carry out the appropriate translations in the manner illustrated above.

QUADRATIC EQUATION

In the previous section, we gave the "standard" formula for the solutions of the quadratic equation. If the discriminant ($b^2 - 4ac$) is positive, the roots are real numbers; if ($b^2 - 4ac$) is negative, the roots are complex; and if the discriminant is zero, the roots are identical and real (double root). In applications, our understanding of the physical situation usually will indicate whether the roots must be real or complex, and this should agree with the predictions of the formula. In other words, we have here an example of the use of physical understanding as a check and verification on our mathematics.

A common source of confusion is the distinction between the pairings real/complex and positive/negative. As stated above, the question of whether the roots are real or complex depends on the sign of the discriminant. The question of the sign of the <u>roots</u>, themselves, can arise only if the roots are real. The sign of a complex root is not a meaningful concept.

Again, physical understanding can serve to help interpret mathematical results. Suppose, for example, that the unknown in a quadratic equation represents a length, and that both roots are real, with one positive and one negative. Since a negative length generally is not possible, we reject the negative root as being extraneous and take the positive root as the desired result. But what if both roots had been negative (or worse yet, complex)? Then we have no acceptable solution and something must be wrong with the derivation!

LOGARITHMS

The log symbol

$$\log_b N = x \quad \text{means} \quad b^x = N$$

In this definition, the symbols b, x, N are all real and b and N must be positive. In particular, note that the log of a negative number is undefined (actually, in advanced applications this case can be defined as a complex number). Thus we have another check point. If our calculations lead to, say $\log_{10}$ (-5.3), an error probably has occurred.

When the logarithm appears in calculus, the base b usually will be the number e; sometimes such a log is written as ln. Conversion formulas from one base to another are available in references. We have to make sure that conversions are carried out in the correct direction. Suppose, for example, we have

$$\log_{10} 2 = 0.301$$

and we wish to convert to the base e (i.e., to find $\log_e 2$). A fool proof method is to rewrite the $\log_{10}$ expression according to the definition of the log,

$$10^{0.301} = 2$$

and then to take the $\log_e$ of both sides. Thus,

$$\log_e 10^{0.301} = \log_e 2, \quad \text{or} \quad \log_e 2 = 0.301 \; (\log_e 10)$$

where the quantity in the parenthesis is a standard conversion number. Fortunately, most hand calculators handle $\log_{10}$ and $\log_e$ computations directly so this tedious approach can be avoided.

TRIGONOMETRIC FUNCTIONS

One source of confusion is how to handle angles not falling in the first quadrant. Say we seek sin 120°. Sketch a unit circle in an

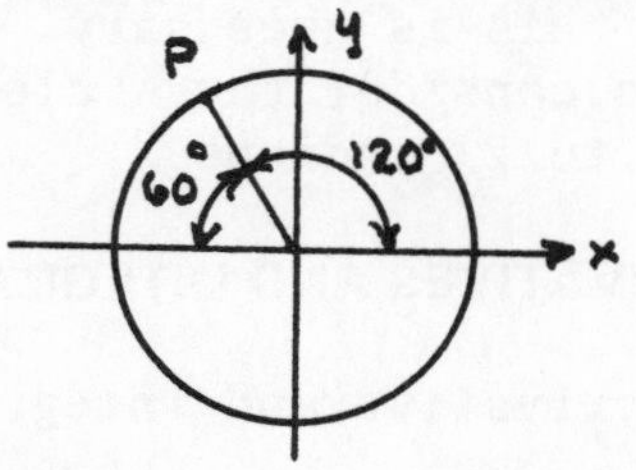

xy-plane with the angle 120° as shown. The sine function is defined as the y coordinate of P. From the sketch we see that this is the same as sin 60° or $\sqrt{3}/2$. Similarly, cos 120° = x(P) = - cos 60° = $-\frac{1}{2}$. Note the appearance of the minus sign because of the negative value of x(P).

Many problems require only a knowledge of the trig functions for the three angles 30°, 45°, 60°. These cases can be remembered easily by constructing two right triangles as shown:

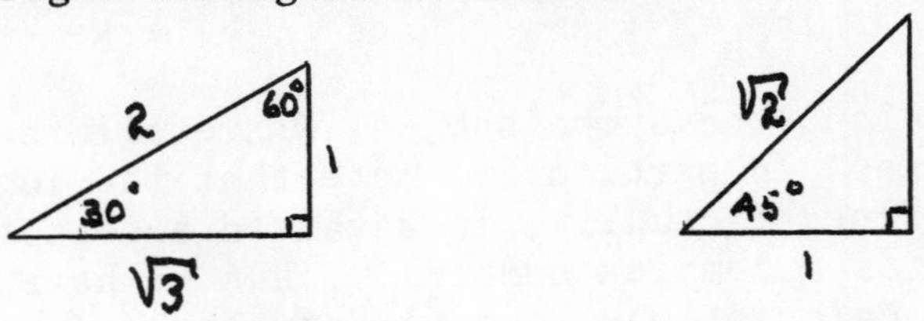

For example, sin 60° = $\sqrt{3}/2$ can be read off the sketch at once. A common error, however, is to mix up the role of the 30° and 60°. In dealing with a particular trigonometric function, such as the sine, it may help to say to oneself "sine is the opposite over the hypotenuse" as one points to the appropriate sides of the triangle.

A frequently occurring situation is that the value of a trigonometric function is given and one seeks the value of another trig function. Suppose, for example, one knows that sin x = 0.2 and that one seeks the value of cos x. Several methods can be used. One possibility is to obtain the value of x from a slide rule or table and then to look up cos x. A second, and perhaps preferable approach, is to sketch a right triangle illustrating the given data, to calculate the third side by the Pythagorean formula, and then to read off the

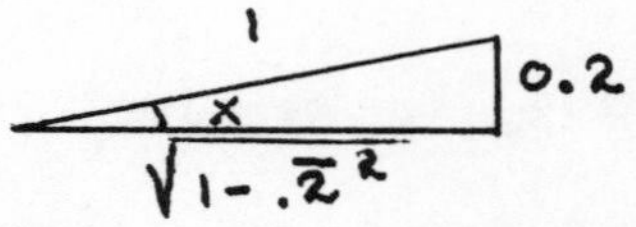

desired result. This second approach remains valid if our starting point is a formula (e.g., sin x = q, then cos x = $\sqrt{1 - q^2}$). This method amounts to the same thing as using appropriate trigonometric identities ($\sin^2 x + \cos^2 x = 1$, in this case). But the triangle method may be easier to remember than the identities.

In calculus applications, angles generally will be expressed in radian measure. Care must be taken to convert properly between radians and degrees when this is necessary. A simple way to remember the proper relation is to consider the whole circle, for which the angle 360° is equivalent to 2π radians.

DERIVATIVES AND INTEGRALS

The most commonly used derivative and integral formulas are set out below.

$$\frac{dx^n}{dx} = nx^{n-1}, \quad \frac{d}{dx}\{f[z(x)]\} = \frac{df}{dz} \cdot \frac{dz}{dx} \quad \text{(chain rule)}$$

$$\frac{d(u \cdot v)}{dx} = u\frac{dv}{dx} + v\frac{du}{dx}, \quad \frac{d(u/v)}{dx} = \frac{v u' - u v'}{v^2},$$

$$\frac{d(\log_e u)}{dx} = \frac{1}{u}\frac{du}{dx}, \quad \frac{d(\sin u)}{dx} = \cos u \frac{du}{dx}, \quad \frac{d(\cos u)}{dx} = -\sin u \frac{du}{dx},$$

$$\int dx = x + C, \quad \int x^n\, dx = \frac{x^{n+1}}{n+1} + C \; (n \neq -1), \quad \int \frac{dx}{x} = \log_e x + C,$$

$$\int \cos x\, dx = \sin x + C, \quad \int \sin x\, dx = -\cos x + C$$

The principal difficulty experienced by engineers in applying these rules arises from the problem of proper substitution. Suppose, for example, we seek

$$\frac{d}{dt}\left[t^a \sin (5t^2)\right]$$

First, we apply the product rule for $(u \cdot v)'$ with $u = t^a$, $v = \sin (5t^2)$ (note that t now replaces the role of x). Thus, our desired derivative becomes

$$t^a \frac{d}{dt}\left[\sin (5t^2)\right] + \sin (5t^2)\frac{d}{dt}(t^a)$$

Next we take care of $\sin (5t^2)$ by using the chain rule together with the special rules for x^n and $\sin x$. Thus:

$$\frac{d}{dt}\left[\sin (5t^2)\right] = \cos (5t^2)\frac{d}{dt}(5t^2) = 10t \cos 5t^2$$

Finally, the sought for derivative becomes

$$10t^{a+1} \cos (5t^2) + at^{a-1} \sin (5t^2)$$

Other points to watch are:

1. The appearance of the "open" constant of integration in the indefinite integral. This constant may be needed in order to meet all the conditions of a particular problem.

2. Look out for the way the algebraic signs appear in the derivatives and integrals of the trigonometric functions. This can be confusing. If in doubt, check with a table of integrals.

3. In the integration of x^n be sure that $n \neq -1$. Otherwise, use the special formula for this case.

In addition to the above there is an occasional mix-up between the formulas

$$\frac{dx^n}{dx} = nx^{n-1} \quad \text{and} \quad \frac{dn^x}{dx} = n^x \log_e n$$

However, the need to find the derivative of n^x is relatively rare.

ACCURACY

Even with a hand calculator accuracy can be a problem in rare situations. This might occur if intermediate values in the computation are not recorded with sufficient accuracy. This arises in the subtraction of two quantities of almost equal magnitude. Typically, this can occur in solving the quadratic equation when both roots are real and b and $\sqrt{b^2 - 4ac}$ are almost of the same magnitude.

Let us look at an example. We wish to find

$$- 10 + \sqrt{101}.$$

Now $\sqrt{101} = 10.05$, so that the overall result is approximately 0.05. However, if we took $\sqrt{101}$ to only three significant figures, we would have 10.0 and a final result of zero instead of 0.05. Problems of this nature which involve a square root can be handled by the approximate formula (based on the binomial expansion):

$$\sqrt{1+\varepsilon} \cong 1 + \frac{1}{2}\varepsilon \qquad |\varepsilon| << |$$

where ε can be either positive or negative, but must be small in magnitude compared with one. We transform to the standard form as follows:

$$\sqrt{101} = \sqrt{100 + 1} = \sqrt{100(1 + .01)}$$

$$= 10\sqrt{1 + .01} \cong 10\ (1 + .005) = 10.05$$

Other commonly used approximation formulas are

$$\sin x \cong x, \quad \cos x \cong 1 - \frac{1}{2}x^2$$

where x must be expressed in radian measure and must be of small magnitude compared with one.

DIFFERENTIAL EQUATIONS

We restrict our attention to two special differential equations which occur often in mechanical engineering.

The d.e. (differential equation)

$$\frac{d^2y}{dx^2} = f(x)$$

is directly solvable by successive integration. Thus if

$$\frac{d^2x}{dt^2} = 6t^2 - 2a$$

we have

$$\frac{dx}{dt} = 2t^3 - 2at + C_1, \quad x = \frac{1}{2}t^4 - at^2 + C_1t + C_2$$

Note the appearance of the integration constants C_1 and C_2 and note also the symbol variation between the standard form of the differential equation and the example.

A more difficult differential equation, which occurs frequently in vibration problems, has the form

$$\frac{d^2x}{dt^2} + p^2 x = \delta$$

where p and δ are constants. The method of solution of this differential equation is set out in standard texts. Here we give only the result:

$$x(t) = \frac{\delta}{p^2} + C_1 \cos pt + C_2 \sin pt$$

or

$$x(t) = \frac{\delta}{p^2} + A \cos (pt-\alpha)$$

The constants C_1, C_2 and A, α are integration constants. The two forms listed above are equivalent, as can be shown by using appropriate trigonometric identities.

Of frequent occurrence is the special case $\delta = 0$, where the solution specializes by the omission of the first term (involving δ).

The presence of trigonometric terms in the above solution leads to the characteristic to-and-fro motion found in vibrating systems. Now suppose that as the result of applying physical principles we arrive at the somewhat different differential equation

$$\frac{d^2x}{dt^2} - p^2 x = \delta$$

Although this differential equation looks almost the same as the previous one its solution has quite a different behavior. In fact, it does not oscillate at all. Thus, if we anticipate oscillatory behavior and end up with the second differential equation, something has gone wrong in the analysis. Here again is a check point for our mathematics.

REVIEW PROBLEMS

Following are 28 simple problems with answers to serve as a partial review and extension of the subject matter outlined above. Try to work the problems before looking at the answers. For each problem try to discern whether some special point is being brought out.

Following the 28 review problems, there are six problems from past professional examinations with answers outlined. Again, try to work the problems before looking at the answers. Note that in many cases there will be more than one correct way to attack a problem.

1. $\log \frac{P}{N} (a - b)^{1/n}$ is equal to:

   (a) $\log \frac{P}{N} + n \log (a - b)$

   (b) $\log P - \log N + \log (a - b) + \log \frac{1}{n}$

   (c) $\log P - \log N + (\frac{1}{n}) \log (a - b)$

2. $\int_0^{10} (x^2 + 1)\, dx$ is equal to:

   (a) 20 (b) 21 (c) 101 (d) 1001 (e) None of these

3. sin 930° is equal to:

   (a) $-\frac{1}{2}$ (b) $+\frac{1}{2}$ (c) $\frac{2}{\sqrt{3}}$ (d) $-\frac{1}{\sqrt{2}}$ (e) None of these

4. $\log \frac{(a + b)}{c}$ is equal to:

   (a) $\log a + \log b - \log c$

   (b) $\frac{\log a + \log b}{\log c}$

   (c) $\log (a + b) - \log c$

   (d) $e^{\frac{(a + b)}{c}}$

   (e) $\log \frac{a}{c} + \log \frac{b}{c}$

5. The first derivative of $\frac{1}{x}$ with respect to x is:

   (a) $-\frac{1}{x^2}$ (b) $-\frac{2}{x^2}$ (c) $\log_e x$ (d) 1 (e) $\frac{dx}{dy}$

6. Find the slope of the line which is tangent to the parabola $y = 8x^2 + 4$ at the point where $x = 2$

7. tan (arc sin 0.5) is equal to:

   (a) $\frac{1}{4}$ (b) $\frac{1}{2}$ (c) $\frac{1}{3^{1/2}}$ (d) 1 (e) $3^{1/2}$

8. $\log_{10}(100)^2 - \ln_e 2.718$ is approximately equal to:

   (a) 2 (b) 3 (c) 4 (d) 5 (e) 6

9. For the position-time function, $x = 3t^2 + 2t$, the velocity in the x direction at $t = 1$ is:

   (a) 4 (b) 5 (c) 6 (d) 7 (e) 8

10. Derive the equation of the largest circle that is tangent to both coordinate axes and has its center on the line,

$$2x + y - 6 = 0$$

11. In the equation, $y = \dfrac{-x^3 + 3x + 2}{x^2 + 2x + 1}$, the limit of y as x approaches a value of minus one is:

   (a) zero (b) 1 (c) 2 (d) 3 (e) infinity

12. The cosine of 120° is equal to:

   (a) $\frac{3^{1/2}}{2}$ (b) $\frac{1}{3^{1/2}}$ (c) $\frac{1}{2}$ (d) $-\frac{1}{2}$ (d) $\frac{-3^{1/2}}{2}$

13. If the sine of angle "A" is given as K, the tangent of angle "A" would be:

   (a) $1 - K$ (b) $\frac{1}{K}$ (c) $(1 - K^2)^{1/2}$ (d) $\frac{1}{(1 - K^2)^{1/2}}$

   (e) $\frac{K}{(1 - K^2)^{1/2}}$

14. $\log_{10}(1000)^3$ is equal to:

   (a) 3 (b) 5 (c) 6 (d) 9 (e) 12

15. $\ln_e (2.718)^{xy}$ is approximately equal to:

   (a) xy (b) exy (c) 2.718 xy (d) x + y (e) 0.434 xy

16. Each interior angle of a regular polygon with five sides is:

   (a) 90 degrees (b) 100 degrees (c) 108 degrees

   (d) 120 degrees (e) 130 degrees

17. The number below that has four significant figures is:

   (a) 1414.0 (b) 1.4140 (c) 0.141 (d) 0.01414

   (e) 0.0014

18. The slope of the curve $y = x^3 - 4x$ as it passes through the origin ($x = 0$, $y = 0$) is equal to:

   (a) + 4 (b) + 2 (c) 0 (d) - 2 (e) - 4

19. Find the area bounded by the parabola, $y^2 = 2x$, and the line, $x = 8$.

20. If $u = e^{3y} \cos 2x$, what is $\frac{du}{dt}$ if both x and y are functions of t?

21. The logarithm of the number, - 0.09, is:

(a) positive
(b) negative
(c) zero
(d) a complex number
(e) none of these

22. The value of sin (A + B) where sin A = 1/3 and cos B = 1/4 is:

(a) 7/12
(b) 1/12
(c) $(1 + 30^{1/2})/12$
(d) $(1 + 2\sqrt{30})/12$
(e) $(8^{1/2} - 1)/12$

23. The derivative of $\frac{x}{(x - 1)}$ is:

(a) $\frac{-1}{(x - 1)^2}$
(b) $\frac{1}{(x - 1)^2}$
(c) $\frac{x}{(x - 1)^2}$
(d) $\frac{1}{(x - 1)}$
(e) $\frac{-x}{(x - 1)^2}$

24. The value of $\int (1 - x)^{1/2}\, dx$ is:

(a) no value
(b) $(2/3)\ (1 - x)^{3/2}$
(c) $(-1/3)\ (1 - x)^{3/2} + C$
(d) $(-2/3)\ (1 - x)^{3/2} + C$
(e) $(-3/2)\ (1 - x)^{3/2} + C$

25. $\int \frac{x\, dx}{x^2 + 1}$ equals:

(a) $\frac{1}{2} \ln (x^2 + 1)$
(b) $\ln (x^2 + 1) + C$
(c) $\ln (x^2 + 1)$
(d) $\frac{(x^2 + 1)^2}{2} + C$
(e) $\frac{1}{2} \ln (x^2 + 1) + C$

26. The integral, $\int_1^2 (x^2 - 1)\,dx$, equals:

(a) 0 (b) 2 (c) 3/4 (d) 4/3 (e) 5/3

27. The area bounded by the x axis, the lines, x = 1 and x = 3, and by $y = x^3$ is equal to:

(a) 81/4 (b) 20 (c) 10 (d) - 20 (e) 41/2

28. The expression $A^3 + B\,X$ written in Fortran is:

(a) A*3 + BX

(b) A*3 + B*X

(c) A**3 + B*X

(d) A**3 + BX

(e) A*3 + B**X

ANSWERS TO PROBLEMS

1. c
2. e
3. a
4. c
5. a
6. 32
7. c
8. b
9. e
10. $(x - 6)^2 + (y + 6)^2 = 36$
11. d
12. d
13. e
14. d
15. a
16. c
17. d
18. e
19. 128/3
20. $\frac{du}{dt} = 3\,e^{3y} \cos 2x \frac{dy}{dt} - 2\,e^{3y} \sin 2x \frac{dx}{dt}$
21. d
22. d
23. a
24. d
25. e
26. d
27. b
28. c

MATH FUNDAMENTALS 1

A cylindrical water tank which is 35 ft. in diameter and 105 ft. in length is placed temporarily on an 18.5° slope as shown in the figure. The filler opening is located flush with the top of the tank at the mid-point.

What is the maximum volume of water which can be placed in the tank?

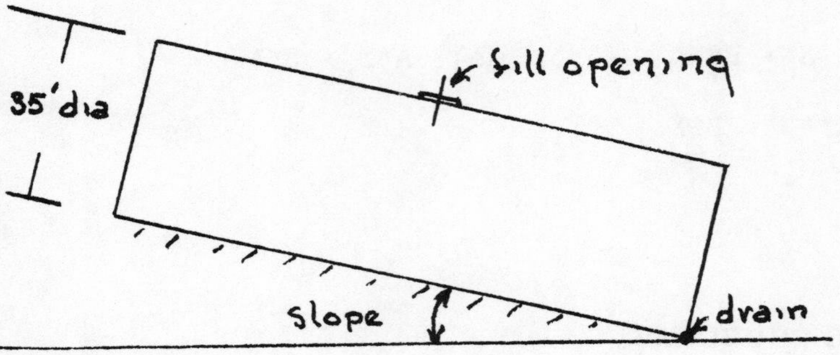

Solution

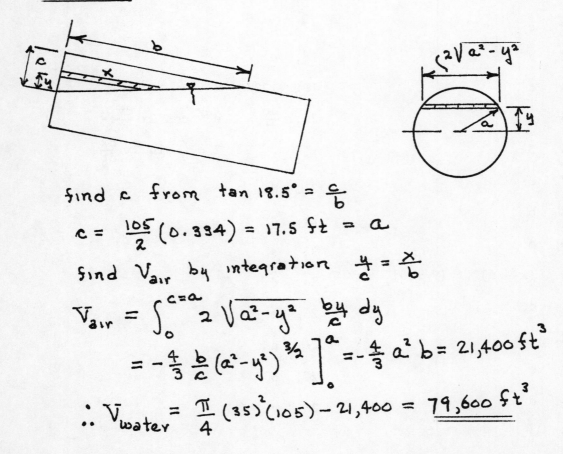

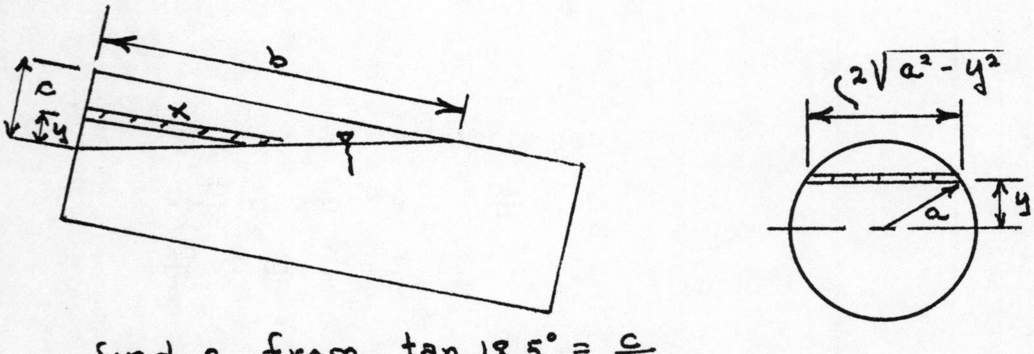

find c from $\tan 18.5° = \frac{c}{b}$

$$c = \frac{105}{2}(0.334) = 17.5 \text{ ft} = a$$

find V_{air} by integration $\frac{y}{c} = \frac{x}{b}$

$$V_{air} = \int_0^{c=a} 2\sqrt{a^2 - y^2}\,\frac{by}{c}\,dy$$

$$= -\frac{4}{3}\frac{b}{c}(a^2 - y^2)^{3/2}\Big]_0^a = -\frac{4}{3}a^2 b = 21{,}400 \text{ ft}^3$$

$$\therefore V_{water} = \frac{\pi}{4}(35)^2(105) - 21{,}400 = \underline{\underline{79{,}600 \text{ ft}^3}}$$

MATH FUNDAMENTALS 2

A bag contains four white balls and six black balls. Three balls are drawn in succession without replacing any ball after it is drawn. Find the probability (chances in a hundred) that the first and second balls will be white and the third ball black.

Solution

1st drawing	4W, 6B	$P(W) = 4/10$
2nd "	3W, 6B	$P(W) = 3/9$
3rd "	2W, 6B	$P(B) = 6/8$

$$P(WWB) = \frac{4}{10} \times \frac{3}{9} \times \frac{6}{8} = \frac{1}{10}$$

or <u>10 chances out of 100</u>

MATH FUNDAMENTALS 3

A solid steel cylinder is three inches in diameter and eighteen inches long. It has been proposed that a concentric hole two inches in diameter be bored into the cylinder from one end and filled with lead so as to move the center of gravity of the cylinder 1/2 inch toward the bored end. Steel weighs 490 pounds per cubic foot and lead weighs 705 pounds per cubic foot. Is this proposal a reasonable one? Explain, verifying by analysis and calculations.

Solution

Without any lead insert, the c.g. is at the center of the large cylinder. Imagine the lead plug growing in length. At first, the c.g. moves toward the bored end. Eventually, the c.g. coincides with the end face of the plug (see figure). Further lengthening of the plug, now moves the c.g. back towards its original position.

Maximum c.g. excursion is for the sketched configuration

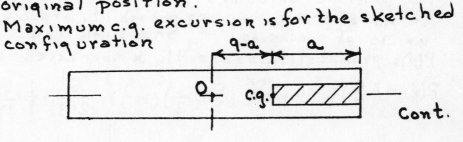

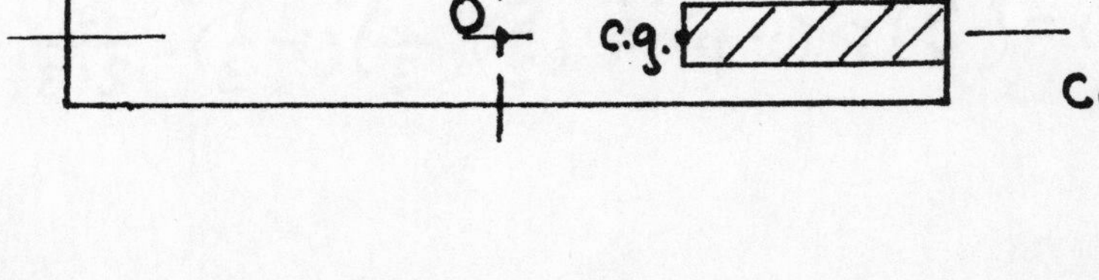

Cont.

(3) Cont

$$\Sigma M_o: (9-a)\left(\frac{\pi}{4}\cdot 3^2\cdot 18\cdot \frac{490}{1728}+\frac{\pi}{4}\cdot 2^2\cdot a\cdot \frac{705-490}{1728}\right)$$

$$= 0+\left(9-a+\tfrac{1}{2}a\right)\left(\frac{\pi}{4}\cdot 2^2\cdot a\right)\frac{705-490}{1728}$$

simplifying

$$a^2+\frac{81\times 98}{43}a-\frac{9\times 81\times 98}{43}=0$$

$$\therefore\ a=-\frac{81\times 49}{43}\pm\sqrt{\left(\frac{81\times 49}{43}\right)^2+\left(\frac{9\times 81\times 98}{43}\right)}$$

for physical meaning we must take the plus sign.

Then: $a=-\dfrac{81\times 49}{43}\left[-1+\sqrt{1+\dfrac{43\times 2}{49\times 9}}\right]$

$$= 9-\frac{43}{98}+\frac{2}{81}\left(\frac{43}{98}\right)^2-\cdots$$

$$= 9-0.438+0.005-\cdots = 9-0.433$$

Maximum displacement of c.g. is

<u>0.433 in so proposal is not reasonable</u>

MATH FUNDAMENTALS 4

If the probability of an event occurring is 1/3, what is the probability that it will occur exactly twice (2) in five (5) times?

<u>Solution</u>

Binomial distribution (independent events with two possible outcomes)

n = no. of trials $\quad$ p = probability of success on a single trial

x = no. of "successes"

$P(x)$ = probability of exactly x successes

$$P(x)=\binom{n}{x}p^x(1-p)^{n-x}=\binom{5}{2}\left(\frac{1}{3}\right)^2\left(1-\frac{1}{3}\right)^3=\frac{5!}{2!3!}\cdot\frac{1}{9}\cdot\frac{8}{27}=\underline{\underline{0.329}}$$

MATH FUNDAMENTALS 5

A system is made up of three black boxes in series having the following reliability:

(a) 0.99 (b) 0.95 (c) 0.70

If the system reliability requirement is 0.85 and redundancy is permitted for one box, can the requirement be achieved?

Solution

Reliability of series $R_s = 0.99 \times 0.95 \times 0.70 = 0.658$

$\therefore$ required R not attainable in series. So, we use redundancy for weakest box (0.70).

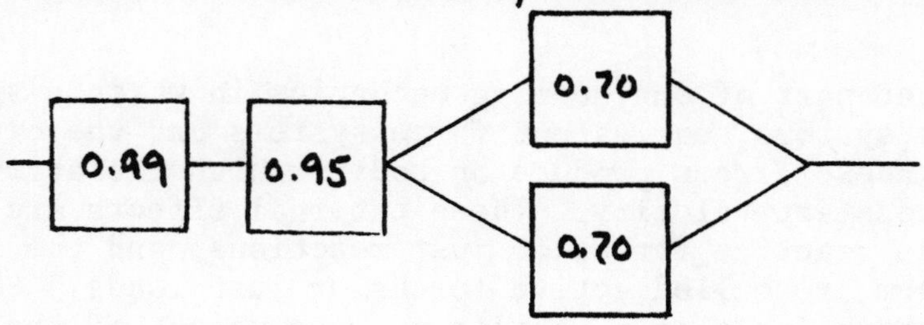

equivalent r for redundant boxes:

$R_{.70/.70}$ = prob. (at least one box ok)

= 1- prob. (both boxes bad)

$= 1 - 0.30 \times 0.30 = 0.91$

$R_s = 0.99 \times 0.95 \times 0.91 = \underline{0.856 > 0.85}$

requirement can be achieved

MATH FUNDAMENTALS 6

A sample of two hubs are sent to a laboratory for testing. The hubs were selected from a lot of six. What is the probability that at least one will be found defective if two of the original six are defective?

Solution

Lot contains 2 bad and 4 good hubs (2B, 4G)

let P_n = prob. of drawing n bad hubs.

hen: $P_0 + P_1 + P_2 = 1 \quad \therefore P_1 + P_2 = 1 - P_0$

$= 1 - \text{prob.}(GG) = 1 - \frac{4}{6} \cdot \frac{3}{5} = \underline{0.6}$

2 Force and Stress Analysis

HAROLD M. TAPAY

STATICS

Statics is that part of engineering mechanics in which a study is made of force systems, equivalent force systems and the external effects that these forces produce on bodies which are at rest or moving with constant velocity. These external effects are usually referred to as reactive forces or just reactions; and the forces that cause them are called active forces or just loads. Some examples of these loads on a vehicle are the weight of the vehicle, the weight of the occupants, and the air resistance. Bodies which are at rest or moving with constant velocity have their active and reactive force systems in equilibrium which means no resultant force or no resultant moment. In solving for the reactive forces a very useful visual aid is a diagram of the structure or a member of the structure in which all the forces applied by contacting bodies plus gravity forces (weight) are shown; this diagram is called a "free body diagram".

The principles used in statics are as listed below:

1. Parallelogram Law -

   Forces F_1 and F_2 can be replaced by F_3, i.e., diagonal of parallelogram through point of concurrency of F_1 and F_2.

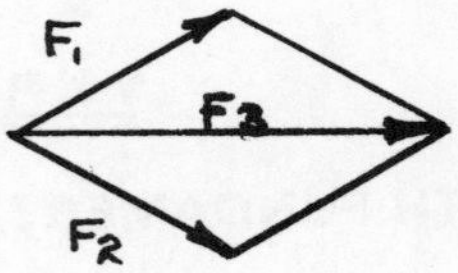

2. Principle of Transmissibility -

   External effects, i.e., reactions R_L and R_R are not changed by moving Force F from A to B or to any point on its line of action.

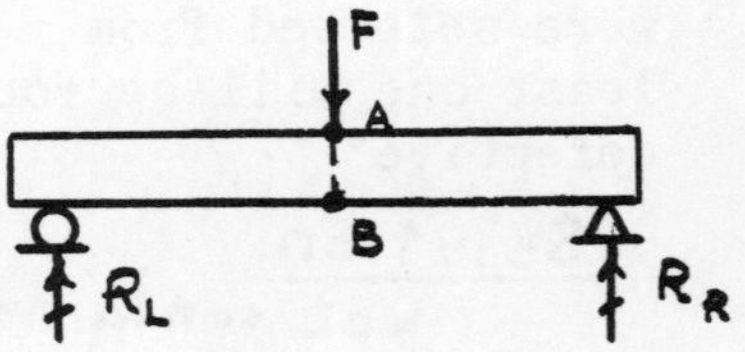

3. Newton's 1st Law -

   Contains principle of equilibrium of forces.

4. Newton's 3rd Law -

   Active and reactive forces are equal in magnitude, opposite in direction and collinear, e.g.,

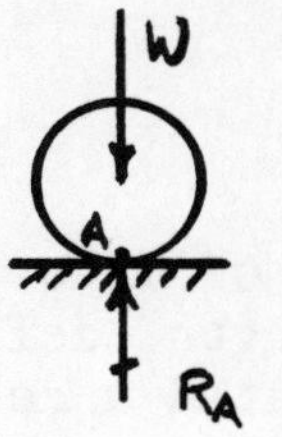

5. Varignon's Theorem - or Principle of Moments -

   Moment of a force about any point is equal to the sum of the moments of its components about the same point, i.e., $r \times F = r \times F_1 + r \times F_2 + r \times F_3$

   where $F = \underbrace{F_1 + F_2 + F_3}_{\text{components of F}}$

For <u>force systems not</u> in <u>equilibrium</u>:

Resultant Force $R = F_1 + F_2 + F_3 + \ldots$

or $R_x = \Sigma F_x$; $R_y = \Sigma F_y$; $R_z = \Sigma F_z$

<u>Equations of equilibrium</u> for a force system are:

$$\Sigma F_x = 0; \quad \Sigma F_y = 0; \quad \Sigma F_z = 0$$

$$\Sigma M_x = 0; \quad \Sigma M_y = 0; \quad \Sigma M_z = 0$$

<u>Very important force system</u> - <u>couple</u>; its resultant is a moment M_z whose magnitude = Fd

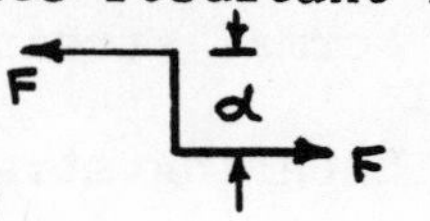

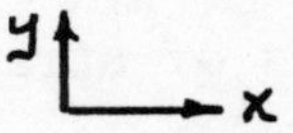

<u>Very important equivalent force system</u>:

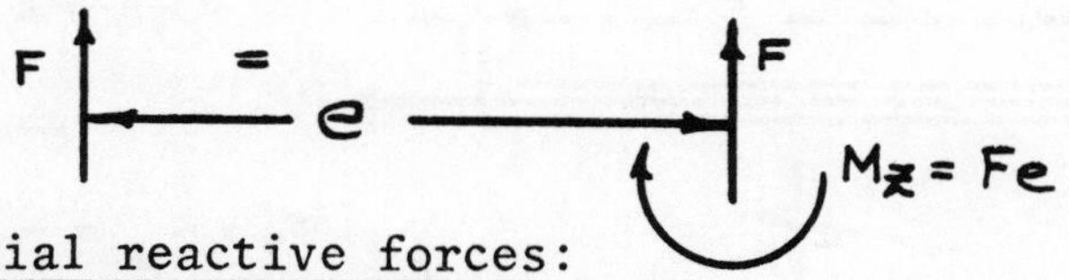

<u>Special reactive forces</u>:

Maximum static friction - $(F_s)_{max}$

Body on verge of moving -

$(F_s)_{max} = \mu_s N$

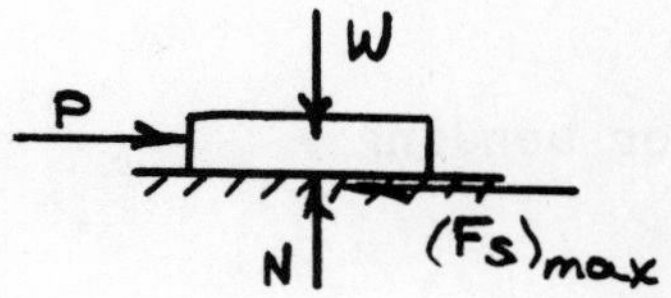

Kinetic friction - F_k

Body moving $F_k = \mu_k N$

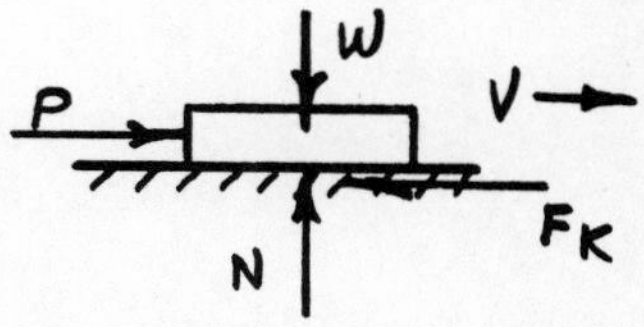

μ_s - coeff. of static friction

μ_k - coeff. of kinetic friction

$\mu_s > \mu_k$

STRENGTH OF MATERIALS

Strength of materials is that part of engineering mechanics in which the strength, deflection and stability of structural members are investigated under different force systems. This investigation results in formulas for stresses, strains, beam deflections, angles of twist, buckling loads, etc. Some of these formulas contain mechanical properties of the material from which the member is made; these properties having been determined previously in a materials testing laboratory. The derivation of the above formulas requires the use of some of the following:

1. Principles of statics.

2. Materials properties, e.g., μ, E, G, etc.

3. Assumed strain distributions.

4. Mathematical properties of members, cross section, e.g., A, I, J.

5. Stress-strain law.

Types of strain: Normal strain $\varepsilon \ \frac{\text{in}}{\text{in}}$ - tensile and compressive; Shearing strain $\gamma \ \frac{\text{in}}{\text{in}}$

For axial loading -

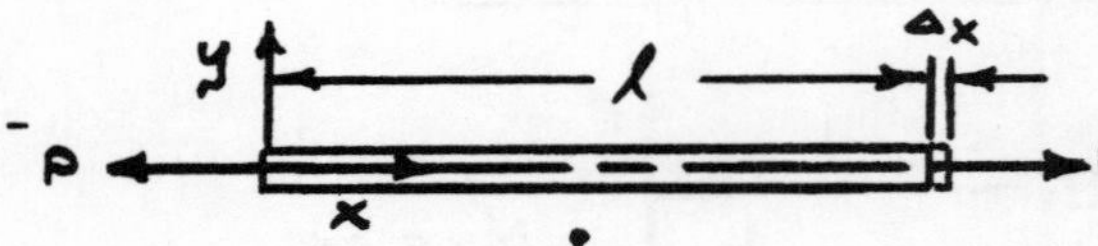

$\Delta x = P\ell/EA$

$\epsilon_x = \Delta x/\ell$

$\epsilon_y = \epsilon_z = -\mu\epsilon_x$

For bending -

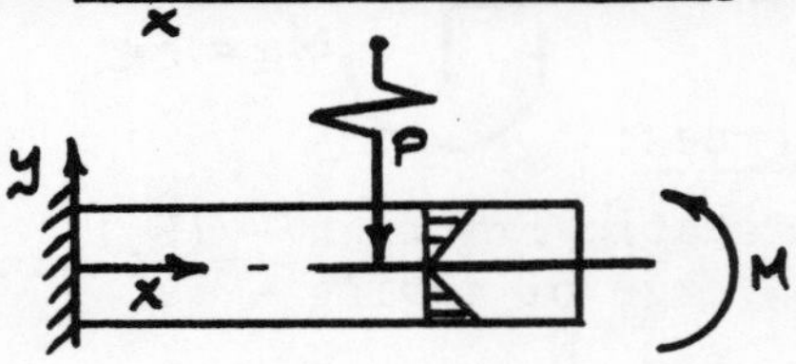

$\epsilon_x \propto y : \epsilon_x = y/\rho$

For torsion

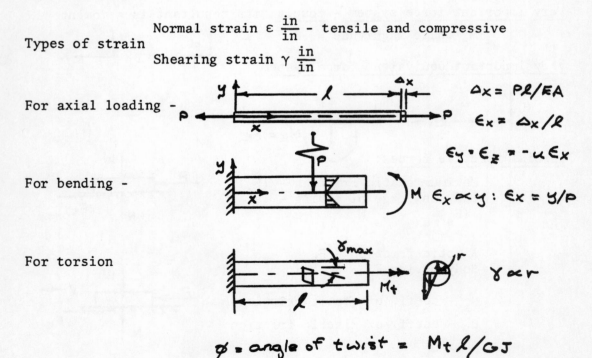

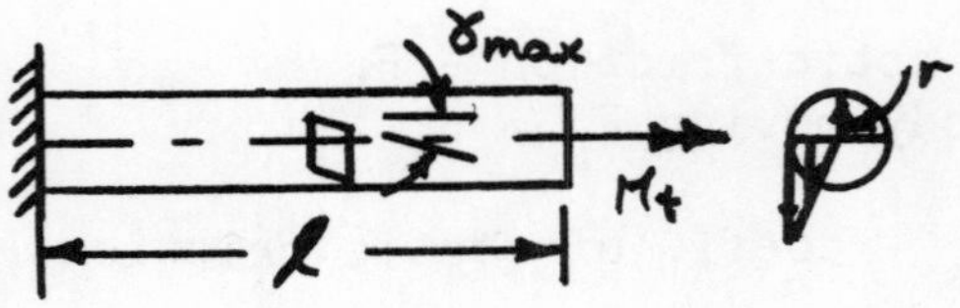

$\gamma \propto r$

ϕ = angle of twist = $M_t \ell/GJ$

Types of stress — Normal stress σ $\frac{\text{lb}}{\text{in}^2}$ - tensile and compressive; Shearing stress τ $\frac{\text{lb}}{\text{in}^2}$

For axial loading

For bending of initially straight beams whose material obeys Hooke's Law

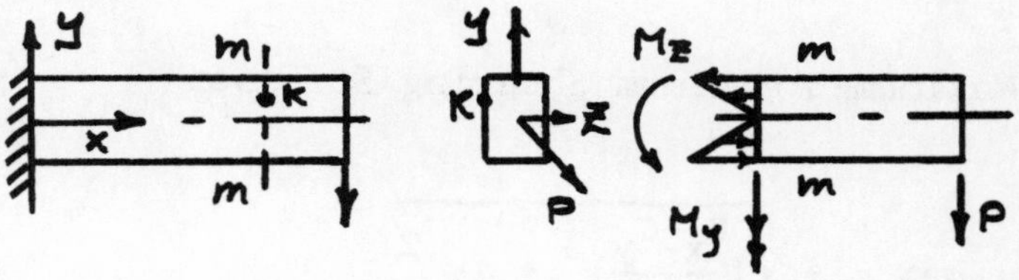

$$\sigma = \pm \frac{M_{zz}\, y}{I_{zz}} \pm \frac{M_{yy}\, z}{I_{yy}}$$

For fiber K both flexure Stresses are tensile.

For torsion of straight members of circular cross section whose material obeys Hooke's Law

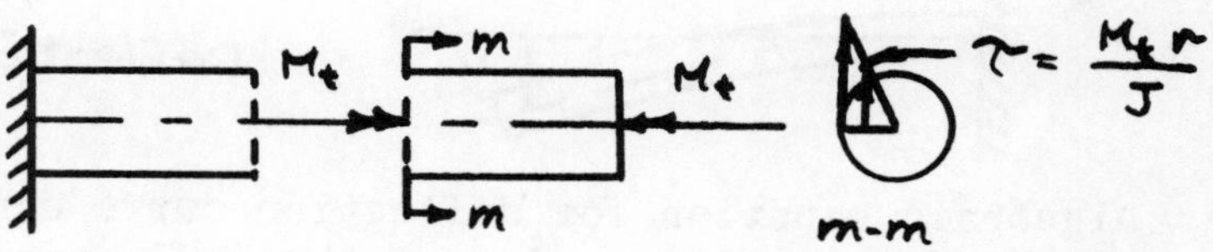

For bending of initially straight beam in one plane

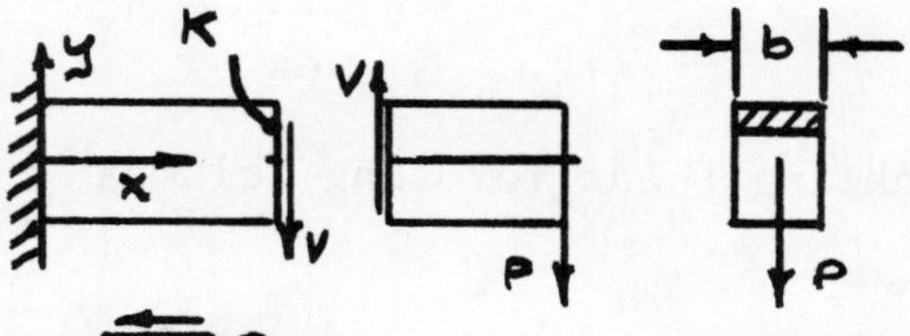

$$\tau_{xy} = \frac{VQ}{bI}$$

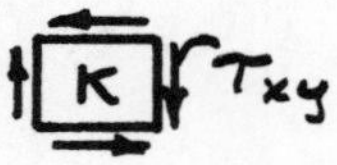

$Q = \int y\, dA$ for Shaded Area

Hooke's Generalized Law

$$\varepsilon_x = \frac{\sigma_x}{E} - \mu \frac{\sigma_y}{E} - \mu \frac{\sigma_z}{E}$$

$$\varepsilon_y = -\mu \frac{\sigma_x}{E} + \frac{\sigma_y}{E} - \mu \frac{\sigma_z}{E}$$

$$\varepsilon_z = -\mu \frac{\sigma_x}{E} - \frac{\sigma_y}{E} + \frac{\sigma_z}{E}$$

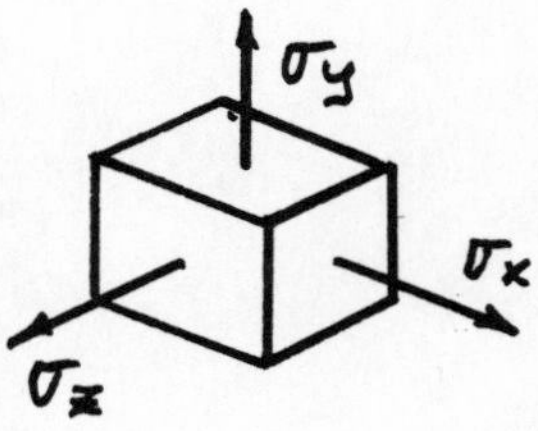

Hooke's Law for shearing stress and shearing strain

$\tau_{xy} = G\, \gamma_{xy}$ $\qquad \tau_{xy} = \tau_{yx}$

$\tau_{yz} = G\, \gamma_{yz}$ $\qquad \tau_{yz} = \tau_{zy}$

$\tau_{zx} = G\, \gamma_{zx}$ $\qquad \tau_{zx} = \tau_{xz}$

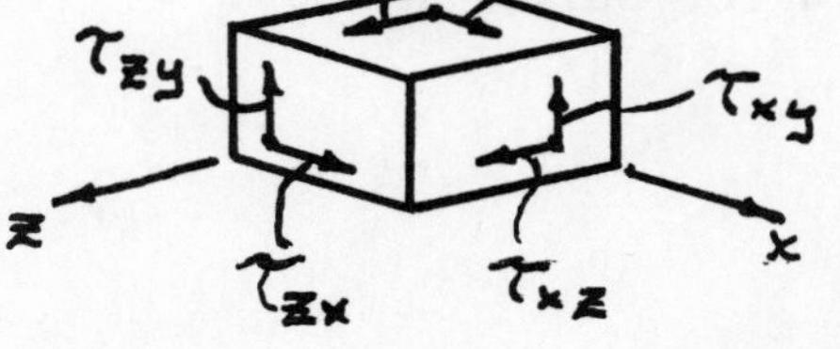

Maximum and Minimum Principal Stresses

$$\sigma_{\substack{max\\min}} = \frac{\sigma_x + \sigma_y}{2} \pm \sqrt{\left(\frac{\sigma_x - \sigma_y}{2}\right)^2 + \tau_{xy}^2}$$

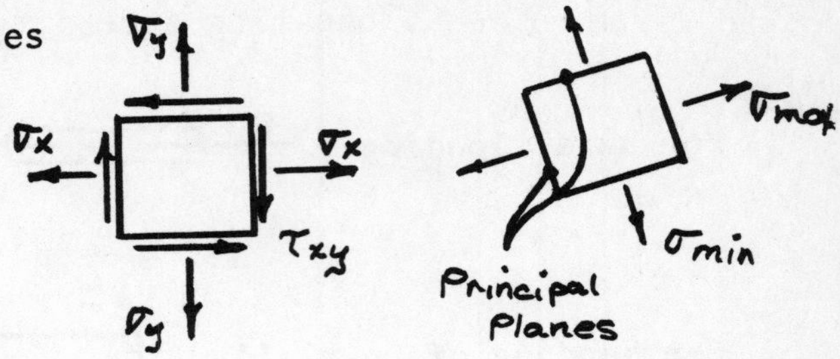

Maximum & Minimum Shearing Stresses

$$\tau_{\substack{max\\min}} = \pm \sqrt{\left(\frac{\sigma_x - \sigma_y}{2}\right)^2 + \tau_{xy}^2}$$

Beam Deflections - y for initially straight beams

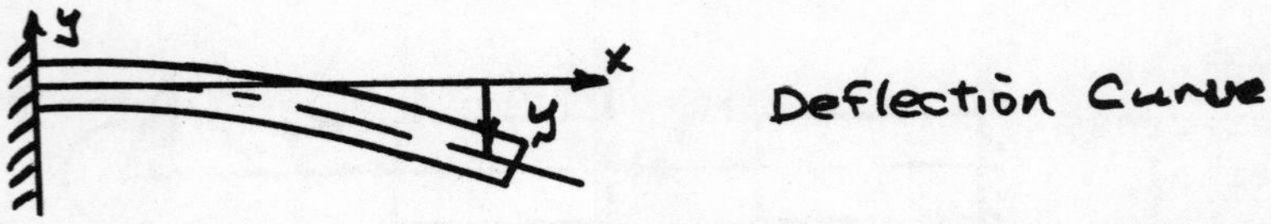

Algebraic equation for deflection curve can be obtained by solving the differential equation of the deflection curve; namely,

$$EI \frac{d^2y}{dx^2} = M$$

Elastic Buckling Loads for Long Columns

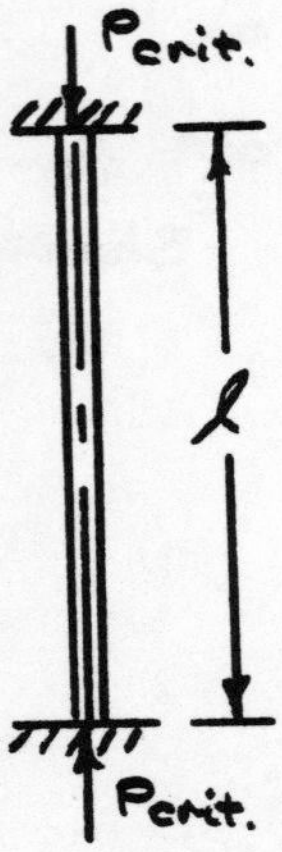

$$P_{crit.} = \frac{n\pi^2 EI}{\ell^2}$$

n - constant - depends on end conditions

e.g., n = 1 for pinned ends.

n = 4 for built in ends.

FORCE AND STRESS ANALYSIS 1

The first section of a rotary well drilling unit has a 6" outside diameter and a 5" inside diameter. If the weight of the drill string above it is 120,000 lbs and it requires a torque of 30,000 lb inch to rotate it, will the stresses in this section be below the allowables of 60,000 psi compression and 20,000 psi shear?

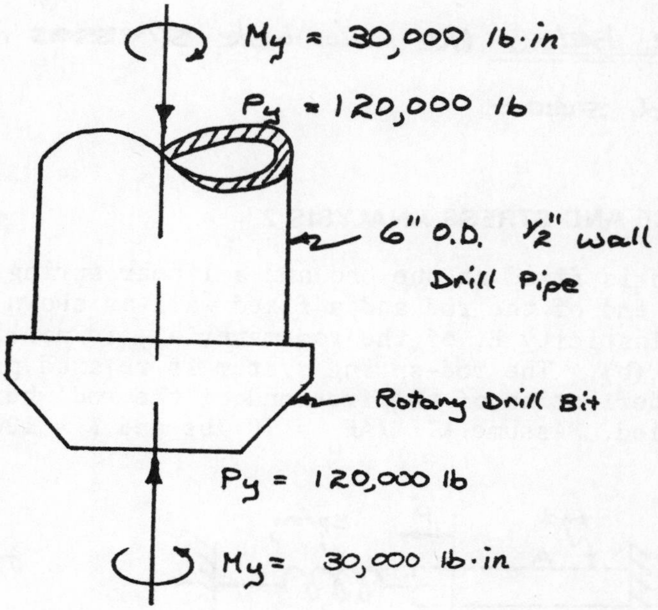

Solution

$$\tau_{xy} = \text{Torsional Shearing Stress} = \frac{(M_y)c}{J} = \frac{30{,}000\,(3)}{\frac{\pi}{32}(6^4-5^4)}$$

$$= \frac{90{,}000}{67.8} = \underline{\underline{1{,}330\ \text{lb/in}^2}}$$

$$\sigma_y = \text{Compressive Stress} = -\frac{P_y}{A} = \frac{-120{,}000}{\frac{\pi}{4}(6^2-5^2)}$$

$$= -\frac{120{,}000}{8.63} = \underline{\underline{-13{,}900\ \text{lb/in}^2}}$$

$$\sigma_{min} = \frac{\sigma_x+\sigma_y}{2} - \sqrt{\left(\frac{\sigma_x-\sigma_y}{2}\right)^2 + \tau_{xy}^2}$$

$$= \frac{-13{,}900}{2} - \sqrt{\left(\frac{13{,}900}{2}\right)^2 + (1{,}330)^2} = -6{,}950 - 7{,}070$$

$$= \underline{\underline{-14{,}020\ \text{lb/in}^2}} \quad \left[\sigma_{allow.} = -60{,}000\ \text{lb/in}^2\right]$$

$$\tau_{max} = \sqrt{\left(\frac{\sigma_x - \sigma_y}{2}\right)^2 + \tau_{xy}^2} = \underline{\underline{7070 \text{ lb/in}^2}}$$

$$\left[\tau_{allow.} = 20{,}000 \text{ lb/in}^2\right]$$

The largest compressive and shearing stresses <u>are below</u> the allowable stresses in compression and shear.

FORCE AND STRESS ANALYSIS 2

A rod is fixed at one end and a linear spring is connected between the free end of the rod and a fixed wall as shown in Fig. (a). The modulus of elasticity E, of the rod material, is nonlinear as indicated in Fig. (b). The rod-spring system is relaxed prior to loading. Determine the deflection of the free end of the rod where a load, P = 20 lbs, is applied. Assume K^2L^2/AE_o = 10 lbs and K = 100 lbs/in.

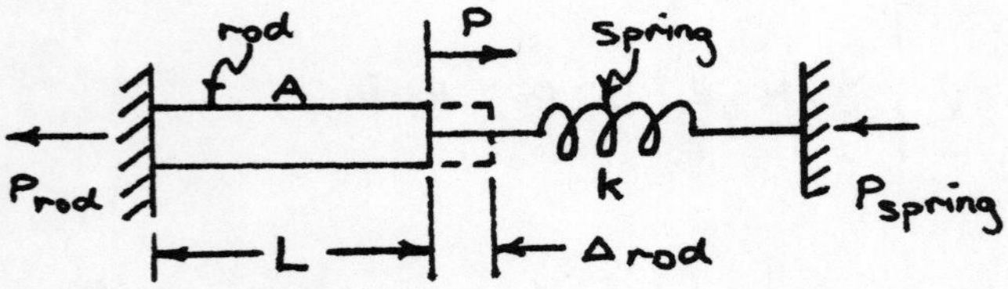

A - Cross sectional area of rod, in^2.
k - Spring constant, lbs/in.

Fig. (a) Spring-Rod System

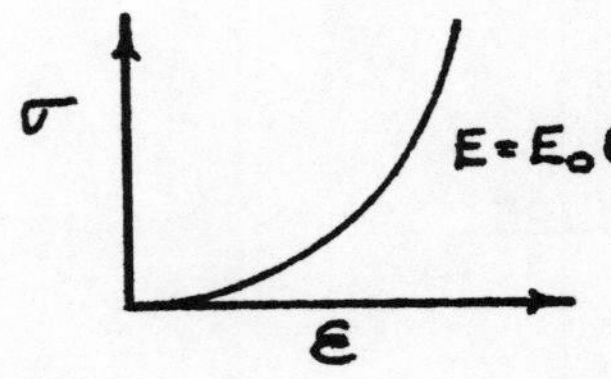

E - Modulus of elasticity, lbs/in^2.
E_o - Constant.
ε - Strain, in/in.

Fig. (b) Stress-Strain Relationship of Rod Material

<u>Solution</u>

1) $\Sigma F_x = 0$ $\qquad P_{rod} + P_{spring} = 20$

2) Since <u>supports</u> are <u>non-yielding</u>, $\Delta rod = \Delta spring$

3) From stress-strain curve for rod material:

$$\frac{d\sigma}{d\epsilon} = E = E_o \epsilon$$

$$\sigma = \frac{E_o}{2} \epsilon^2$$

$$P_{rod} = \sigma A = (E_o A/2)\epsilon^2$$

For rod, $\epsilon = \Delta_{rod}/L$

For spring, $\Delta_{spring} = P_{spring}/K$

$$\therefore \quad \epsilon = P_{spring}/KL \quad : \quad \epsilon^2 = (P_{spring})^2/K^2L^2$$

$$P_{rod} = \frac{E_oA}{2}\,\frac{(P_{spring})^2}{K^2L^2} = \frac{(P_{spring})^2}{2\,\dfrac{K^2L^2}{AE_o}} = \frac{(P_{spring})^2}{20}$$

Substituting into Eq. (1) gives:

$$\frac{1}{20}(P_{spring})^2 + P_{spring} = 20$$

$$P_{spring}^2 + 20\,P_{spring} - 400 = 0$$

$$P_{spring} = \frac{-20 \pm \sqrt{20^2 + 4(400)}}{2} = \frac{-20 \pm 44.8}{2}$$

$$= \underline{\underline{12.4 \text{ lb}}}$$

$$\Delta_{rod} = \Delta_{spring} = \frac{P_{spring}}{K} = \frac{12.4}{100} = 0.124 \text{ in}$$

$$\longrightarrow \approx \underline{\underline{1/8 \text{ in}}}$$

FORCE AND STRESS ANALYSIS 3

The solid circular shaft shown below of diameter 3.0 inches is simply supported by two bearings at A and D. The shaft carrier two pulleys at B and C. Pulley B is 15.0 inches in diameter and carries the vertical **loads** P_{z1} = 4000 lbs and P_{z1} = 2000 lbs. Pulley C is 12.0 inches in diameter and carries two horizontal loads in the X direction, P_{x1} = 5750 lbs and P_{x2} = 3250lbs.

REQUIRED: Find the maximum principal stress for element E halfway along the shaft and at the outside fibre: Element E is at X = 1.5, Y = 10.0, Z = 0.0 inches. Neglect the weight of the shaft and pulleys in your calculations.

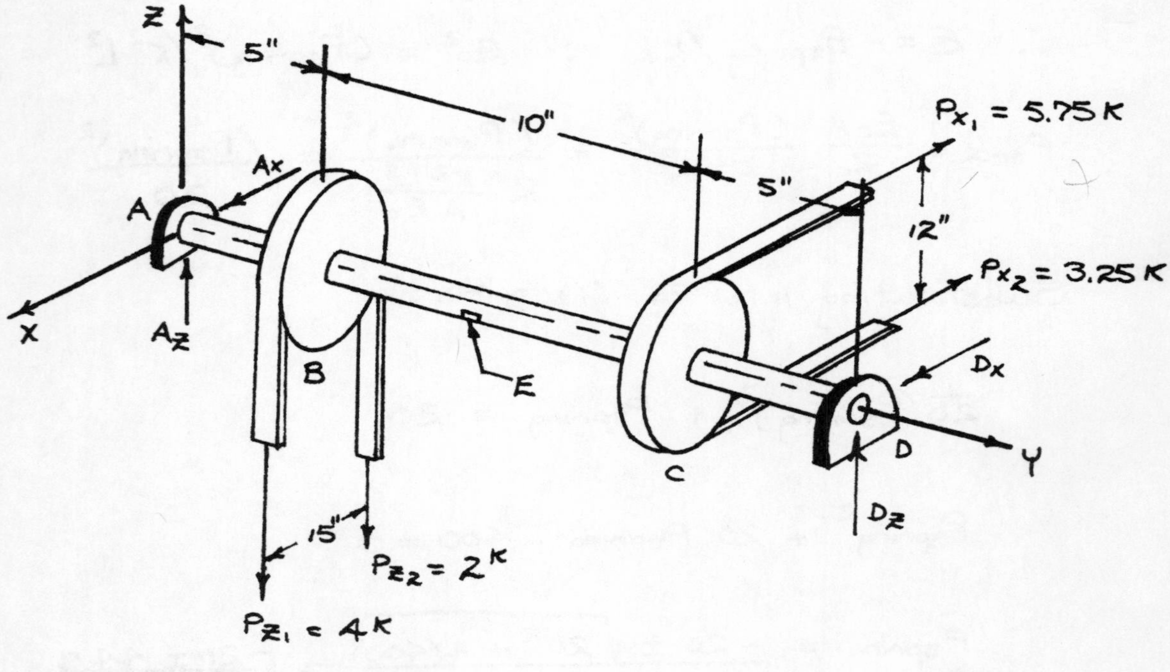

Solution

I. Determination of Bending Moment "M_z", Shear "U_z", and Twisting Moment "M_y" at section containing Element E.

$\Sigma(M_z)_D = 0$ $\quad 20\,A_x = 9(5)$ $\quad A_x = 2.25K$

$\Sigma(M_x)_D = 0$ $\quad 20\,A_z = 15(6)$ $\quad A_z = 4.5K$

$$M_z = 10(2250) = \underline{\underline{22500 \text{ lb. in}}}$$

$$U_z = 6K - A_z = \underline{\underline{1.5K}} \quad \Sigma F_z = 0$$

$$M_y = 4(7.5) - 2(7.5) = 15 \text{ Kip in} = \underline{\underline{15{,}000 \text{ lb.in}}}$$

II. For Element E

$$\sigma_y = \text{Bending Stress due to } M_z = \frac{M_z c}{I} = \frac{32\, M_z}{\pi d^3}$$

$$= \frac{32\,(22{,}500)}{\pi\,(3)^3} = \underline{\underline{-8500 \text{ psi}}}$$

$$\tau_{yz} = \text{Torsional shearing stress due to } M_y = \frac{M_y c}{J}$$

$$= \frac{16\, M_y}{\pi d^3} = \frac{16\,(15{,}000)}{\pi\,(3)^3} = \underline{\underline{2830 \text{ psi}}}$$

$$\tau_{yz} = \text{Direct Shearing Stress due to } U_z = \frac{Q\,U_z}{I\,d}$$

$$= \frac{4}{3}\frac{U_z}{A} \;;\; A = \frac{\pi d^2}{4} = 7.08 \text{ in}^2$$

$$= \frac{4}{3}\,\frac{1500}{7.08} = 280 \text{ psi}$$

$$(\tau_{yz})_{Total} = 2830 + 280 = \underline{\underline{3110 \text{ psi}}}$$

$$\sigma_{max} = \text{Max. principal stress} = \frac{\sigma_y}{2} + \sqrt{\left(\frac{\sigma_y}{2}\right)^2 + (\tau_{yz})^2}$$

$$= -4250 + \sqrt{-4250^2 + 3110^2}$$

$$= -4250 + 5250$$

$$= \underline{\underline{1000 \text{ psi}}}$$

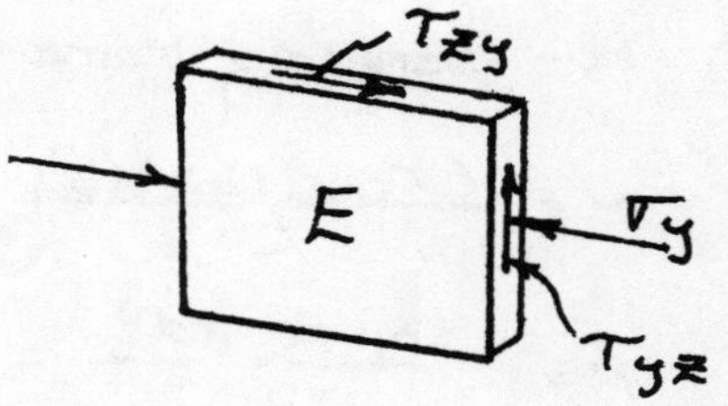

FORCE AND STRESS ANALYSIS 4

A rotating beam fatigue test was run which yielded the test results shown plotted on semi-log coordinates on the following page. Under pure bending, the specimens failed at 52 KSI and 10^5 cycles. The endurance limit for the material was at 30 KSI (10^6 cycles).

REQUIRED: (a) Determine the number of cycles to failure at 41 KSI.

(b) For a test specimen minimum diameter of 0.3 inches, what is the bending moment carried at the endurance limit?

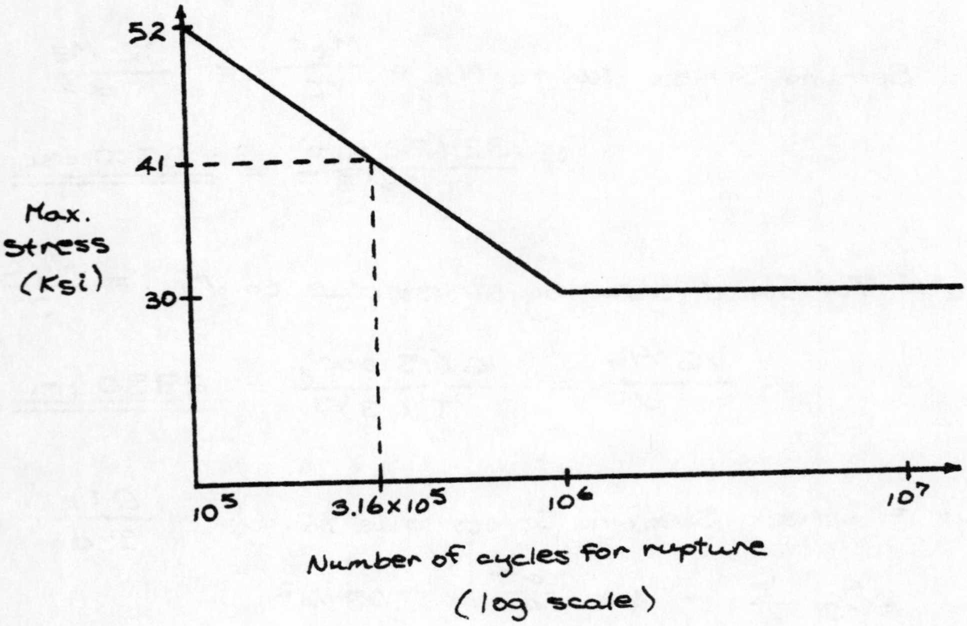

Solution

a) From the above graph for a fatigue strength of 41 Ksi, the number of cycles for rupture is

$\approx$ **320,000**

b) M - Bending Moment carried at Endurance limit

$$M = \frac{(\sigma_{End.\,limit})(I)}{c} = \frac{30{,}000}{d/2}\left(\frac{\pi d^4}{64}\right) \quad ; \quad d = 0.3 \text{ in.}$$

$$M = \frac{30{,}000\,\pi d^3}{32} = \frac{30{,}000\,\pi (0.3)^3}{32} = \underline{79.3 \text{ lb}\cdot\text{in}}$$

FORCE AND STRESS ANALYSIS 5

A boat trailer frame is being designed as shown. The support points carry the weight of the boat equally, with the axle attachment directly below the center support. A 4 x 2 in channel (area = 1.57 in^2 and I = 3.83 in^4) has been chosen as the primary member with a round bar held out from the flange by vertical spacers for additional strength. The center of the bar will be 6 in from the flange. If the maximum stress is to be limited to 10,000 psi, what is the diameter and location of the bar? Boat weighs 1500 lbs.

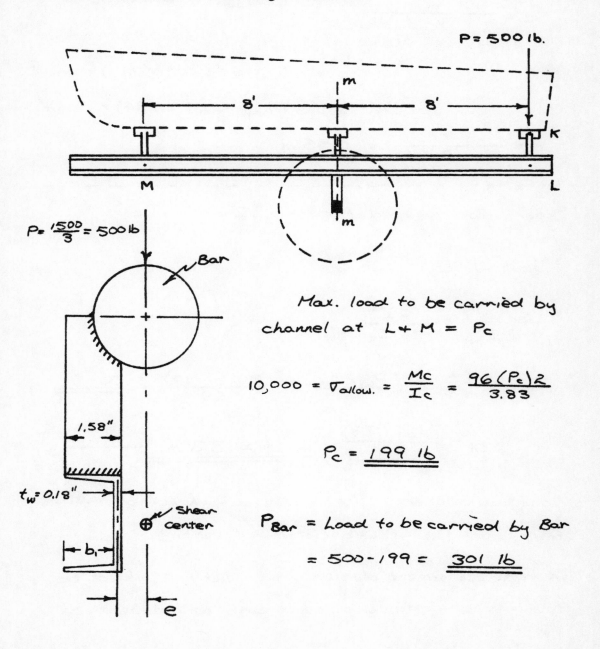

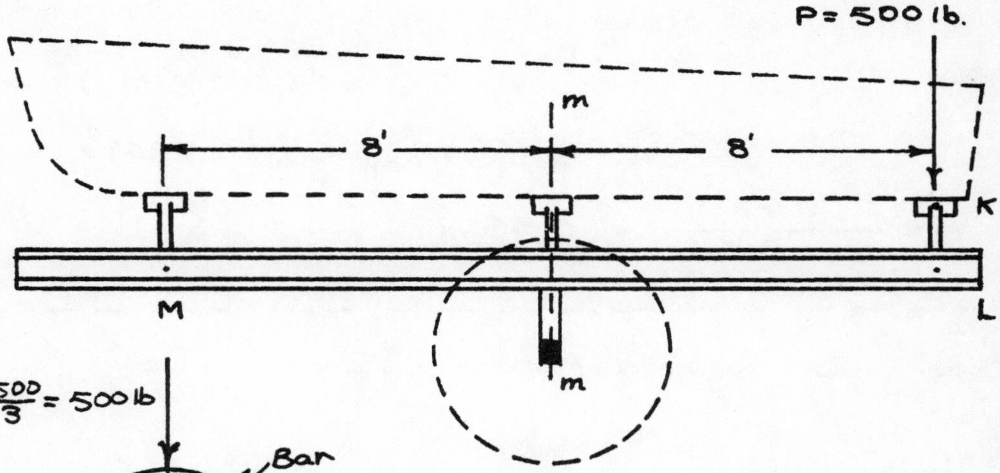

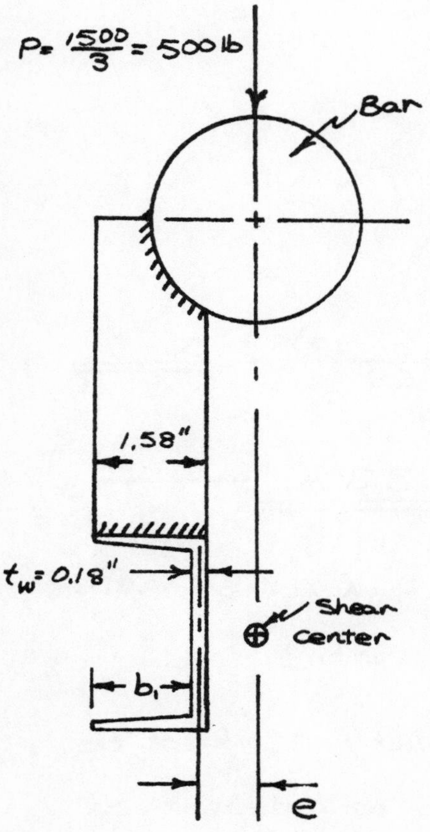

Max. load to be carried by channel at L + M = P_c

$$10{,}000 = \sigma_{allow.} = \frac{Mc}{I_c} = \frac{96(P_c)2}{3.83}$$

$$P_c = \underline{\underline{199 \text{ lb}}}$$

P_{Bar} = Load to be carried by Bar

$$= 500 - 199 = \underline{\underline{301 \text{ lb}}}$$

For no torsional shearing stresses in channel, line of action of load P applied by boat at each support should pass through shear center of channel.

Location of Shear Center

$$e = \frac{\frac{1}{2} b_1}{1 + \frac{1}{6}\frac{A_{web}}{A_{flange}}}$$

$(t_f)_{ave.} = 0.296$ in.

$b_1 = 1.58 - 0.18 = 1.4$ in.

$A_{web} \approx (4 - 0.3)(0.18) = .66 in^2$

$A_{flange} \approx 1.4(0.296) = .41 in^2$

$$e = \frac{\frac{1}{2}(1.4)}{1 + \frac{0.66}{6(0.41)}} = \underline{\underline{0.55 \text{ in}}}$$

Solving for required Moment of Inertia of bar.

$$y_K = y_L$$

$$\frac{P_{Bar} \cancel{X}^3}{\cancel{3E} I_{Bar}} = \frac{P_c \cancel{X}^3}{\cancel{3E} I_c}$$

$$I_{Bar} = \frac{\pi d^4}{64} = \frac{P_{bar} I_c}{P_c} = \frac{301(3.83)}{199} = \underline{\underline{5.8 \text{ in}^4}}$$

$$d^4 = \frac{64(5.8)}{\pi} \quad ; \quad \underline{\underline{d = 3.3 \text{ in.}}} \longleftarrow$$

Longitudinal center line of this bar should be 0.55 in. from center line of web (lateral direction).

A more economical design (much lighter) would be to have a continuous shear web as the vertical spacer. This web would be welded to the channel and to a much lighter round bar or tubing.

FORCE AND STRESS ANALYSIS 6

Compute the minimum web thickness required to prevent shear buckling in the beam shown below.

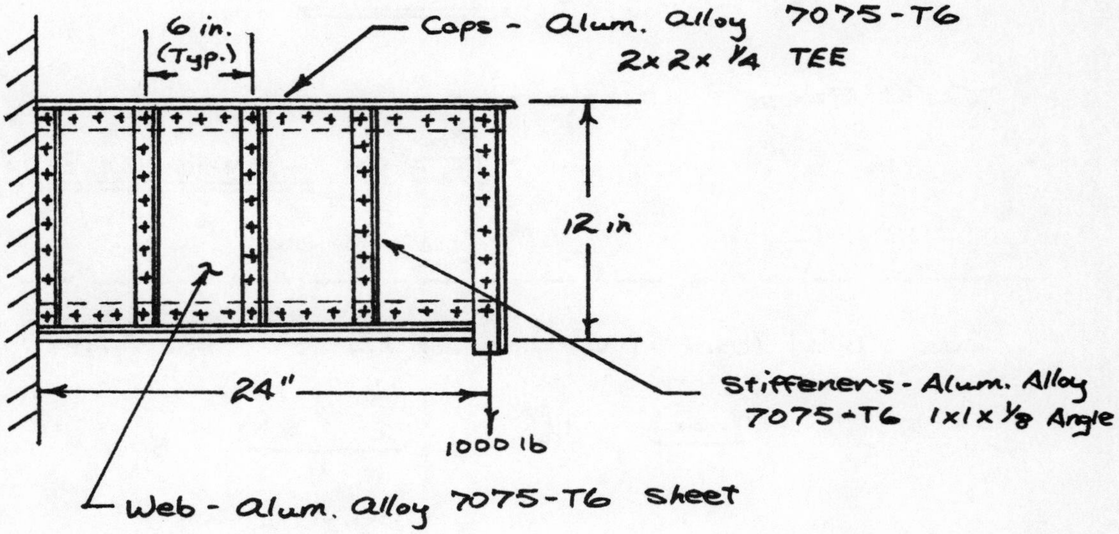

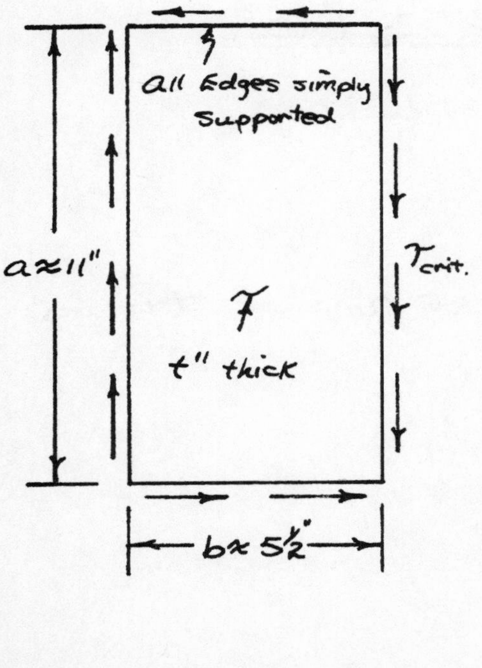

$$\tau_{vert.} \approx \frac{1000}{12t} \quad ; \quad \mu \approx 1/3$$

$$E = 10^7 \text{ lb/in}^2$$

$$\frac{1000}{12t} \approx \tau_{crit.} = \frac{K_s \pi^2 E}{12(1-\mu^2)}\left(\frac{t}{b}\right)^{2*}$$

For $a/b \approx 2$

$$K_s \approx 5.8$$

$$1000(1-\mu^2)(b)^2 = K_s \pi^2 E t^3$$

$$t^3 = \frac{1000(8/9)(5.5)^2}{5.8(\pi^2)(10^7)}$$

$$t^3 = 47\,(10^{-6})$$

$$t = 3.6\times10^{-2} = \underline{\underline{0.036 \text{ in}}} \longleftarrow$$

Axial load in stiffeners $\quad P_s \approx \dfrac{Vd}{h}$ * $\quad ; \; A_s = 0.23 \text{ in}^2$

$$V = 1000 \text{ lb} \quad ; \; h = 11 \text{ in.} \quad ; \; d \approx 6 \text{ in.}$$

$$P_s = \frac{1000(6)}{11} = \underline{\underline{545 \text{ lb. C}}}$$

Axial stress in stiffener

$$\sigma = \frac{P_s}{A_s} = -\frac{545}{0.23} = \underline{\underline{-2380 \text{ lb/in}^2}}$$

(low, no buckling)

Max. axial load in compression flange $\quad A_c = 1.07 \text{ in}^2$

$$F_c = \frac{M_{max}}{h} + \frac{V}{2}^{*} \approx \frac{24(1000)}{11} + 500$$

$$= 2680 \text{ lb.}$$

$$\sigma = \frac{F_c}{A_c} = \frac{2680}{1.07} = \underline{\underline{-2500 \text{ lb/in}^2}}$$

(low, no buckling)

* Bruhn – Analysis and Design of Airplane Structures

FORCE AND STRESS ANALYSIS 7

The equation governing beam deflection is $\frac{d^2y}{dx^2} = \frac{M}{EI}$.

Using the finite difference method, set up the necessary equations with boundary conditions included, to establish the deflection curve for the cantilever beam shown. Use the increment h = L/4. Do not solve the equations. (Assume EI constant for the beam.)

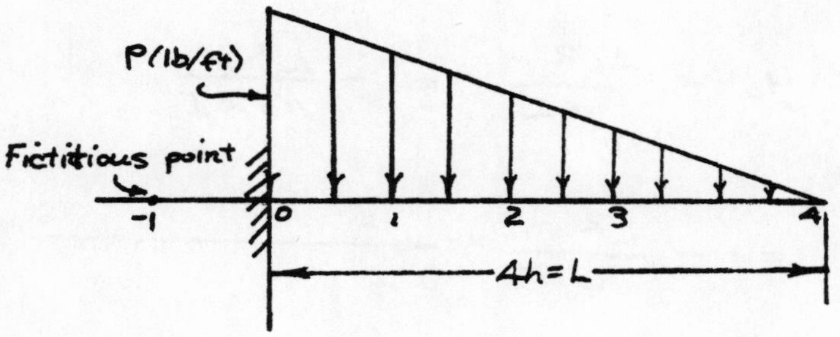

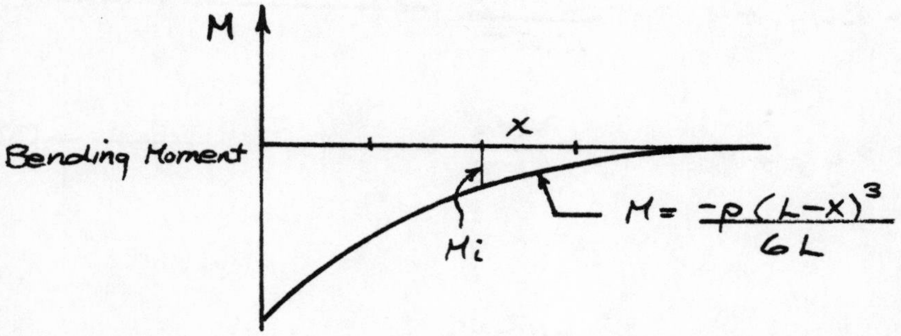

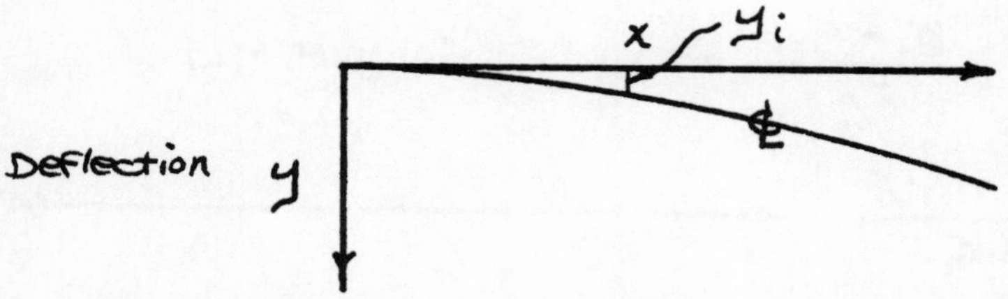

From Finite Difference Method

$$\left(\frac{d^2y}{dx^2}\right)_i \approx \frac{y_{i+1} - 2y_i + y_{i-1}}{h^2}$$

From differential equation of elastic curve

$$\left(\frac{d^2y}{dx^2}\right)_i = -\frac{M_i}{(EI)_i}$$

$$\therefore\ y_{i+1} - 2y_i + y_{i-1} = \frac{-h^2 M_i}{(EI)_i}$$

For $i = 1$:

$$y_2 - 2y_1 + y_0 = \frac{-h^2 M_1}{EI} = -\frac{L^2 M_1}{16EI}$$

$$y_0 = 0$$

$$\therefore\ y_2 - 2y_1 = -\frac{L^2 M_1}{16EI} \longleftarrow (1)$$

Similarly:

$$y_3 - 2y_2 + y_1 = -\frac{L^2 M_2}{16EI} \longleftarrow (2)$$

$$y_4 - 2y_3 + y_2 = -\frac{L^2 M_3}{16EI} \longleftarrow (3)$$

For point 0

$$y_1 - 2y_0 + y_{-1} = \frac{-L^2 M_0}{16EI}$$

and $\left(\frac{dy}{dx}\right)_0 \approx \frac{y_1 - y_{-1}}{2h} = 0 \qquad \therefore\ y_1 = y_{-1}$

$$\therefore\ 2y_1 = -\frac{L^2 M_0}{16EI} \longleftarrow (4)$$

Now have 4 equations and 4 unknowns.

FORCE AND STRESS ANALYSIS 8

A pin-connected tripod frame consisting of three round steel bars AB, AC and AD supports a load of 2000 lb at A. Each bar is 30 inches long and has a cross-sectional area of 0.5 in^2. The radius of the base circle BCD is 12 inches.

If no buckling occurs, find the stress in each of the bars in psi and state whether tension or compression.

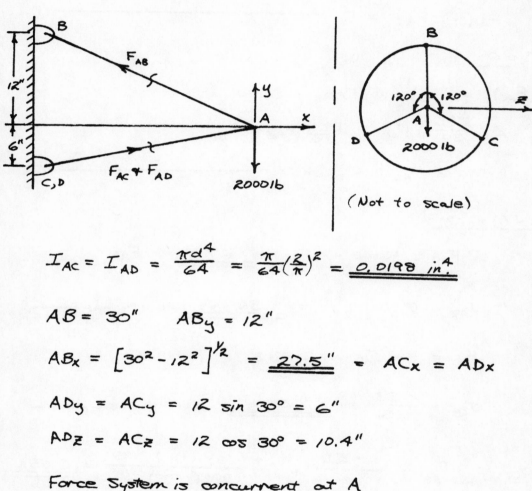

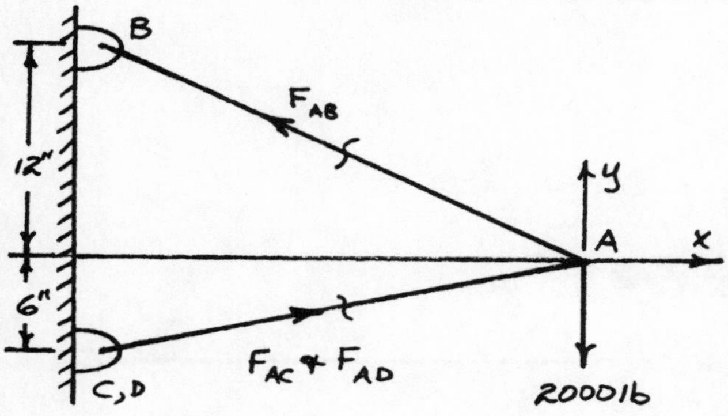

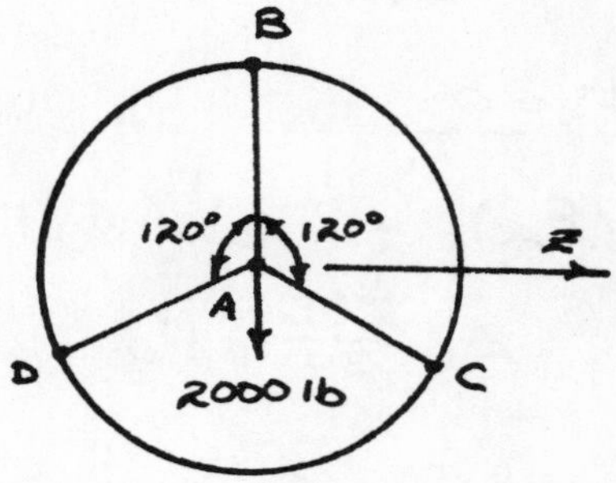

(Not to scale)

$$I_{AC} = I_{AD} = \frac{\pi d^4}{64} = \frac{\pi}{64}\left(\frac{2}{\pi}\right)^2 = \underline{\underline{0.0198\ in^4}}$$

$$AB = 30'' \qquad AB_y = 12''$$

$$AB_x = \left[30^2 - 12^2\right]^{1/2} = \underline{\underline{27.5''}} = AC_x = AD_x$$

$$AD_y = AC_y = 12 \sin 30° = 6''$$

$$AD_z = AC_z = 12 \cos 30° = 10.4''$$

Force System is concurrent at A

$\underline{\Sigma F_z = 0}$

$$+(F_{AD})_z - (F_{AC})_z = 0$$

$$\frac{10.4}{30}(F_{AD}) - \frac{10.4}{30}(F_{AC}) = 0$$

$F_{AD} = F_{AC}$ ←

$\Sigma F_y = 0$

$$(F_{AB})_y + (F_{AC})_y + (F_{AD})_y - 2 = 0$$

$$\frac{12}{30} F_{AB} + \frac{2(6)}{30} F_{AC} - 2 = 0$$

$$12\ F_{AB} + 12\ F_{AC} = 60$$

$$F_{AB} + F_{AC} = 5 \longleftarrow (1)$$

$\Sigma F_x = 0$

$$(F_{AC})_x + (F_{AD})_x - (F_{AB})_x = 0$$

$$2\left(\frac{27.5}{30}\right) F_{AC} - \frac{27.5}{30} F_{AB} = 0$$

$$2\ F_{AC} - F_{AB} = 0 \longleftarrow (2)$$

Add (1) to (2)

$$3\ F_{AC} = 5 \qquad F_{AC} = \underline{\underline{1.67\ K}} = F_{AD}$$

$$\sigma_{AC} = \sigma_{AD} = \frac{1.67}{0.5} = \underline{\underline{3.34\ \text{Ksi}}}\ \text{compr.} \longleftarrow$$

$$F_{AB} = 5 - F_{AC} = 3.33\ K$$

$$\sigma_{AB} = \frac{3.33}{0.5} = \underline{\underline{6.66\ \text{Ksi}}}\ \text{tensile} \longleftarrow$$

$$(P_{crit.})_{AC\ or\ AD} = \frac{\pi^2 EI}{\ell^2} = \frac{\pi^2 (3)(10^7)(0.0198)}{900(1000)} = \underline{\underline{6.55K}}$$

No Buckling Occurs, $6.55\ K > 1.67\ K$

FORCE AND STRESS ANALYSIS 9

What will be the vertical deflection of rigid member BC under force P?

AB and CD have length L, area A, modulus of elasticity E, and moment of inertia I.

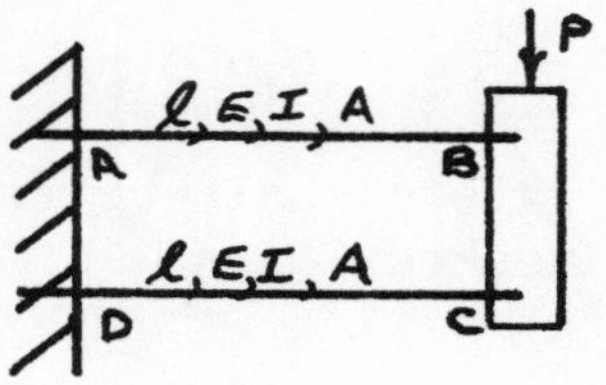

Solution

Deflected Shape

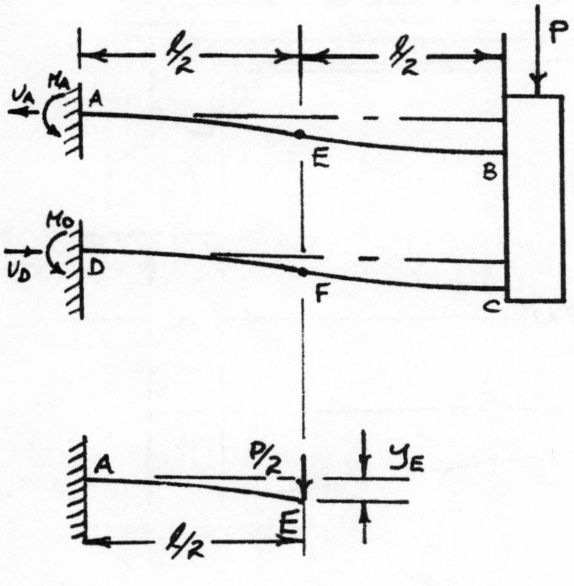

* Points E + F are points of contraflexure. ($M_E = M_F = 0$)

* Since members AB + DC have same properties, length, etc.

$U_A = U_D = P/2$ ←

$M_A = M_D = \frac{P}{2}\frac{\ell}{2} = \frac{P\ell}{4}$ ←

$$y_B = 2y_E = y_C = 2y_F$$

$$= 2\left[\frac{(P/2)(\ell/2)^3}{3EI}\right] = \frac{P\ell^3}{24EI}$$ ←

Vertical Deflection of Rigid Member $= \dfrac{P\ell^3}{24EI}$

FORCE AND STRESS ANALYSIS 10

A truck axle is to be designed to withstand a torque load of 2880 ft-lbs and a wheel load of 2000 lbs.

(a) For the configuration shown, what is the axle diameter required to the nearest 1/16 inch?

Assume $S_{t\ max} = 100{,}000$ psi

$S_{s\ max} = 50{,}000$ psi

(b) The minimum factor of safety shall be 2.50 in tension. What is the factor of safety in shear?

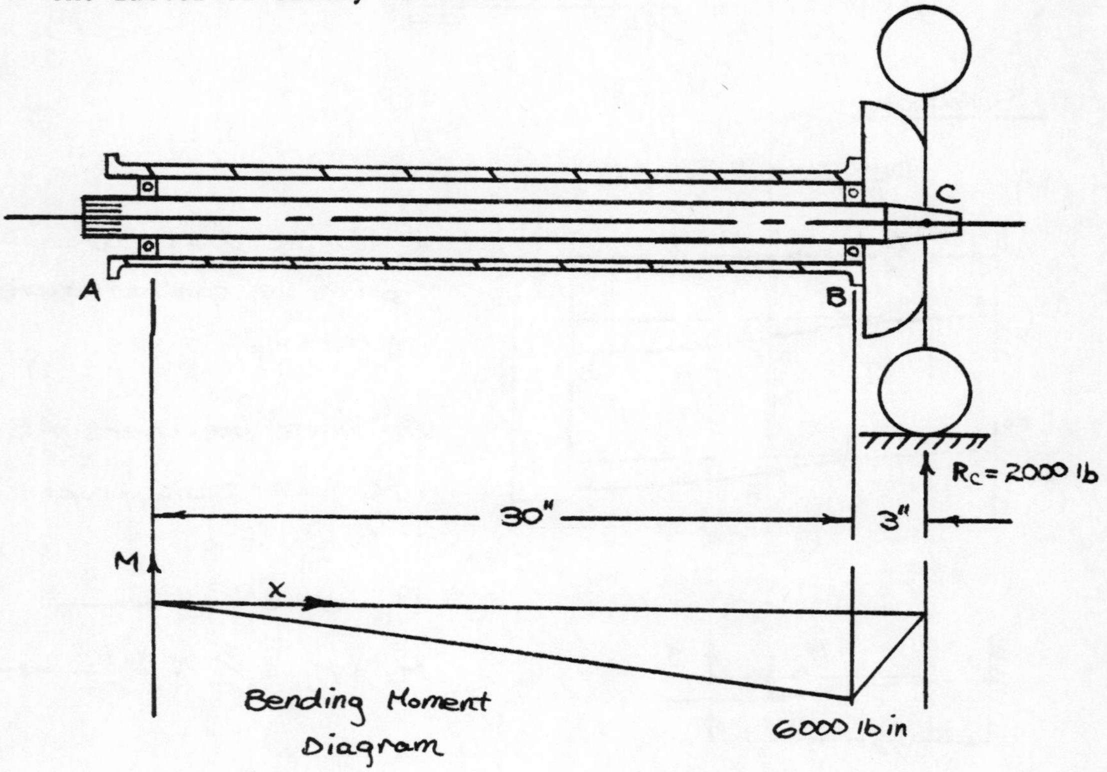

$$\sigma_x = \frac{Mc}{I} = \frac{6000(32)}{\pi d^3} \; ; \; \tau_{xy} = \frac{2880(12)(16)}{\pi d^3}$$

$$= \frac{34600(16)}{\pi d^3}$$

$$a)\ (S_s)_{max} = 50{,}000 = \sqrt{\left(\frac{\sigma_x}{2}\right)^2 + \tau_{xy}^2}$$

$$= \sqrt{\left[\frac{6000(16)}{\pi d^3}\right]^2 + \left[\frac{34600(16)}{\pi d^3}\right]^2}$$

$$50{,}000 = \frac{16}{\pi d^3}\sqrt{6000^2 + 34{,}600^2} = \frac{16}{\pi d^3}(35{,}100)$$

$$d^3 = \frac{16(35{,}100)}{50{,}000\,(\pi)} = 3.57$$

$$d = \underline{\underline{1.53\ in}} \longleftarrow$$

$$(S_t)_{max} = 100{,}000 = \frac{\sigma_x}{2} + \sqrt{\left(\frac{\sigma_x}{2}\right)^2 + \tau_{xy}^2}$$

$$= \frac{6000(16)}{\pi d^3} + \frac{16}{\pi d^3}\sqrt{6000^2 + 34{,}600^2}$$

$$= \frac{16}{\pi d^3}\left[6000 + 35{,}100\right] = \frac{16}{\pi d^3}(41{,}100)$$

$$d^3 = \frac{16(41{,}100)}{\pi(100{,}000)} = 2.09\ in^3$$

$$d = \underline{\underline{1.28\ in}} \longleftarrow$$

Required Diameter = $\underline{\underline{1\,^9\!/_{16}''}} \longleftarrow$

---

b) For a factor of Safety of $\underline{\underline{2.5}}$ in tension

$$d^3 = \frac{16(41{,}000)}{\pi\left(\frac{100{,}000}{2.5}\right)} = 5.21 \qquad d = \underline{\underline{1.74\ in}}$$

$$(S_s)_{max} = \frac{16(35{,}100)}{\pi(1.74)^3} = 33{,}900\ lb/in^2$$

$$\text{Safety Factor in Shear} = \frac{50{,}000}{33{,}900} = \underline{\underline{1.47}}$$

FORCE AND STRESS ANALYSIS 11

During a static loading test, a 45 degree strain rosette attached to the test object indicated the following strain outputs: $\varepsilon_x = 600$, $\varepsilon_m = -83$, and $\varepsilon_y = 100$ microinches. The material's modulus of elasticity is 30 x 106 psi, while Poisson's ratio is 0.3.

From this data draw Mohr's circle of stress for this point indicating the maximum and minimum stress values and the stresses along the x and y directions. Find also the angle between the x axis and the axis of maximum principal stress.

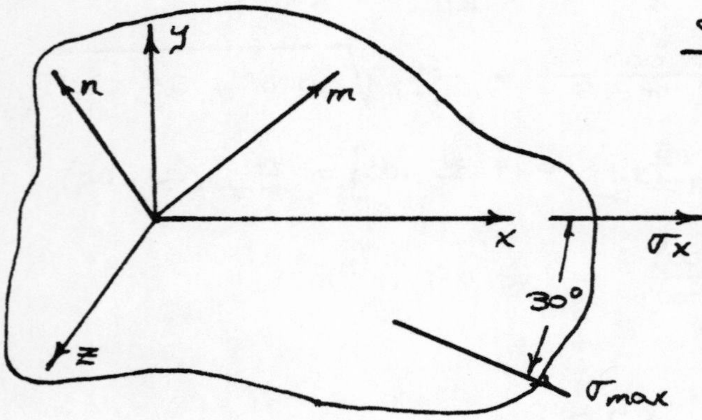

Solution

From Mohr's Circle Strain Theory

Location of circle center

$$OA = \frac{\varepsilon_x + \varepsilon_y}{2}$$

$$R = \text{radius} = \sqrt{AH^2 + AB^2}$$

$$BC = AH = \frac{\gamma_{xy}}{2}$$

$$R = \sqrt{\left(\frac{\varepsilon_x - \varepsilon_y}{2}\right)^2 + \left(\frac{\gamma_{xy}}{2}\right)^2}$$

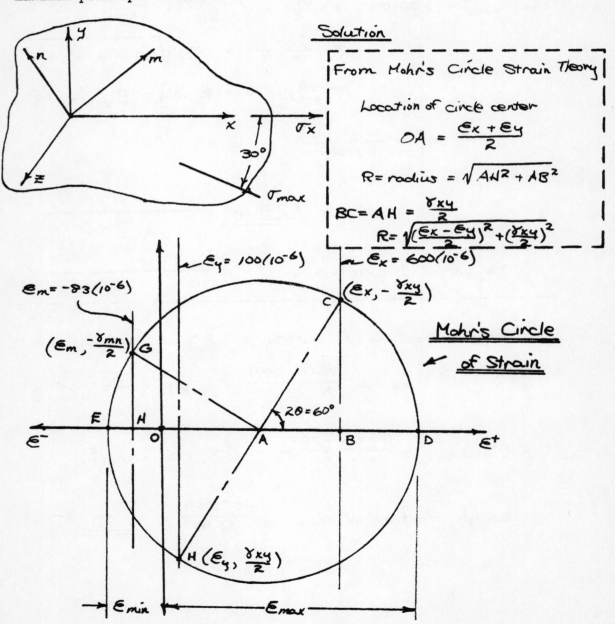

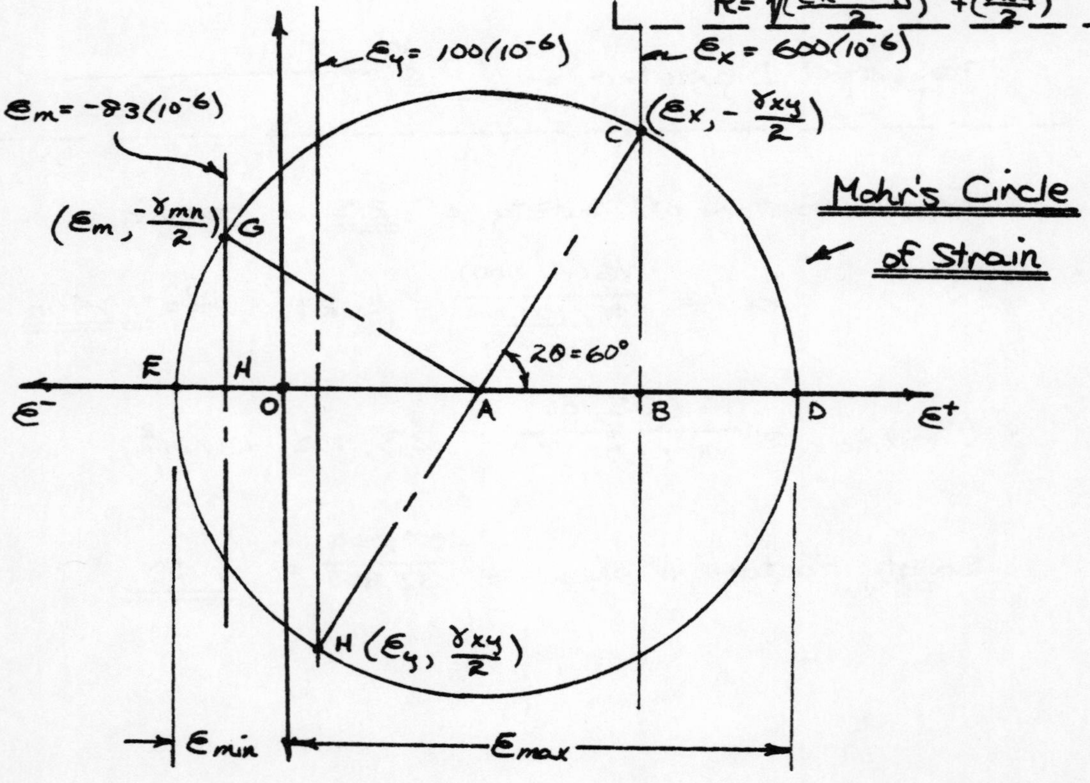

Radius of Mohrs Circle of Strain

$$= AC = (AB^2 + BC^2)^{1/2} = \left[(250\times10^{-6})^2 + (433\times10^{-6})^2\right]^{1/2}$$

$$= 500\times10^{-6} \text{ in/in}$$

$$\epsilon_{max} = OD = 850\times10^{-6} \text{ in/in}$$

$$\epsilon_{min.} = OE = -150\times10^{-6} \text{ in/in}$$

$$\sigma_{max} = \frac{E}{1-u^2}\left[\epsilon_{max} + u\,\epsilon_{min}\right]$$

$$= \frac{30\times10^6}{1-0.3^2}\left[850\times10^{-6} - 0.3(150)(10^{-6})\right]$$

$$= 33\left[850-45\right] = \underline{\underline{26{,}500 \text{ lb/in}^2}} \longleftarrow$$

$$\sigma_{min} = \frac{E}{1-u^2}\left[\epsilon_{min} + u\,\epsilon_{max}\right]$$

$$= 33\left[-150+255\right] = \underline{\underline{3{,}460 \text{ lb/in}^2}} \longleftarrow$$

$$\sigma_x = \frac{E}{1-u^2}\left[\epsilon_x + u\,\epsilon_y\right] = 33\left[600 + 0.3(100)\right] = \underline{\underline{20{,}800 \text{ lb/in}^2}}$$

$$\sigma_y = \frac{E}{1-u^2}\left[\epsilon_y + u\,\epsilon_x\right] = 33\left[100 + 0.3(600)\right] = \underline{\underline{9{,}250 \text{ lb/in}^2}}$$

* Note - above stress are also found from Mohr's circle of stress on the following page.

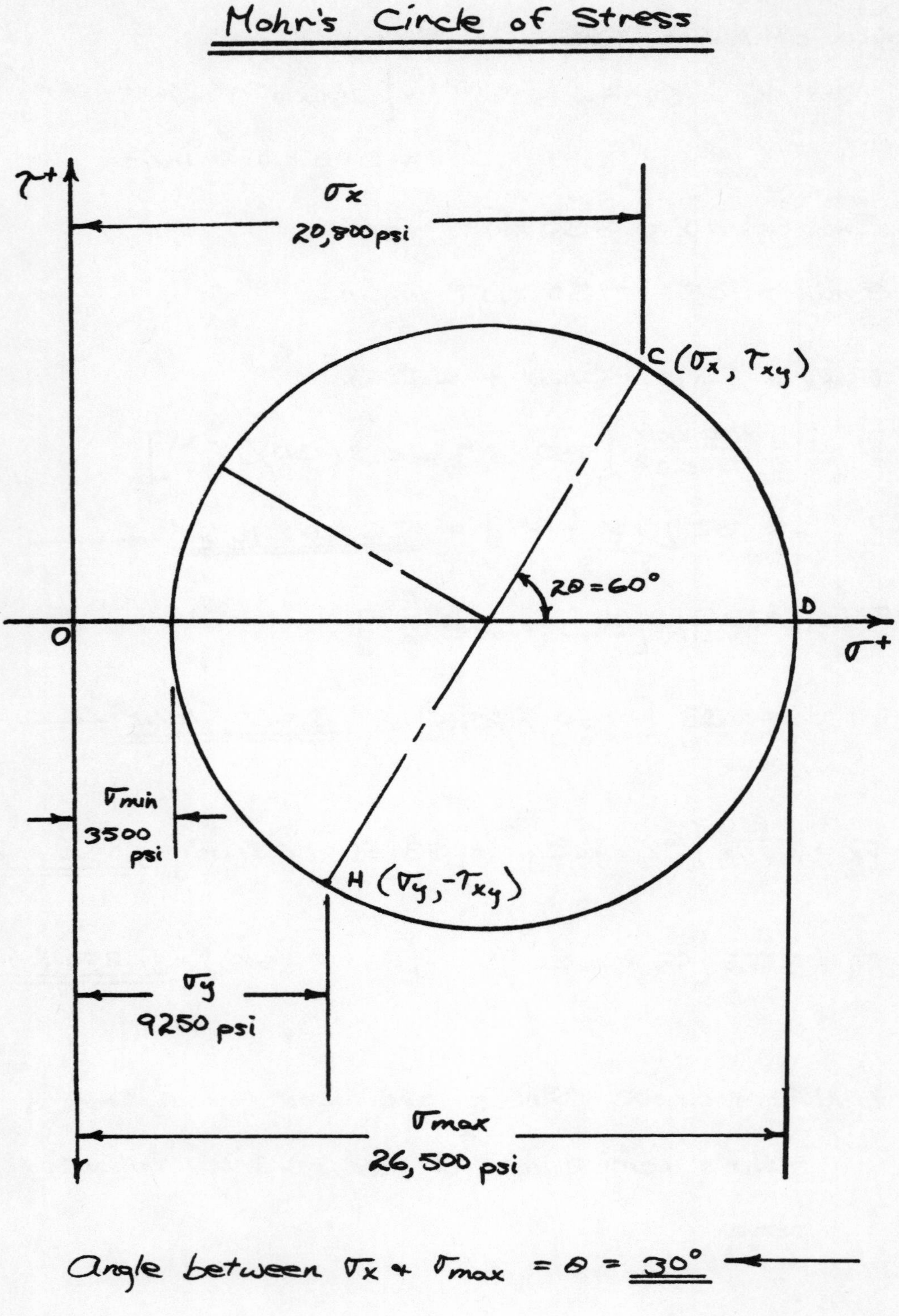
Mohr's Circle of Stress
τ+
σx
20,800 psi
C (σx, τxy)
2θ=60°
O
D
σ+
σmin
3500
psi
H (σy, -τxy)
σy
9250 psi
σmax
26,500 psi
angle between σx + σmax = θ = 30°

FORCE AND STRESS ANALYSIS 12

A uniformly-loaded cantilever beam is supported at the free end, but the support is not at the same level as the fixed end. It was jacked into place when the beam was fully loaded, thus reducing the deflection of the free end by 50%.

What are the reactions at each end? What is the bending moment at the fixed end?

Solution

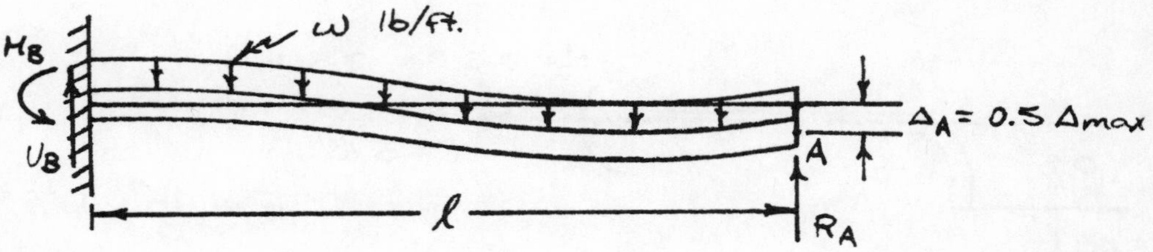

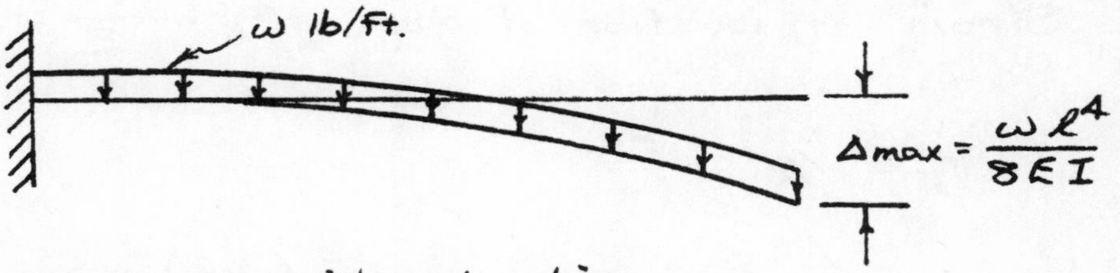

For cantilever with w loading

$$\Delta_{max} = \frac{wl^4}{8EI} = \Delta_A$$

For cantilever with R_A loading $\quad \Delta_A = -\frac{R_A l^3}{3EI}$

$$\frac{wl^4}{8EI} - \frac{R_A l^3}{3EI} = \frac{0.5\,wl^4}{8EI} = \frac{wl^4}{16EI}$$

$$R_A = \frac{3wl}{16} \longleftarrow$$

$$M_B = -\frac{wl^2}{2} + R_A(l) = -\frac{wl^2}{2} + \frac{3}{16}wl(l)$$

$$= -\frac{5}{16}wl^2 \longleftarrow$$

FORCE AND STRESS ANALYSIS 13

A load cell in the form of a bent aluminum strip is shown in the diagram. The strip is one-quarter inch wide and one-tenth inch thick. Determine the strain gage reading for a 5 lb load.

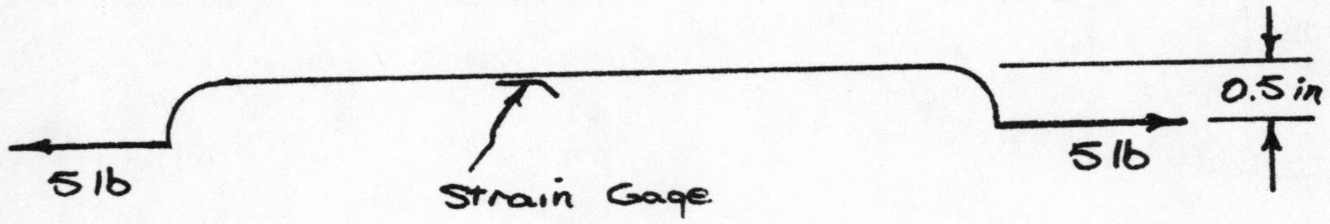

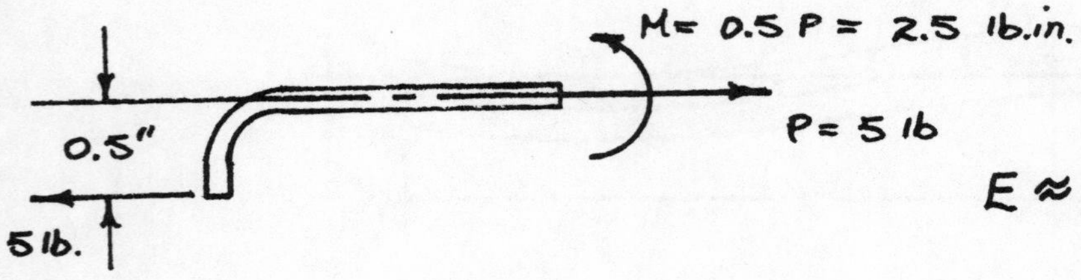

$E \approx 10^7 \text{ lb/in}^2$

Strain at location of gage $\epsilon_K = \dfrac{\sigma_K}{E}$

$$\sigma_K = P/A + \frac{Mc}{I}$$

$$= \frac{5}{(1/4)(1/10)} + \frac{2.5\left(\frac{1}{20}\right)}{\frac{1/4\left(\frac{1}{10}\right)^3}{12}}$$

$$= 200 + 6000 = \underline{\underline{6200 \text{ lb/in}^2}}$$

$$\epsilon_K = \frac{6200}{10 \times 10^6} = \underline{\underline{620 \times 10^{-6} \text{ in/in}}} \longleftarrow$$

FORCE AND STRESS ANALYSIS 14

A square steel bar of 1 x 1 inch cross section and 6 feet long is to be used as a column. The ends of the bar are rounded so as to be free to rotate, but are placed in sockets and may not be displaced. If an average unit stress of 30,000 psi is allowable, find the maximum load that this column may support.

Solution

For a long column with rounded ends

Elastic Buckling load $P_{crit.} = \frac{\pi^2 E I}{\ell^2}$; r = gyration radius

For a long column the minimum ℓ/r for above formula to be valid is obtained as follows:

$$\frac{P_{crit.}}{A} = \frac{\pi^2 E \overbrace{(Ar^2)}^{I}}{A \ell^2} = \frac{\pi^2 E}{(\ell/r)^2} = \sigma_{crit} \approx \sigma_{Prop.\,Limit}$$

For elastic Buckling with a $\sigma_{prop.\,limit} \approx 45$ Ksi,

$$\left(\frac{\ell}{r}\right)^2_{min} = \frac{\pi^2 E}{45{,}000} = \frac{\pi^2 (30)(10^6)}{45{,}000} = 6570$$

$$\left(\frac{\ell}{r}\right)_{min} = \underline{\underline{81}} \longleftarrow \quad b = h = 1 \text{ in.}$$

For a column with a 1" x 1" cross section + ℓ = 72"

$$r = \left(\frac{I}{A}\right)^{1/2} = \left[\frac{bh^3}{12(bh)}\right]^{1/2} = \frac{h}{\sqrt{12}} = \frac{1}{\sqrt{12}} = 0.28 \text{ in.}$$

$$\frac{\ell}{r} = \frac{72}{0.28} = 257 \longleftarrow$$

The above column is definitely a long column so

the maximum load that the column can carry is not governed by strength but by stiffness

$$P_{max} = P_{critical} = \frac{\pi^2 EI}{\ell^2} = \frac{\pi^2 (3)(10^7)(1^4)}{12\,(72)^2}$$

$$P_{max} = \underline{\underline{4760 \text{ lb}}} \longleftarrow$$

assuming a safety factor of 3,

$$P_{allow} = \frac{4760}{3} \approx \underline{\underline{1600 \text{ lb}}} \longleftarrow$$

3

Dynamics and Vibrations

HAROLD M. TAPAY

DYNAMICS

Dynamics is that part of engineering mechanics in which a study is made of particles and bodies in motion; it is divided into branches--namely, kinematics and kinetics. Kinematics consists essentially of deriving equations relating to displacement, velocity, acceleration and time. Kinetics consists essentially of deriving equations relating to the unbalanced force or the unbalanced moment to the change in motion.

Kinematics:

Types of particle motion -

(1) Rectilinear motion

$$V = \text{absolute velocity} = \frac{ds}{dt} \qquad (1)$$

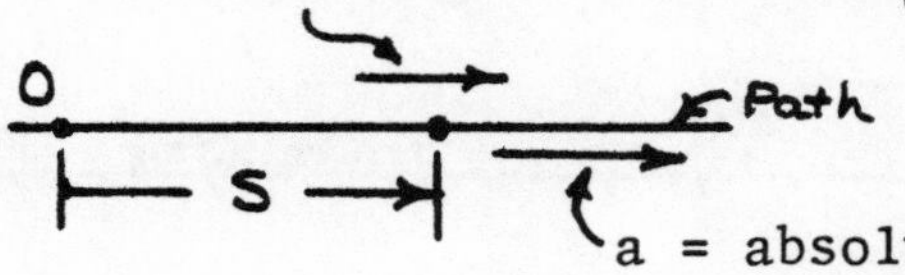

$$a = \text{absolute accel.} = \frac{dV}{dt} \qquad (2)$$

S = displacement $\quad VdV = ads \qquad (3)$

(2) Curvilinear motion

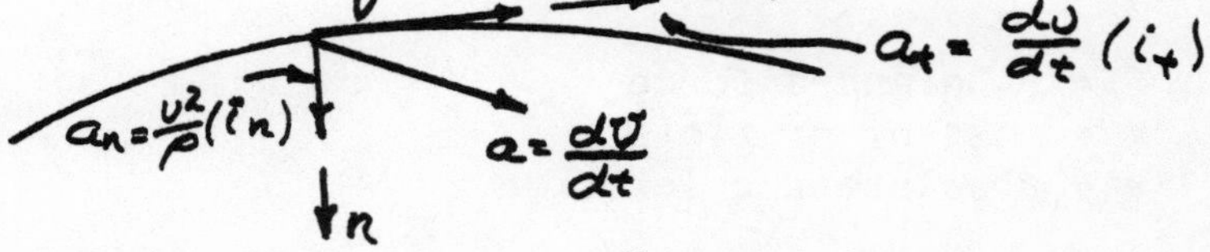

Tangential component of "a" = $a_t = \frac{dv}{dt}\ (i_t)$

Normal component of "a" = $a_n = \frac{v^2}{P}\ (i_n)$ $\qquad$ v = speed

Types of Rigid Body Plane Motion:

(1) Rectilinear Translation - all particles move on parallel straight lines and have same displacements, velocities and accelerations.

(2) Curvilinear Translation - same as above except that all particles move on congruent curves.

(3) Fixed Axis Rotation - all particles move on concentric circular paths

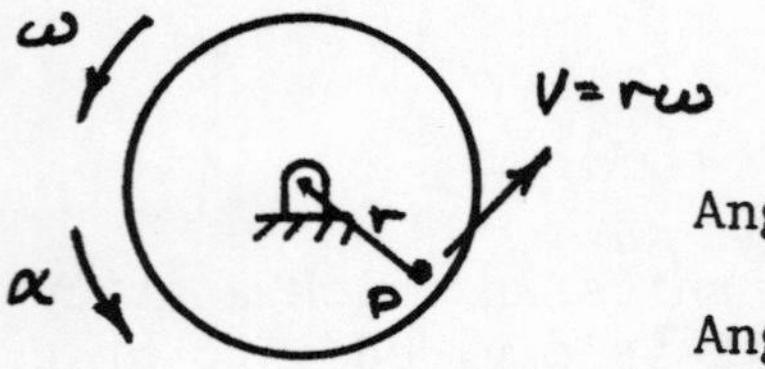

Angular Velocity $\omega = \frac{d\theta}{dt}$ —— (1)

Angular Acceleration $\propto = \frac{d\omega}{dt}$ —— (2)

$\omega d\omega = \propto d\theta$ —— (3)

For Particle P -

Tangential acceleration of particle $a_t = \frac{r d\omega}{dt} = r\propto$

Normal acceleration of particle $a_n = \frac{v^2}{r} = r\omega^2$

(4) General Plane Motion - a motion which can be resolved into translation plus fixed axis rotation

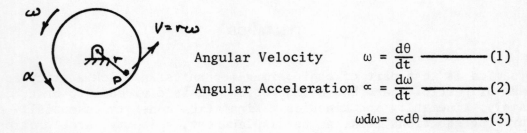

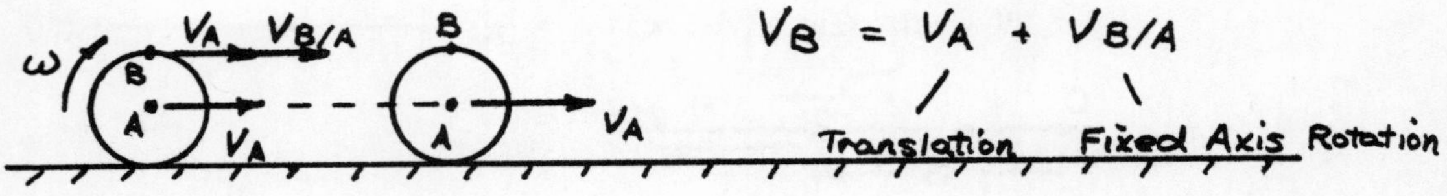

Kinetics:

Newton's 2nd Law of Motion for a particle -

$$F = ma \quad \text{—— (1)}$$

F = unbalanced force
m = mass of particle
a = absolute acceleration

For English Gravitation System of Units -

F - lb; m - slugs = $\frac{W}{g}$; a - ft/sec^2

For a system of particles (rigid or non-rigid body) -

$$F = (\Sigma m)(\bar{a}) = M\bar{a} \qquad (2)$$

F = unbalanced force
M = mass of system of particles
a = absolute acceleration of center of mass

For rigid-body rotation about fixed axis

$$(M_o)_z = \int (r^2 dm)\alpha \qquad (3)$$

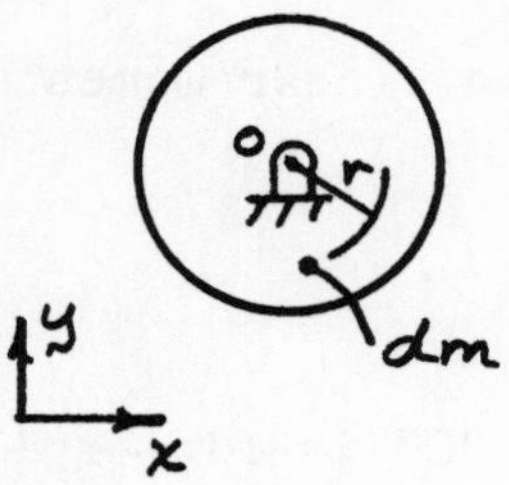

where $\int r^2 dm$ = mass moment of inertia about axis of rotation = $(I_o)_z$

$$(M_o)_z = (I_o)_z \alpha$$

where $\Sigma(M_o)_z$ is unbalanced moment about axis of rotation

Work Energy Principle

By taking Newton's 2nd Law of Motion and using components of "F" and "a" in the tangential direction and multiplying each side of equation by ds and integrating, the work energy principle is obtained:

$$F_t = m\frac{dv}{dt}; \quad F_t ds = m\, dv\, \frac{ds}{dt}; \quad F_t ds = mv\, dv = dU$$

$$\Delta U_{1\to 2} = \int_1^2 dU = m\int_1^2 v\, dv = \frac{1}{2}\, mv_2^2 - \frac{1}{2}\, mv_1^2 = \Delta K.E. \qquad (4)$$

where $\frac{1}{2} mv^2$ is kinetic energy of a particle.

$\Delta U_{1\to 2}$ = work done by F_t which is tangential component of unbalanced force acting on the particle

Work Done by a Force on a Spring

$\Delta U_{1\to 2}$ = work done by a force P slowly applied to a linear elastic spring = 1/2 Px = 1/2 kx^2

k = spring constant = P/x
l = free length of spring

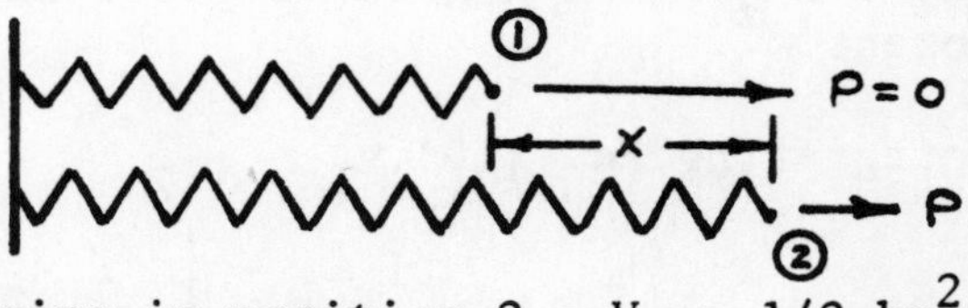

Potential energy of strain for spring in position 2 = V_e = 1/2 kx^2

$$\Delta V_e = (V_e)_2 - (V_e)_1 = 1/2\ kx^2$$

Potential energy of position in gravitational field = V_g

Second form of work energy equation

$$\Delta U_{1\to 2} = \Delta K.E. + \Delta V_g + \Delta V_e \text{ ——— } (5)$$

If no work is done then K.E. + V_g + V_e = constant, i.e., mechanical energy is conserved.

Linear Impulse - Linear Momentum Principle

By rearranging Newton's 2nd Law of Motion the linear impulse-linear momentum principle is obtained; G = linear momentum -

$$F = m\frac{dV}{dt} \quad Fdt = d(mv) = dG \text{ ——— } (6)$$

where Fdt is linear impulse and d(mv) is change in linear momentum.

Integrating Eq. (6) for a particle,

$$mV_1 + \int Fdt = mV_2$$

If no unbalanced force acts then linear momentum "G" is conserved, i.e.,

$$G = \text{constant} \text{ ——— } (7)$$

Angular Impulse - Angular Momentum Principle

(Angular Momentum of a Particle)$_z$ = (Moment of Linear Momentum)$_z$

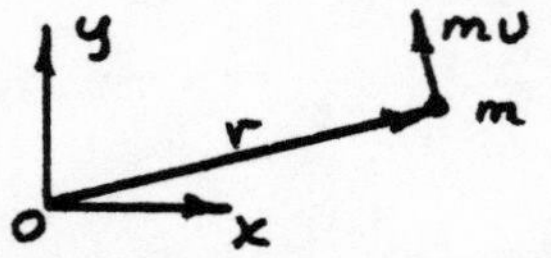

$$H_z = (mV)(r) = \text{Angular Momentum of Particle}$$

For a rotating rigid body

$$H_z = \int (dm)(r\omega)\, r = \omega \int r^2 dm = I_z\omega \text{ ——— } (8)$$

$$\Sigma M_z = I_z\alpha = I_z\frac{d\omega}{dt} = \frac{d}{dt}(I_z\omega) = \frac{dH_z}{dt} \text{ ——— } (9)$$

$$\Sigma M_z\, dt = d(H_z) = d(I_z\omega) \text{ ——— } (10)$$

where ΣM_z dt is angular impulse and $d(I_z\omega)$ is change in angular momentum.

Integrating Eq. (10),

$$I_z\omega_1 + \int_1^2 \Sigma M_z\, dt = I_z\omega_2 \text{ ——— } (11)$$

From Eq. (10) if no unbalanced moment acts then angular momentum H_z is conserved, i.e.,

$$H_z = \text{constant} \text{ ——— } (12)$$

Gyroscopic Motion

When the axis of rotation of a rotating body itself rotates, i.e., precesses, a gyroscopic moment is set up; this moment is obtained from angular momentum-angular impulse principle as follows:

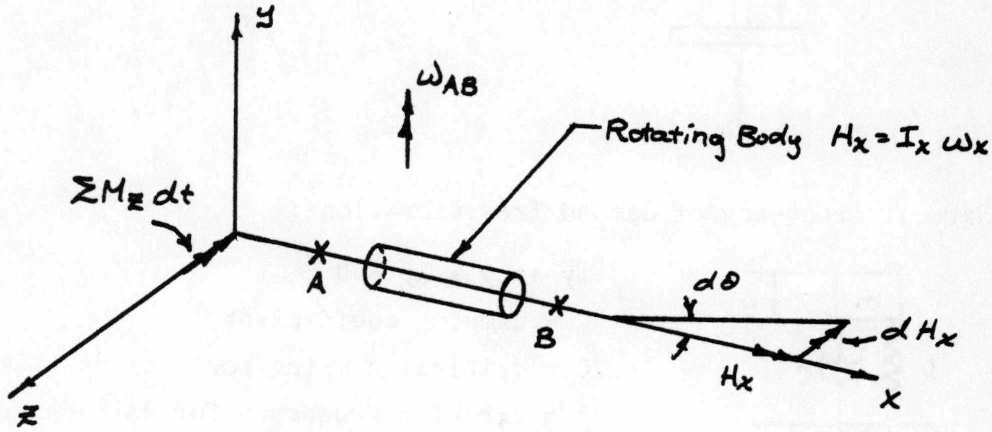

$$\text{Gyroscopic Moment} = \Sigma M_z = \frac{d(H_x)}{dt} = \frac{(H_x d\theta)}{dt} = I_x \left(\frac{d\theta}{dt}\right) \omega_x \text{——(13)}$$

$$\frac{d\theta}{dt} = \omega_{AB} = \text{precession velocity}$$

VIBRATIONS

Natural frequency "f" of free vibrations for systems shown below are obtained by solving equations of motion.

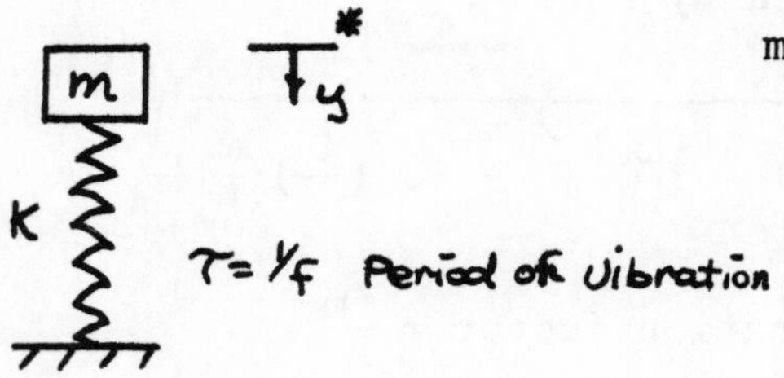

$$m\ddot{y} + ky = 0$$

$$\ddot{y} + \frac{k}{m} y = 0$$

$$p^2 = \frac{k}{m}$$

p = circular frequency

$$f = \frac{p}{2\pi} = \frac{1}{2\pi}\sqrt{\frac{k}{m}}$$

*Static Equilibrium Position

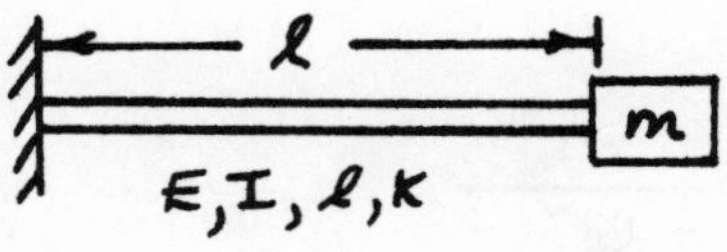

$$k = \frac{3EI}{\ell^3}$$

$$m\ddot{y} + ky = 0$$

$$\ddot{y} + \frac{k}{m} y = 0$$

$$f = \frac{p}{2\pi} = \frac{1}{2\pi}\sqrt{\frac{k}{m}}$$

$$\tau = \frac{1}{f}$$

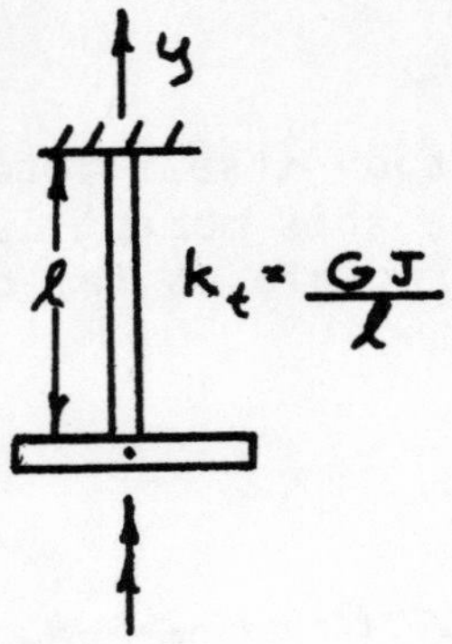

$$\Sigma M_y = I_y \frac{d^2\theta}{dt^2}$$

$$\frac{d^2\theta}{dt^2} + \frac{k_t\theta}{I_y} = 0$$

$$f = \frac{1}{2\pi}\sqrt{\frac{k_t}{I_y}}$$

$$\tau = \frac{1}{f}$$

Natural frequency of damped free vibrations:

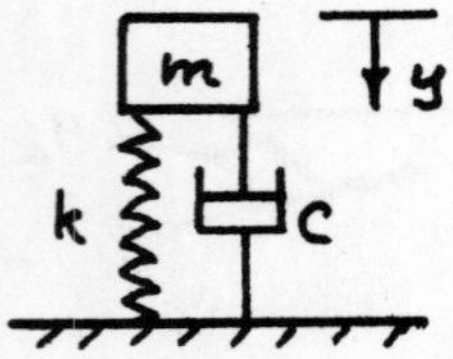

$m\ddot{y} + c\dot{y} + ky = 0$

C = damping coefficient

C_c= critical damping coefficient = 2mp

p = circular frequency for no damping

$f = \frac{\omega}{2\pi}$ where ω is circular frequency of damped free vibrations

$$\omega = p\sqrt{1 - \left(\frac{C}{C_c}\right)^2} \qquad f = \left(\frac{p}{2\pi}\right)\sqrt{1 - \left(\frac{C}{C_c}\right)^2}$$

Damped forced vibrations:

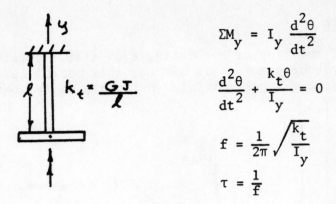

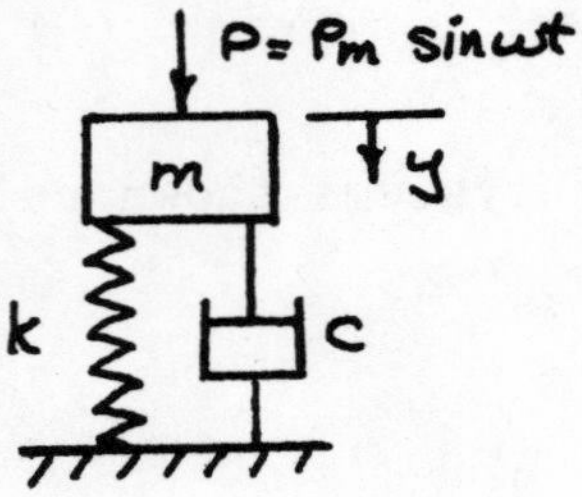

Equation of motion -

$m y + c\ddot{y} + k\dot{y} = P_m \sin \omega t$

Amplitude of steady-state vibration produced by P is $\frac{P_m}{k}$

$$y_m = \frac{\frac{P_m}{k}}{\sqrt{\left[1 - \left(\frac{\omega}{p}\right)^2\right]^2 + \left[2\left(\frac{C}{C_c}\right)\left(\frac{\omega}{p}\right)\right]^2}}$$

$$\text{Magnification factor} = \frac{y_m}{\frac{P_m}{k}}$$

For support vibrating - $\delta = \delta_m \sin \omega t$

$$y_m = \frac{\delta_m}{\sqrt{\left[1 - \left(\frac{\omega}{p}\right)^2\right]^2 + \left[2\left(\frac{C}{C_c}\right)\left(\frac{\omega}{p}\right)\right]^2}}$$

$$\text{Magnification factor} = \frac{y_m}{\delta_m}$$

DYNAMICS AND VIBRATIONS 1

In a recent edition of Mechanical Engineering, published by ASME, an inventor described a patented skid he had developed for towing heavy objects. The device is shown below in an idealized representation. A mass M oscillates relative to a wedge shaped form M_o. The wedge is of angle α, the spring constant is K and has unstretched length X_o. There is no friction acting between mass M and the wedge, but the coefficient of friction between the wedge and ground is μ. When the mass M is close to its minimum displacement relative to the wedge ($X_{r\,min}$), it creates a normal force and friction large enough so the skid does not move backwards. When the mass M is close to its maximum displacement relative to the wedge ($X_{r\,max}$), the inertia force overcomes friction; and the wedge and mass go forward a distance X. In this way the skid can tow an object intermittently forward if it is attached to the wedge.

REQUIRED: Analyze the following two unloaded situations:

(a) When X_r is minimum find the minimum coefficient of friction needed to prevent the skid from going backward. $\ddot{X}$ is + 100 ft/sec^2 for that position.

(b) If μ = 0.5, find the $\ddot{X}$ acceleration of the skid forward when X_r is maximum. $\ddot{X}_r$ is - 100 ft/sec^2 for that position.

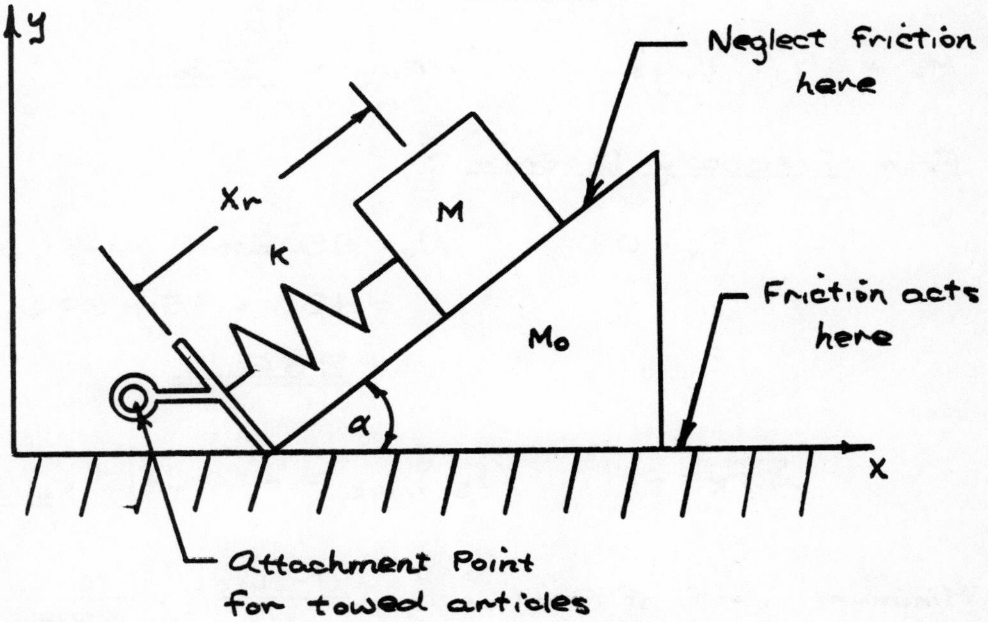

M = 4.5 Slugs
M_o = 9.5 slugs
K = 90 lbs/in
X_o = 12 in
α_o = 75°

Solution

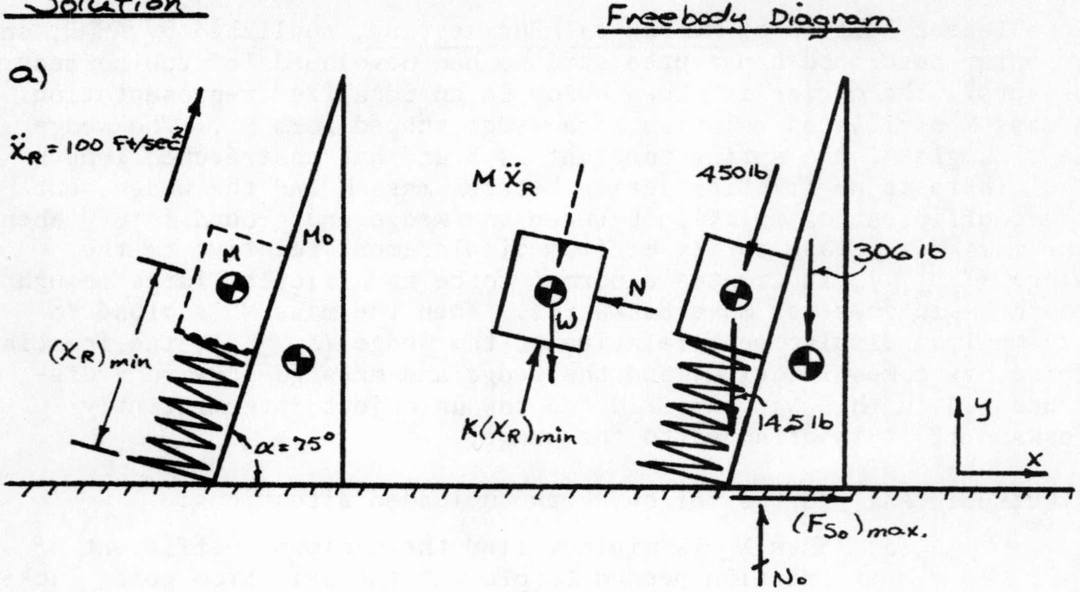

$M = 4.5$ slugs : $W = mg = 4.5(32.2) = \underline{\underline{145 \text{ lb}}}$

$M\ddot{x}_R = 4.5(100) = \underline{\underline{450 \text{ lb}}}$

$M_o = 9.5$ slugs : $W_o = M_o g = \underline{\underline{306 \text{ lb}}}$

From Freebody Diagram :

$\Sigma F_y = 0 \qquad N_o = 450 \cos 15° + 145 + 306$

$= 435 + 145 + 306$

$= \underline{\underline{886 \text{ lb}}}$

$\Sigma F_x = 0 \qquad (F_{So})_{max.} = 450 \sin 15° = \underline{\underline{116.5 \text{ lb}}}$

Minimum coeff. of friction $\mu = \dfrac{(F_{So})_{max}}{N_o} = \dfrac{116.5}{886}$

$\approx \underline{\underline{0.13}}$ ⟵

b) $\ddot{x}_R = -100 \text{ ft/sec}^2$: $M\ddot{x}_R = 4.5(100) = \underline{\underline{450 \text{ lb}}}$

Freebody Diagram of assembly

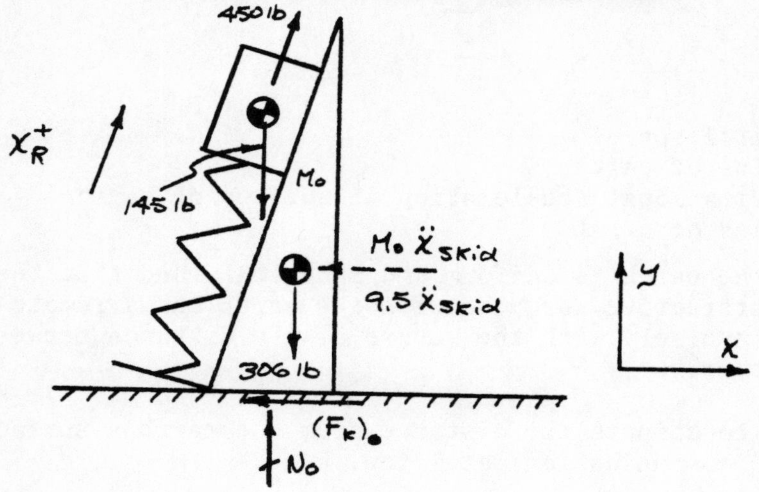

$\Sigma F_y = 0$

$N_o = 306 + 145 - 450 \cos 15° = 451 - 450(0.966)$

$= \underline{\underline{16 \text{ lb}}}$

$(F_k)_o = \mu_k N_o = 0.5(16) = \underline{\underline{8 \text{ lb}}}$

$\Sigma F_x = 0$

$+450 \cos 75° = 9.5 \ddot{x}_{skid} - 8 = 0$

$\ddot{x}_{skid} = \frac{116.5 - 8}{9.5} = \underline{\underline{11.4 \text{ ft/sec}^2}} \longleftarrow$

Acceleration of skid is 11.4 ft/sec² to the right.

DYNAMICS AND VIBRATIONS 2

(a) Show, by equating "centrifugal" and "gravitational" forces, that speed of a satellite in a circular orbit about the earth is given by:

$$V = R_e\sqrt{\frac{g_e}{R}}$$

where V = orbital speed
R_e = radius of earth
g_e = gravitational acceleration at surface of earth
R = radius of orbit

(Note: Assume the earth is uniform and spherical, and that the gravitational attractive force between the earth and a remote object varies inversely with the square of the distance between their mass centers.)

(b) Use the formula to compute the distance from the earth's surface to a synchronous communication satellite.

Solution

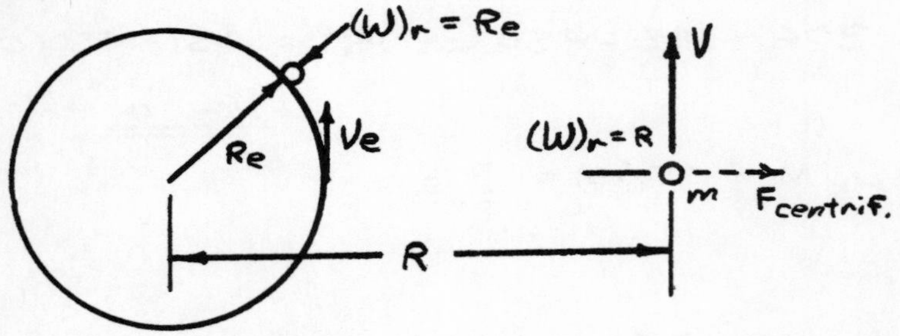

a) $$F_{centrif.} = \frac{mv^2}{R} = (W)_{r=R} = \frac{m\, m_{earth}\, G}{R^2}$$

$$v^2 = \frac{G m_e}{R}$$

$$(W)_{r=R_e} = m g_e = \frac{m\, m_e\, G}{R_e^2}$$

$$G m_e = g_e R_e^2$$

$$v^2 = \frac{g_e R_e^2}{R}$$

$$v = R_e\sqrt{\frac{g_e}{R}}$$

b) For a synchronus satellite

$$\frac{U_e}{U} = \frac{R_e}{R} \qquad U = \frac{R}{R_e} U_e$$

$$U_e = R_e \omega \qquad \omega = \text{Earth's Angular Velocity}$$

$$= \frac{1(2\pi)}{24(60)(60)} = 7.28(10^{-5})\ \text{sec}^{-1}$$

$$U_e = (3960)(5280)(7.28)(10^{-5}) = \underline{\underline{1520\ \text{ft/sec}}}$$

$$U = \frac{R}{R_e} U_e = R_e \sqrt{\frac{g_e}{R}}$$

$$R^3 U_e^2 = R_e^4 g_e \qquad R^3 = \frac{R_e^4 g_e}{U_e^2}$$

$$R^3 = \frac{(3960)^4 (5280)^4 (32.17)}{(1520)^2} = 2.67 \times 10^{24}$$

$$R = 1.39 \times 10^8\ \text{ft} = \underline{\underline{26{,}400\ \text{miles}}}$$

Distance from Earth's surface to satellite:

$$26{,}400 - 3960 = \underline{\underline{22440\ \text{miles}}}$$

DYNAMICS AND VIBRATIONS 3

A straight piece of 1/8 inch diameter steel rod is clamped at one end in a vise; the other end is free.

Find the length in inches to make the fundamental bending frequency 264 Hz.

Assume steel weighs 0.284 lb/in^3.

Solution

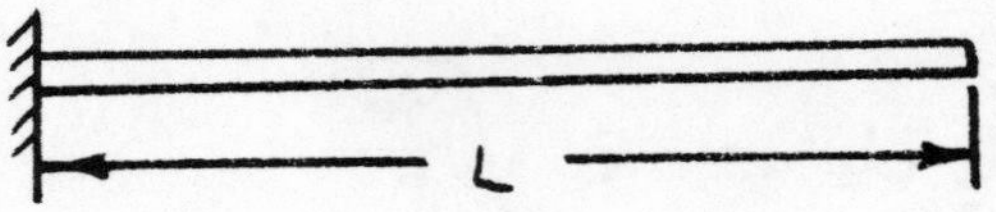

For cantilever beam:

f = fundamental bending freq. = 264 cycles/sec.

$$f \overset{*}{=} \frac{CK(10^5)}{L^2}$$

K = gyration radius $= \left(\frac{I}{A}\right)^{1/2}$

$K = R/2$

C = 1.13 for steel cantilever beam

$$f = \frac{1.13}{L^2}\left(\frac{1}{32}\right)(10^5) = 264$$

$$L^2 = \frac{1.13(10^5)}{264(32)} = 13.4$$

$$L = \underline{\underline{3.67 \text{ in.}}}$$

* This equation comes from

$$f = \frac{a}{2\pi}\left(\frac{1.875}{L}\right)^2 \quad \text{where } a^2 = \frac{EIg}{A\gamma}$$

DYNAMICS AND VIBRATIONS 4

A child on a "pogo stick" can be idealized (for engineering purposes) as a mass on a spring, as shown in the diagram. At the top of a jump the C.G. of the child is above the ground by an amount H_o.

Compute the maximum axial load in the spring (stiffness K, unloaded length L_o) at the bottom of the jump.

Given: Child's weight: 100 lbs
K = spring stiffness: 100 lbs/in
H_o: 2 ft
L_o: 4 in

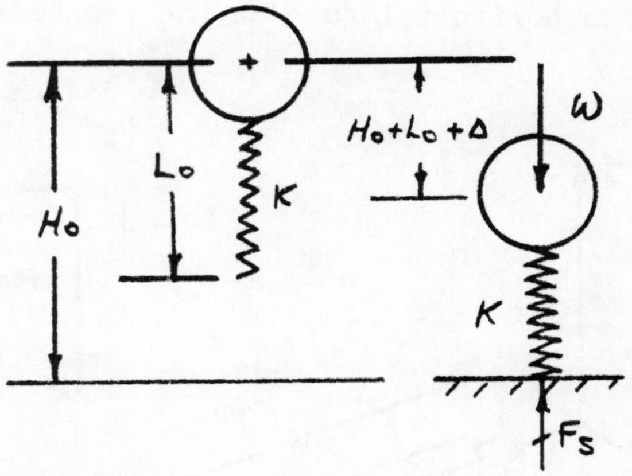

Solution

Let Δ = max. compressive deformation in spring at impact

Using Work-energy principle :

$$\Delta U = \Delta K.E. + \Delta U_g + \Delta U_e$$

$$0 = 0 - (W)(H_o - L_o + \Delta) + \tfrac{1}{2} K \Delta^2$$

$$0 = 0 - 100(24 - 4 + \Delta) + 50\Delta^2$$

$$0 = -2000 - 100\Delta + 50\Delta^2$$

$$\Delta = \frac{100 \pm \sqrt{(-100)^2 + 4(2000)(50)}}{100}$$

$$= \frac{100 \pm}{100} = \underline{\underline{7.4 \text{ in}}}$$

THIS SPRINGS FREE LENGTH Lo IS ONLY 4 IN., ∴ IT CAN'T ACCOMODATE A LINEAR ELASTIC DEFLECTION OF 7.4 IN, SPRING WILL BOTTOM OUT.

DYNAMICS AND VIBRATIONS 5

The exhaust turbine of an aircraft directly drives a small supercharger at a constant speed of 18,000 rpm. The shaft connecting the two is parallel to the aircraft's horizontal center line. The turbine weighs 20 lbs; the blower weighs 10 lbs. Their radii of gyration are 4.0 and 4.5 inches, respectively.

If the bearings supporting the shaft are spaced 10 inches apart, determine the dynamic bearing loads created when the plane makes a 100 ft radius horizontal turn on the ground at a constant speed of 15 mph.

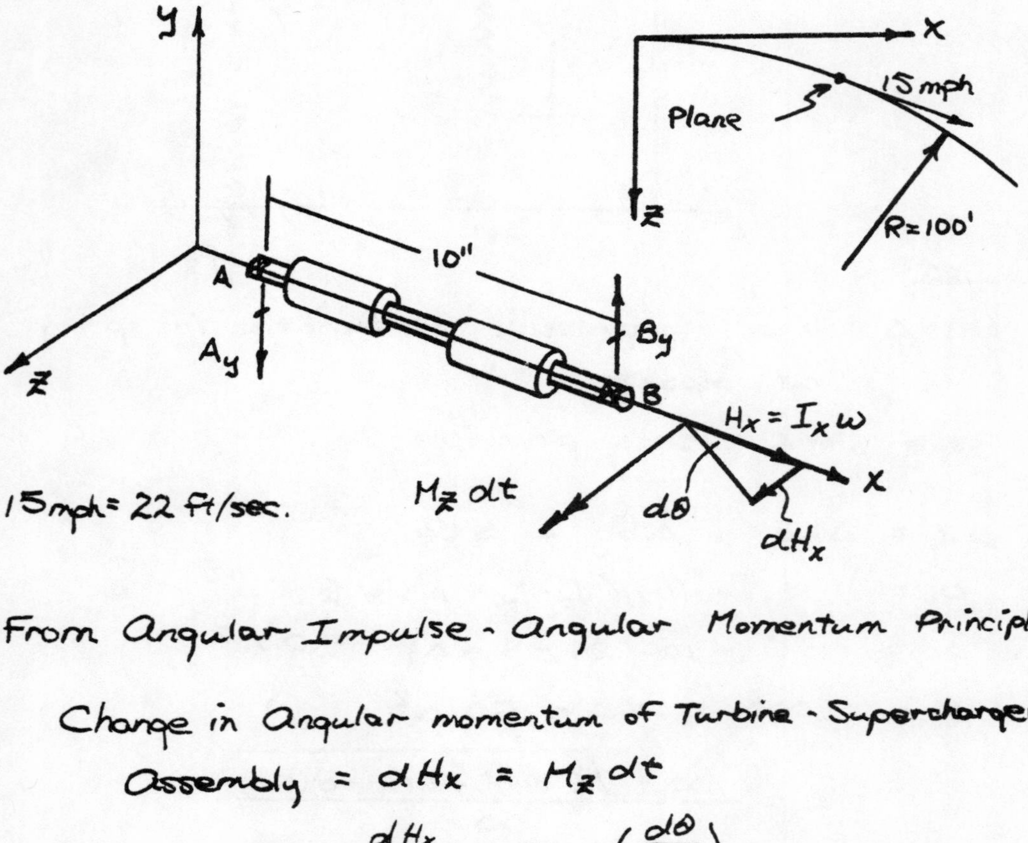

From Angular Impulse - Angular Momentum Principle

Change in Angular momentum of Turbine - Supercharger

Assembly $= dH_x = M_z\,dt$

$$M_z = \frac{dH_x}{dt} = I_x \omega \left(\frac{d\theta}{dt}\right)$$

$\frac{d\theta}{dt}$ = Precession velocity of Shaft axis

$= V/R = 22/100 = 0.22$ rad/sec.

$$I_x = \frac{20}{g}\left(\frac{4}{12}\right)^2 + \frac{10}{g}\left(\frac{4.5}{12}\right)^2$$

$$= \frac{2.22}{g} + \frac{1.41}{g} = \frac{3.63}{g} \text{ slug}\cdot\text{ft}^3$$

$$\omega = \frac{18{,}000}{60}(2\pi) = \underline{\underline{600\pi \text{ rad/sec}}}$$

Dynamic Bearing loads $A_y = B_y = \frac{M_z}{10}(12)$

$$= 1.2\, M_z$$

$$M_z = \frac{3.63}{g}(600\pi)(0.22) = 46.7 \text{ lb}\cdot\text{ft}$$

$$A_y = B_y = (1.2)(46.7) = \underline{\underline{56 \text{ lb}}} \leftarrow$$

DYNAMICS AND VIBRATIONS 6

In launching a torpedo from a submarine, the turbulence within a torpedo tube produces a torque on the torpedo's propeller. The propeller is directly attached to the armature of an electric motor, which is free to turn.

Through how many revolutions does the propeller turn if the average torque is 3,000 in-lb and it takes 500 millisec for the torpedo to leave the tube?

Assume the armature can be idealized as a solid copper cylinder 6 inches long and 6 inches in diameter. Neglect the moment of inertia of the propeller.

Solution

$$\bar{I}_{arm} = \tfrac{1}{2} m R^2 \qquad \gamma_{copper} = 556 \text{ lb/ft}^3$$

$$\bar{I}_{arm} = \tfrac{1}{2} \frac{(556)}{g} \frac{\pi}{4}\left(\tfrac{1}{2}\right)^2\left(\tfrac{1}{2}\right) = \underline{\underline{0.85 \text{ slug ft}^2}}$$

Angular Acceleration of Armature $= \alpha = \dfrac{\Sigma M}{I}$

$$= \frac{3000}{12(0.85)} = \underline{\underline{294 \text{ rad/sec}^2}}$$

Angular Movement of Propeller $= \theta$

$$\theta = \tfrac{1}{2}\alpha t^2 = \tfrac{1}{2}(294)\left(\tfrac{1}{2}\right)^2 = 36.8 \text{ rad.}$$

$$= \underline{\underline{5.87 \text{ rev.}}}$$

DYNAMICS AND VIBRATIONS 7

The torsion pendulum shown in the figure, a thin reçtangular aluminum plate suspended by massless vertical wire and rigid support rods, is used to measure moment of inertia of any given object. The plate has the dimensions 18" x 1" x 24"; the pendulum's period of free oscillation is 1.1 second. When an object was placed on the plate with its center of mass coincident with the plate's center, the observed period of oscillation changed to 1.2 seconds.

What is the moment of inertia of this object?

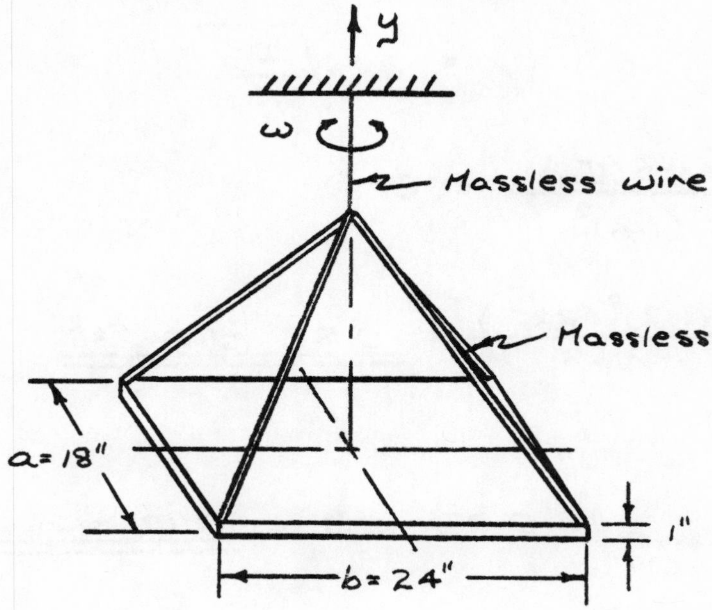

Solution

$$\text{Period} = \tau = 2\pi\sqrt{\frac{I_y}{K}}$$

where K = torsional spring const. of wire

$$I_y = (I_y)_{Pl} = \frac{1}{12} m_{Pl}(a^2+b^2) \qquad \gamma_{Al.} = .1 \text{ lb/in}^3$$

$$m_{Pl} = \frac{18(24)(1)(0.1)}{g} = 1.34 \text{ slugs}$$

$$\tau^2 = \frac{4\pi^2 (I_y)_{Pl}}{K}$$

$$(I_y)_{pl} = \frac{1.34}{12}(1.5^2 + 2^2) = \underline{\underline{0.697 \text{ slug ft}^2}}$$

$$K = \frac{4\pi^2(0.697)}{(1.1)^2} = 22.8 \text{ lb·ft./rad.}$$

$$\text{New Period} = \tau_{new} = 2\pi\sqrt{\frac{(I_y)_{pl} + (I_y)_{object}}{K}}$$

$$= 2\pi\sqrt{\frac{I_y}{K}}$$

$$K = \frac{4\pi^2(I_y)}{(1.2)^2} = 22.8$$

$$I_y = \frac{(1.2)^2(22.8)}{4\pi^2} = \underline{\underline{0.83 \text{ slug ft}^2}}$$

$$(I_y)_{object} = 0.83 - 0.7 = \underline{\underline{0.13 \text{ slug ft}^2}} \longleftarrow$$

DYNAMICS AND VIBRATIONS 8

The figure represents resilient packing of a rigid instrument weighing 15 lbs. The box is dropped 5 ft from rest and strikes a hard surface. Assume that, at the instant of impact, there is no relative velocity between the instrument W and the box, and the force in the packing is zero.

(a) Find the spring constant k, in pounds per inch, required to limit the motion of instrument W to 2 inches relative to the box.

(b) Find the maximum deceleration of the instrument in g's.

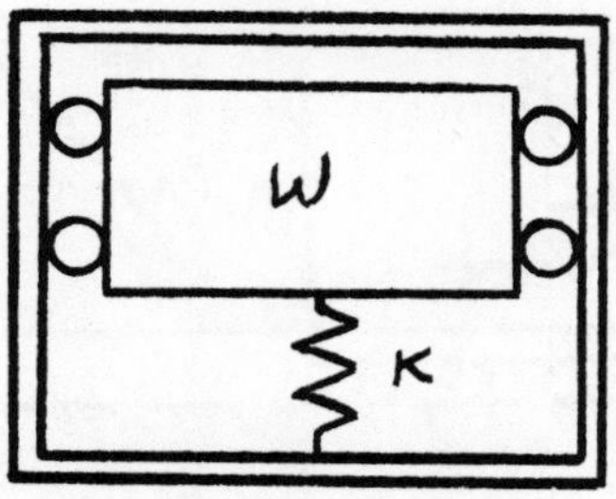

a) Assuming elastic behavior and no friction between Instrument and box.

$$\Delta U = \Delta K.E. + \Delta V_g + \Delta V_e$$

Between position shown and position where spring experiences max. deformation.

$$0 = 0 - 15\left[60 + y\right] + \tfrac{1}{2} K y^2$$

y = max. deformation in spring = 2 in.

$$0 = 0 - 15(62) + \tfrac{1}{2} K(2)^2$$

$$K = 930/2 = \underline{\underline{465\ lb/in}}$$

b) F_s = spring force $= Ky = 465(2) = 930$ lb.

$$\Sigma F_y = ma_y = 930 - 15 = \frac{15}{g} a_y$$

$$a_y = \frac{915g}{15} = \underline{\underline{61g}}$$

DYNAMICS AND VIBRATIONS 9

A large hangar door is required to be stopped at the end of its travel by a constant-force snubber. The door weighs 10,000 lbs and its speed is 1 ft per second when it strikes the snubber.

What constant force is required by the snubber to stop the door within a distance of 1 ft?

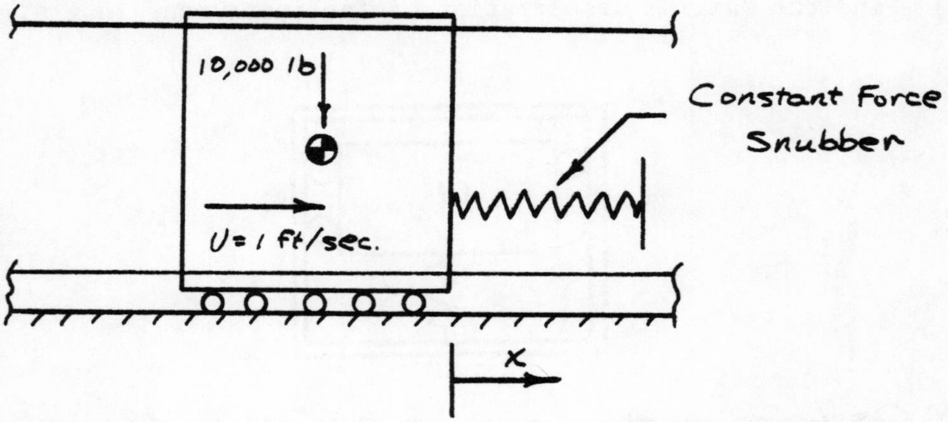

Assuming no friction and using Work Energy Principle

$$\Delta U = \Delta K.E. \qquad U - \text{work}$$

K.E. – Kinetic Energy

$$-(\text{Snubber force})(1') = -\tfrac{1}{2}\,\frac{10{,}000}{g}(1)^2 = -\frac{5000}{g}$$

$$\text{Snubber force} = \underline{\underline{155 \text{ lb}}} \longleftarrow$$

DYNAMICS AND VIBRATIONS 10

A centrifuge in the form of an axially symmetric dish (cross-section illustrated) used for dynamic testing of transistors, etc., is to be accelerated to a speed of 12,000 rpm in 30 seconds. Assuming a constant acceleration over the entire speed range, determine the horsepower requirement for the drive motor.

What is the transistor acceleration?

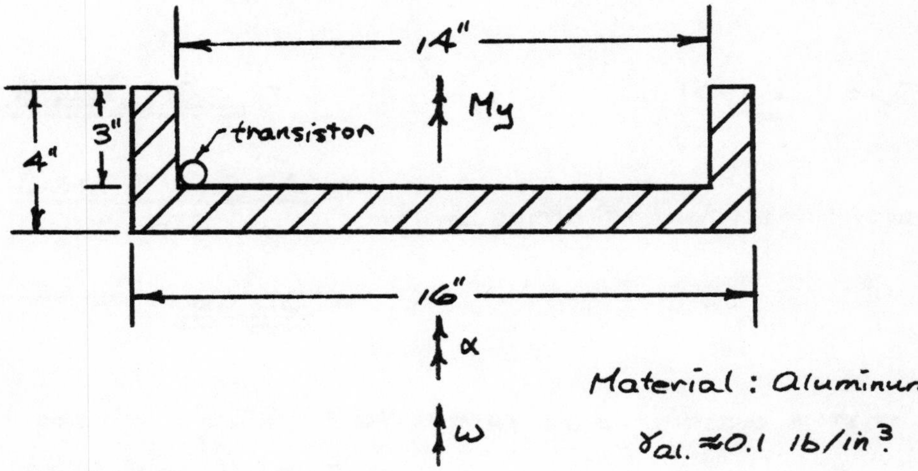

$$\text{Horsepower Output of drive motor} = \frac{(M_y)(\omega_{max})}{550}$$

$$\omega_{max} - \text{max. angular velocity} = \frac{12000}{60}(2\pi)$$

$$\omega_{max} = 1256 \text{ rad/sec.}$$

Neglecting friction, mass of transistor

M_y = motor torque required to acc. centrifuge

$$= (I_y)_{centrifuge}(\alpha)$$

$$\alpha = \omega/t = 1256/30 = \underline{\underline{41.9 \text{ rad/sec}^2}}$$

Mass Moment of Inertia of Centrifuge

$$I_y = \frac{1}{2}\frac{W_1 R_o^2}{g} - \frac{1}{2}\frac{W_2}{g} R_i^2$$

W_1 = wt. of 16" Dia. alum. disk, 4" thick

$= \pi(8)^2(4)(0.1) = 81$ lb.

W_2 = wt. of 14" Dia. alum. disk 3" thick

$= \pi(7)^2(3)(0.1) = 46.2$ lb.

$$I_y = \frac{1}{2g}\left[81\left(\frac{8}{12}\right)^2 - 46.2\left(\frac{7}{12}\right)^2\right] = \underline{\underline{0.314 \text{ slug ft}^2}}$$

Horsepower output of drive motor $= \dfrac{(0.314)(41.9)(1256)}{550}$

$= \underline{\underline{30 \text{ hp}}}$ ←

Transistor acceleration when $\omega = 1256$ rad/sec.

$\alpha = 41.9$ rad/sec²

$$a_n \approx R_i \omega^2 \approx \frac{7(1256)^2}{12} = \underline{\underline{920{,}000 \text{ ft/sec}^2}}$$

$$a_t \approx R_i \alpha \approx \frac{7(41.9)}{12} = \underline{\underline{24.4 \text{ ft/sec}^2}}$$

Transistor Acceleration $\approx 920{,}000$ ft/sec²

$= \underline{\underline{28{,}600 \text{ g}}}$ ←

DYNAMICS AND VIBRATIONS 11

The uniform bar ABC weighs 5 lbs, and a concentrated mass of 3 lbs is attached at C.

Find the natural frequency in cycles per second for angular motion about A. The static deflection of the spring under the conditions shown is 0.55 inches.

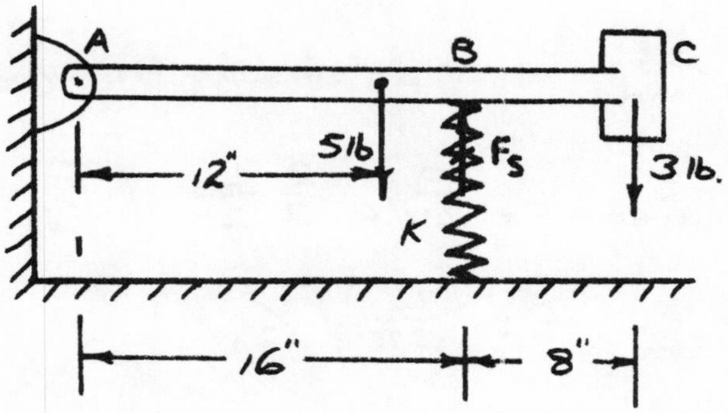

Fig. 1

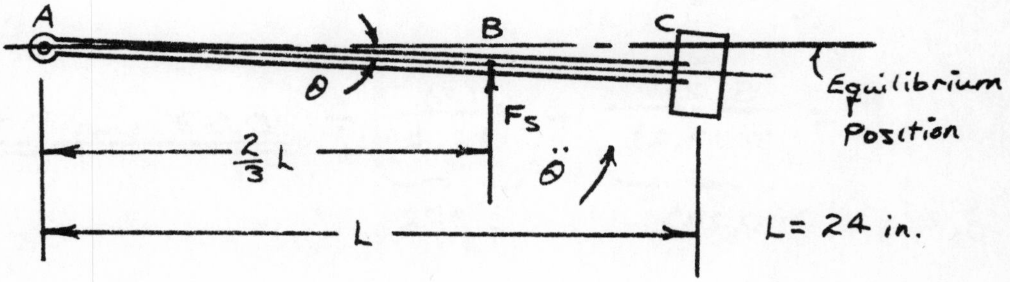

Fig. 2

Determination of K

From Fig. 1 :

$$16\,F_s = 5(12) + 3(24) \qquad F_s = 8.25 \text{ lb}$$

$$F_s = Ky \qquad K = \frac{8.25}{0.55} = 15 \text{ lb/in.}$$

From Fig. 2 :

$$\Sigma M_A = I_A \ddot{\theta} = F_S(2/3 L) = -K(2/3 L\theta)(2/3 L)$$

$$I_A \ddot{\theta} + 4/9\, KL^2 \theta = 0$$

$$\ddot{\theta} + \left(4/9\ \frac{KL^2}{I_A}\right)\theta = 0$$

Let $p^2 = 4/9 \dfrac{KL^2}{I_A}$ $\quad$ p = circular freq. (rad/sec.)

natural freq. $f = \dfrac{p}{2\pi} = \dfrac{2}{3}\dfrac{L}{2\pi}\sqrt{K/I_A}$

$$f = \frac{L}{3\pi}\sqrt{K/I_A}$$

$$I_A = \frac{W_{bar} L^2}{3g} + \frac{W_c L^2}{g}$$

$$= \underbrace{\frac{5(2)^2}{3(32.2)}}_{0.207} + \underbrace{\frac{3(2)^2}{32.2}}_{0.372} = \underline{\underline{0.58 \text{ slug ft}^2}}$$

$$f = \frac{2}{3\pi}\sqrt{\frac{15\,(12)}{0.58}} = \underline{\underline{3.7 \text{ cps}}}$$

(for small values of θ)

DYNAMICS AND VIBRATIONS 12

A machine weighing 25 lbs is placed on rubber mounts which deflect 0.08 inches under its weight. Assume that the mounts have 10% of critical damping.

(a) Find the damped natural frequency for vertical motion, in cycles per second.

(b) When the machine speed is run slowly through resonance, the maximum double amplitude of vibration is observed to be 0.3 inches. Find the amplitude of the disturbing force in pounds.

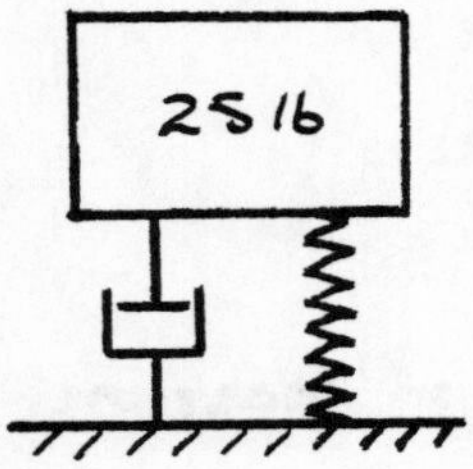

a) $K = \frac{25}{0.08} = 312 \text{ lb/in}$

$\zeta = 0.1 = \frac{\text{coeff. of damping}}{\text{coeff. of critical damping}} = \frac{C}{C_c}$

Damped Circular freq. $q = p\sqrt{1 - \left(\frac{C}{C_c}\right)^2}$

where p = undamped circular freq. $= \sqrt{\frac{K}{m}}$

$= \sqrt{\frac{312(386)}{25}} = 69.6 \text{ rad/sec.}$

$q = 69.6\sqrt{1 - 0.01} = 69.5 \text{ rad/sec}$

Damped natural freq. $= q/2\pi = \underline{\underline{11.1 \text{ cps}}}$ ←

b) Amplitude of vibration for $q/p = 1$ is

$$y_0 = \frac{F_0/K}{\sqrt{\left[1-(q/p)^2\right]^2 + \left[2\zeta\, q/p\right]^2}}$$

where F_0 is the amplitude of disturbing Force.

$$0.15 = \frac{F_0/K}{\sqrt{(1-1)^2 + (0.2)^2}}$$

$F_0 = K(0.15)(0.2) = 0.03K = 0.03(312) = \underline{9.36 \text{ lb}}$ ←

DYNAMICS AND VIBRATIONS 13

A mass traveling at 15.0 miles/sec is in a circular orbit about a second mass.

If the radius of this orbit is 70.0 x 10^6 miles, what specific potential energy change is required to change this orbit to another circular orbit with a 10 x 10^7 mile radius?

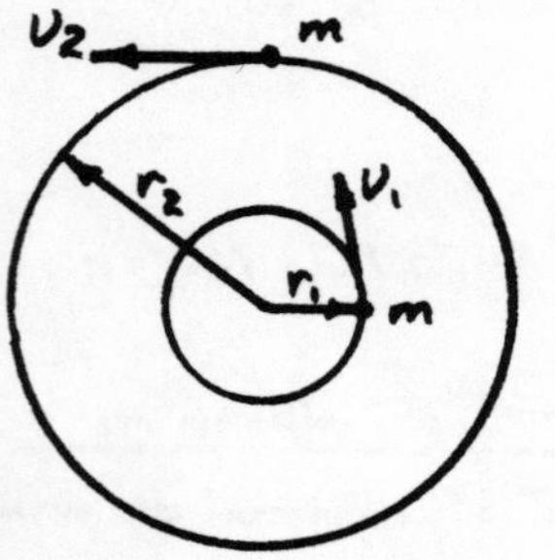

$$v_1 = 15 \text{ miles/sec.}$$

$$r_1 = 70 \times 10^6 \text{ miles}$$

Potential Energy of mass m with respect to mass M

$$(U_g)_1 = -\frac{GMm}{r_1}$$

$$\text{Force between } m + M = F = ma = \frac{GMm}{r_1^2}$$

$$\frac{mv_1^2}{r_1} = \frac{GMm}{r_1^2} \quad ; \quad M = \frac{v_1^2 r_1}{G} = \frac{v_2^2 r_2}{G}$$

$$v_2^2 = v_1^2 \frac{r_1}{r_2}$$

$$(U_g)_1 = -\frac{Gm}{r_1}\left[\frac{v_1^2 r_1}{G}\right] = -mv_1^2$$

$$\text{Specific Potential Energy Change} = -\frac{m}{m}\left[v_2^2 - v_1^2\right]$$

$$= v_1^2 - v_1^2 \frac{r_1}{r_2} = v_1^2\left[1 - \frac{r_1}{r_2}\right]$$

$$= (15)^2(5280)^2\left[1 - \frac{70(10)^6}{100(10)^6}\right]$$

$$= \underline{\underline{1.87 \times 10^9 \text{ ft}\cdot\text{lb/slug}}} \longleftarrow$$

DYNAMICS AND VIBRATIONS 14

A spacecraft is spin-stabilized at a speed of 150 rpm. It is desired to reduce this spin speed, after boost, to 10 rpm. This is to be accomplished by the radial displacement of two masses as indicated in the accompanying diagram. The moment of inertia of the spacecraft about its spin axis is 10 slug-ft^2, excluding the de-spin masses. The mass of each of the de-spin masses is 0.5 slug.

Determine the required radial displacement R.

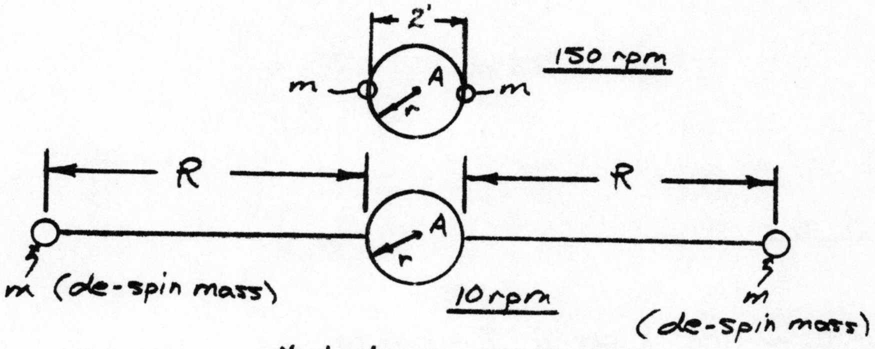

* Let $r_f = r + R$

Solution

Initial Angular Momentum $(H_A)_i = I_A \omega_i + 2(m r \omega_i) r$

$$\omega_i = \frac{150}{60}(2\pi) = 5\pi \text{ rad/sec.}$$

$$H_A = 10(5\pi) + 2(0.5)(1)(5\pi)(1)$$
$$= 55\pi \frac{\text{slug ft}^2}{\text{sec.}}$$

$(H_A)_f$ = Final angular momentum = $(H_A)_i$ since moment about A i.e. $\Sigma M_A = 0$

$$\omega_f = (10/60)(2\pi) = \pi/3 \text{ rad/sec}$$

$$(H_A)_f = I_A \omega_f + 2(m r_f \omega_f)(r_f) = 10(\pi/3) + \pi/3\, r_f^2 = 55\pi$$

$$1/3\, r_f^2 = 55 - 3.33 = 51.67 \;:\; r_f = 12.45 \text{ ft.}$$

$$R = 12.45 - 1 = \underline{\underline{11.45 \text{ ft}}} \longleftarrow$$

DYNAMICS AND VIBRATIONS 15

A piece of putty weighing 25 lbs and traveling in a straight line at 15 mph is struck by another piece of putty weighing 35 lbs and traveling at 20 mph in a straight line in the opposite direction. How much kinetic energy is lost in the impact?

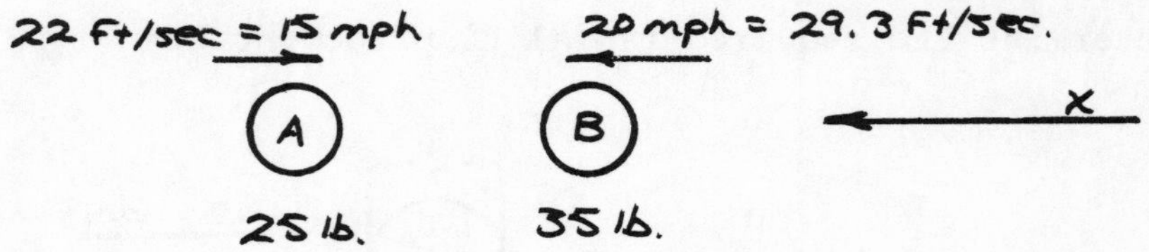

Solution

Assuming Plastic Impact

$$(U_A)_{final} = (U_B)_{final} = U$$

Initial linear Momentum of system

$$G_i = \frac{35(29.3)}{g} - \frac{25(22)}{g} = \frac{475}{g} \text{ lb. sec.}$$

$G_{final} = G_i$ since $\Sigma F_x = 0$

$$G_f = \frac{(35+25)}{g} U = \frac{475}{g} \quad ; \; U = \underline{\underline{7.92 \text{ ft/sec}}}$$

$$(K.E.)_i = \tfrac{1}{2} m_A U_A^2 + \tfrac{1}{2} m_B U_B^2$$

$$= \tfrac{1}{2}\,\frac{25}{g}(22)^2 + \tfrac{1}{2}\,\frac{35}{g}(29.3)^2 = 188 + 466 = \underline{\underline{654 \text{ ft·lb}}}$$

$$(K.E.)_f = \tfrac{1}{2}(m_A + m_B)U^2 = \frac{60}{2g}(7.92)^2 = \underline{\underline{58.5 \text{ ft·lb.}}}$$

$$\Delta K.E. = 654 - 58.5 \approx \underline{\underline{596 \text{ ft·lb}}} \longleftarrow$$

DYNAMICS AND VIBRATIONS 16

Ship A of 4,000 tons weight is towing Ship B of 2,000 tons weight by means of a cable. At the instant the cable becomes taut, the velocity of Ship A is 12 ft/sec and that of Ship B is 8 ft/sec. The cable is constructed so that a pull of 2,000 lbs will elongate it 0.0024 in per foot of length and its safe load is 200,000 lbs. What is the least permissible length of the cable?

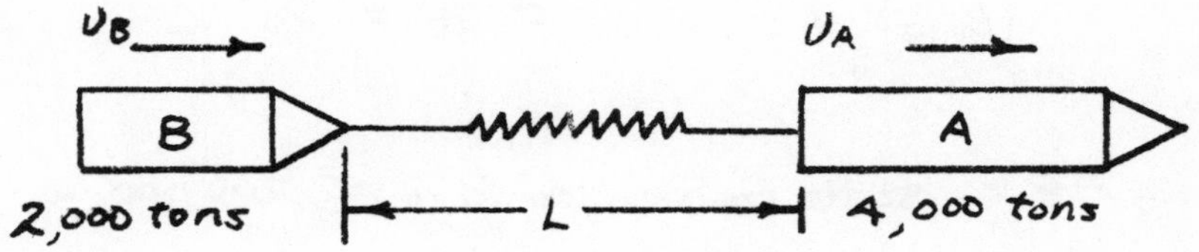

Solution

G_i = Initial Momentum of System

$$= m_A U_A + m_B U_B = \frac{4000(2000)(12)}{g} + \frac{2000(2000)8}{g}$$

$$= \frac{128 \times 10^6}{g} \text{ lb. sec.}$$

From Law of conservation of Linear Momentum

$$G_F = m_A U + m_B U = \frac{(4000+2000)(2000)}{g} U = \frac{128(10)^6}{g}$$

$U = \underline{10.7 \text{ ft/sec}}$ = vel. of each boat

From Law of conservation of Mechanical Energy

$$\tfrac{1}{2} m_A U_A^2 + \tfrac{1}{2} m_B U_B^2 = \tfrac{1}{2}(m_A + m_B) U^2 + \text{Strain Energy in cable}$$

– continued –

$$W_A v_A^2 + W_B v_B^2 = (W_A + W_B)v^2 + 2g(U_e)$$

$$8(10^6)(12)^2 + 4(10^6)(8)^2 = 12(10^6)(10.7)^2 + 2g\,U_e$$

$$1150(10^6) + 256(10^6) = 1380(10^6) + 2g\,U_e$$

$$U_e = 4.05(10)^5 \text{ ft}\cdot\text{lb} = \frac{P}{2}\Delta$$

Where: P = safe cable tension = 200,000 lb.

Δ = total elongation

$$U_e = 100{,}000\,\Delta = 100{,}000(\epsilon)(L) = 4.05\times10^5 \text{ ft}\cdot\text{lb.}$$

For cable tension = 2000 lb $\quad \epsilon = \frac{0.0024}{12}$

For cable tension = 200,000 lb $\quad \epsilon = \frac{0.24}{12} = \underline{\underline{0.02}}$

$$L = \frac{4.05\times10^5}{100{,}000(0.02)} = \underline{\underline{202.5 \text{ ft}}} \longleftarrow$$

DYNAMICS AND VIBRATIONS 17

A weight W and a coil spring with modulus K are attached to the end of a wooden cantilever beam of length L as shown.

W = 100 lbs
E = 2,000,000 psi for beam
I = 0.1 in^4 for beam
K = 100 lb/in

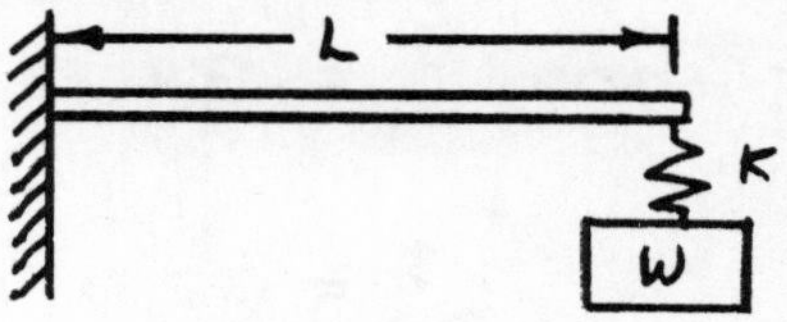

Determine the length L of the beam so that the natural frequency of the system will be 2 cycles per second. Neglect the mass of the spring and beam.

Solution

Weight W is supported by two springs in series, namely, the beam and the coil spring

$$K_{c.s.} = 100 \text{ lb/in} \quad ; \quad K_{beam} = \frac{3EI}{L^3} = \frac{3(2)(10^6)(\frac{1}{10})}{L^3}$$

$$= \frac{6(10^5)}{L^3} \text{ lb/in.}$$

Resultant Spring Constant

$$K = \frac{K_{c.s.}\, K_{Beam}}{K_{c.s.} + K_{Beam}} = \frac{6(10^7)}{L^3}\left[\frac{L^3}{100L^3 + 6(10^5)}\right]$$

$$= \frac{6(10^7)}{100L^3 + 6(10)^5}$$

Natural freq. $f = \frac{1}{2\pi}\left(\frac{gK}{W}\right)^{1/2}$

$$f^2 = \frac{gK}{4\pi^2 W} = \frac{g}{4\pi^2}\,\frac{6(10^7)}{100}\left[\frac{1}{100L^3 + 6(10)^5}\right]$$

$$(2)^2 = 58.8(10^5)\left[\frac{1}{100L^3 + 6(10)^5}\right]$$

$$400L^3 + 24(10)^5 = 58.8(10)^5$$

$$400L^3 = 34.8(10)^5 = 3480(10^3)$$

$$L^3 = 8.7(10^3)$$

$$L = 2.06(10) = \underline{\underline{20.6\ in}}$$

Length of beam $\approx \underline{\underline{20\ 5/8\ in}}$ ←

4

Machine Design

EUGENE J. FISHER

This chapter is arranged to present typical Professional Engineering examination problems. Some will be preceded with an explanation, while in other cases only the problem and solution will appear. If there are solutions that are not clear, an excellent text on this subject is Mechanical Engineering Design by Shigley (published by McGraw-Hill). The references in this chapter are to the Fourth Edition.

The order of the chapter will be by problem type: Bolted Joints, Bearing and Shaft Loads, Interference Fits, Size of Shafts and Critical Speeds, Mechanisms, Gear Trains, Miscellaneous Problems.

Bolted joints that are loaded eccentrically may be analyzed by dividing the load per bolt into two components, one the direct shear and the second the moment load or secondary shear. The assumptions for bolted joints are that (a) none of the load is taken in friction between the two members, and (b) that the moment load on a bolt is directly proportional to its distance from the centroid of the bolt pattern.

To solve an eccentrically loaded bolted problem one may divide the load by the number of bolts to obtain the direct shear load per bolt, then determine the proportionality factor for obtaining the secondary shear force per bolt. This proportionality factor may be determined by solving the equations relating to the force per bolt and the distance to the centroid of the bolt pattern. Thus, $F_b = rk$, where F_b is the force on the bolt, r is the distance from the bolt centroid to this particular bolt and k is the unknown proportionality constant. Another equation may relate the total secondary shear load to each bolt:

$$Pe = F_1r_1 + F_2r_2 + F_3r_3 + F_4r_4 \;\ldots\ldots\; F_ur_u$$

In this equation the total eccentric load is P and the eccentricity from the center line of the bolt pattern is the distance e.

Substituting the first equation into the second equation for the force per bolt:

$$Pe = k(r_1^2 + r_2^2 + r_3^2 + r_4^2 \ldots . r_u^2)$$

Thus, the only unknown is k.

MACHINE DESIGN 1

A removable bracket must be attached to the frame of a machine tool as a support. The need for interchangeability dictated bolting. Clearances and accessibility to the bolts required a pattern as shown. Mild steel bolts (shear strength 30,000 psi, bearing strength 120,000 psi) are to be used.

For a safety factor of three, what diameter standard bolt should be used?

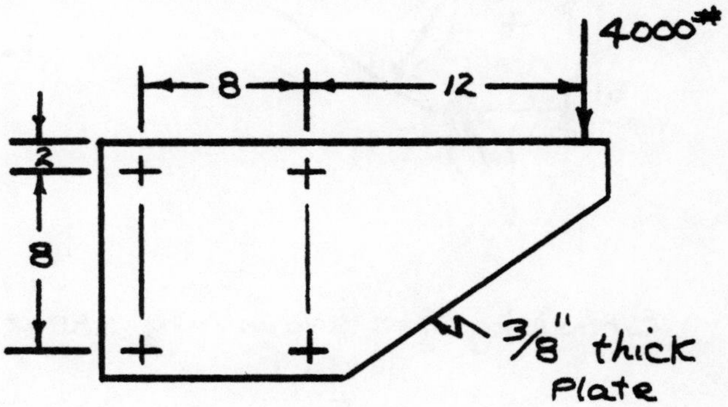

Solution

Determine the centroid of the bolt pattern to be 6" below the top edge and 4" to the right of the top left bolt.

Primary Shear

$$\frac{\text{Load}}{\#\text{ of bolts}} = \frac{4000}{4} = 1000\#/\text{bolt}$$

Secondary Shear

$$P_e = K(r_1^2 + r_2^2 + r_3^2 + r_4^2) = 4000(12+4)$$
$$= K(4)(4^2+4^2)^{1/2}$$

(4 = # of bolts)

$$K = 500\#/\text{in.}$$

$$F_{secondary} = 500\sqrt{32} = \underline{\underline{2840\#/\text{bolt}}}$$

—continued—

These two shear loads will add vectorially, and by sketching these reactions one should see that the most highly loaded bolt is at position c or d.

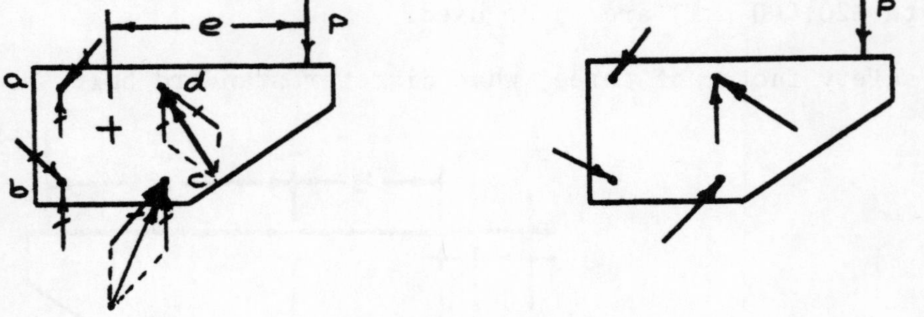

Breaking the secondary shear into its two components

$$F_{Horiz.} = F_{vert.} = 0.707(2840) = 2{,}010^{\#}$$

Thus,

$$F_{Total} = \sqrt{(2{,}010)^2 + (2010 + 1000)^2} = 3620^{\#}$$

Shearing Stress $\tau = P/A$

$$A = P/\tau = \frac{3620}{(30{,}000/3)}$$ ← factor of Safety

$$A = \underline{\underline{0.362 \text{ in}^2/\text{Bolt}}}$$

__Bolt Size (net)__ - (Ref. P 361 Shigley)

3/4 Dia 0.302 in^2 for coarse series UNC

7/8 Dia 0.419 in^2

Therefore select 7/8" bolts

MACHINE DESIGN 2

The combustion chamber of a small attitude control rocket engine is attached to the injector head with four 8-32 bolts as shown. During lift off the engine is subjected to peak accelerations of 200 g's combined with high transverse air loads.

What is the minimum preload for the bolts to prevent fatigue stresses in the bolts and separation of the chamber from injector head?

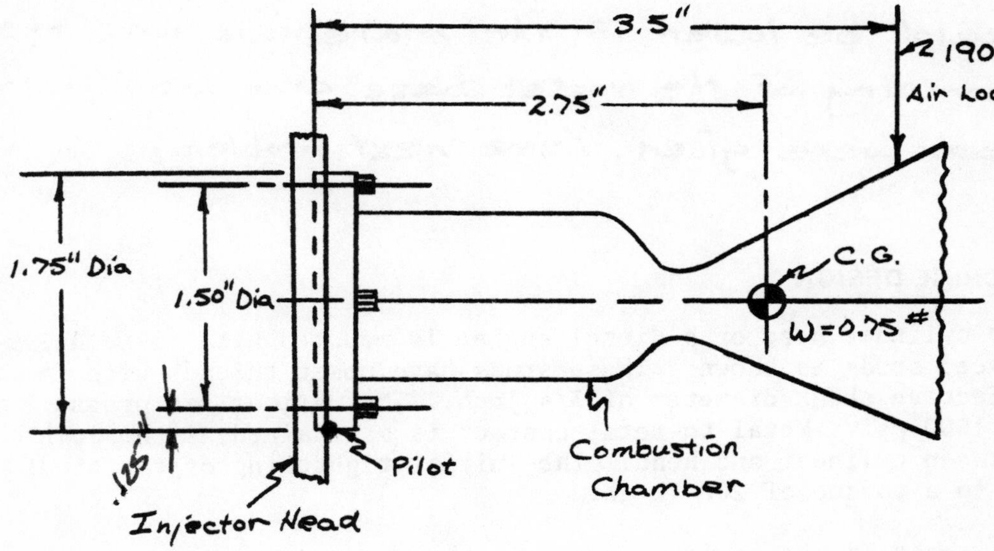

Solution

Assume that the pilot will support the primary shear, so that tension is the only load that the bolts will support. Assume also that the injector head is rigid so that the attitude control rocket will tend to rotate about the lower portion of the pilot. The load on the upper bolt would be:

$$P_1 e_1 + P_2 e_2 = K(r_1^2 + 2r_2^2 + r_3^2)$$

$$(190)(3.5) + [(0.75^{\#})(200)]\,2.75 = K(0.125^2 + 2(0.75 + .125)^2 + 1.625^2)$$

$$K = 257 \ \#/in$$

$$F_{Top\ bolt} = K r_{top} = 257(1.625) = \underline{\underline{418\ lb}}$$

This value becomes equal to the minimum preload in order to prevent separation of the chamber from the injector head. This value would be lower if the materials and the geometry of the rocket base and the injector head were given. (see next problem)

MACHINE DESIGN 3

The cylinder head on a diesel engine is mounted with six equally-spaced studs as shown. These studs have upset threads with an effective shank diameter of 3/4 inch. The maximum gas pressure is 1500 psi. Metal-to-metal contact is maintained as the joint between cylinder and head. The initial tightening of the studs is to a torque of 200 lb-in.

(a) What is the maximum stress in the studs during operation?

(b) What is the effect of a gasket between the metal surfaces?

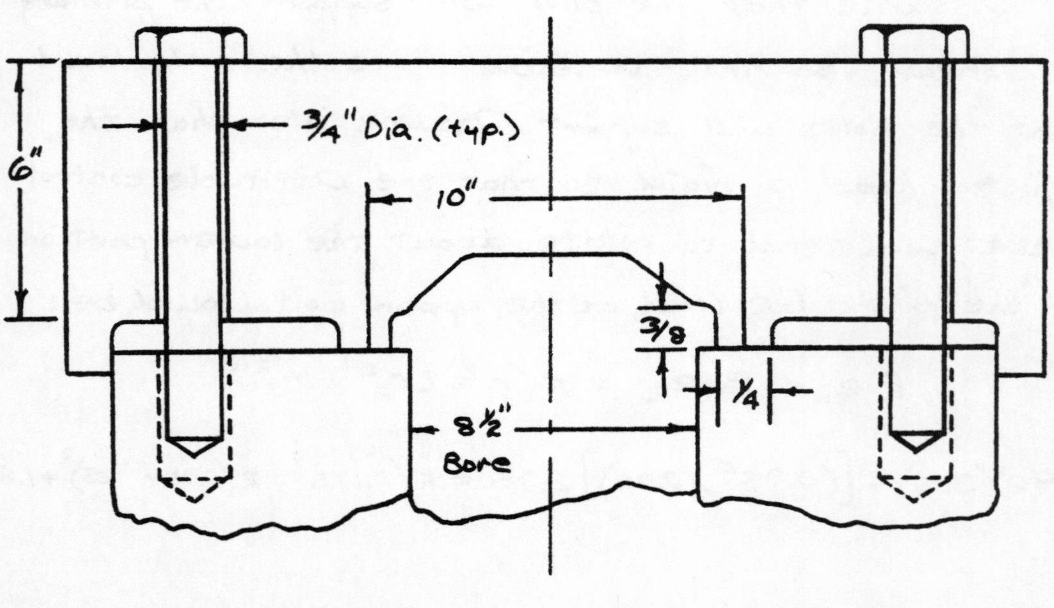

Solution

* Initial force on bolts

$$Torque = .2 F_{initial}\, d_m \quad \text{(Ref. - P. 378 Shigley)}$$

$$F_i = \frac{T}{0.2(3/4)} = \frac{200 \text{ in.lb}}{0.2(.75)} = 1{,}333 \text{ lb.}$$

(bolt preload should be equal to 90% of bolt proof load)

* Deflection of Parts (Ref. - P. 370 Shigley)

$$\delta = Fl/AE \qquad \text{If } \delta = 1'', \quad F = k$$

$$\text{Thus } k = AE/l$$

$$k_b = AE/l$$

$$k_b = \frac{\left(\frac{\pi d^2}{4}\right)(30 \times 10^6 \text{ psi})(6)}{6\ 3/8} = 12.5 \times 10^6 \text{ lb/in.}$$

$$k_m = AE/l = \frac{\frac{\pi}{4}\left[(10\tfrac{1}{4})^2 - (9.75)^2\right] 12 \times 10^6 \text{ psi.}}{3/8}$$

E_{ASTM} Grade 25 cast iron (Ref. - P. 828 Shigley)

$k_m = 252 \times 10^6$ lb/in (considering only 3/8 x 1/4 ring because the rest of the head would be much stiffer since there is more area.)

$$c = \frac{k_b}{k_b + k_m} = \frac{12.5 \times 10^6}{(12.5 + 252)10^6} = 0.047$$

c may vary from 0 to 0.5.

-continued-

$$F_{\text{Total per bolt}} = CF_t + F_i = 0.047\left[\frac{1500\text{ psi}\left(\frac{\pi}{4}(10)^2\right)}{6\text{ Bolts}}\right] + 1333\text{ lb}$$

$$= 930 + 1333 = \underline{\underline{2263\text{ lb/bolt}}}\ \text{(During Operation)}$$

* Max. Stress

$$\sigma = P/A = 2263/0.334 = \underline{\underline{6{,}775\text{ psi}}}$$

(max. stress on bolt during operation)

* <u>Part B</u>

Gasket is used to seal surfaces.

MACHINE DESIGN 4

The force on a pivot bracket casting bolted to the frame of a back hoe can be applied at any point on the 180° arc as shown. The design is such that the bolts "B" are not subjected to shear.

What is the maximum force on each of those bolts?

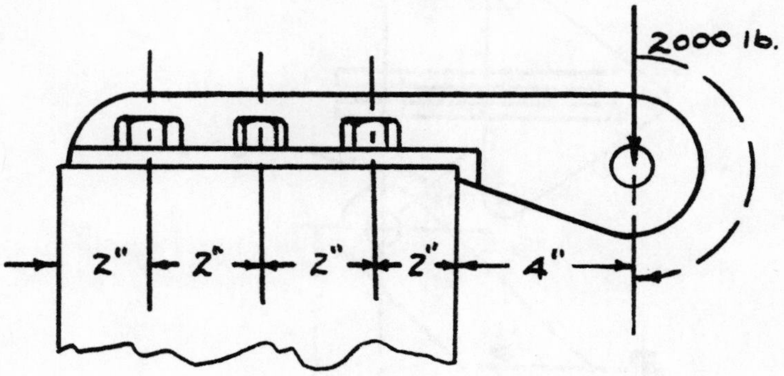

Solution

Assume that the structure is rigid and that it pivots about the point at the right or left side of the bracket. The longer lever arm will occur when the force is in the vertically upward direction, the shear will be absorbed by the pilot.

$$Pe = K(r_1^2 + r_2^2 + r_3^2) = 2000(12)$$
$$= K(2^2 + 4^2 + 6^2)$$

$$K = \frac{24000}{56} = 430 \text{ lb/in}$$

Thus, the max. force of each bolt is :

$F_1 = 430(2) = 860$ lb.

$F_2 = 430(4) = 1720$ lb.

$F_3 = 430(6) = 2580$ lb.

MACHINE DESIGN 5

A hydraulic actuator on a skip loader is mounted to the frame by means of a rigid cast steel bracket. The actuator under one condition exerts a force of 1250 lb as shown.

What are the loads on the cap screws?

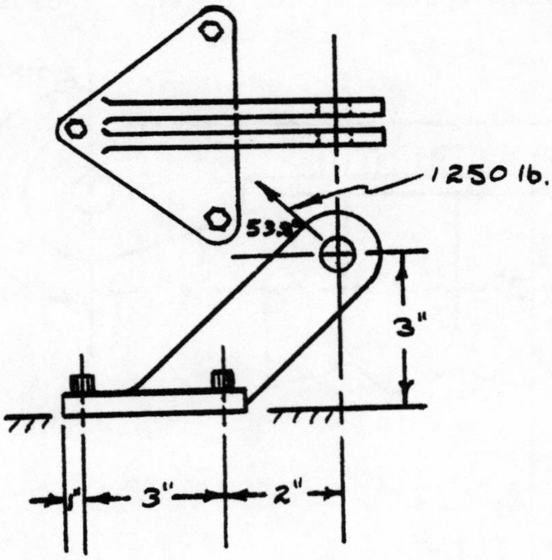

Solution

1) Determine the Horizontal and Vertical components of the force.

$$F_{Horiz.} = 1250(\cos 53.8°) = 750^{\#}$$

$$F_{vert.} = 1250(\sin 53.8°) = 1000^{\#}$$

2) take moment about left hand edge, assuming both bracket and mounting pad to be rigid.

$$F_{\substack{right\\bolt}} = \frac{[1000(6) + 750(3)]}{4} = \frac{8250}{4}$$

$$= 2{,}060 \text{ lb/right two bolts}$$

$$F/\text{bolt} = 2{,}060/2 = 1030 \text{ lb/bolt}$$

(tensile load)

$$F_{Shear} = \frac{750}{3} = 250 \text{ lb/bolt}$$

The loads on these bolts would have been larger if one assumed the bracket to yield across the single left bolt hole. These are only the External loads, The other loads may be initial loads, residual and torsional.

MACHINE DESIGN 6

Under certain conditions a wheel and axle is subjected to the loading shown.

REQUIRED: (a) What are the loads acting on the axle at A-A?

(b) What maximum direct stresses are developed at that section?

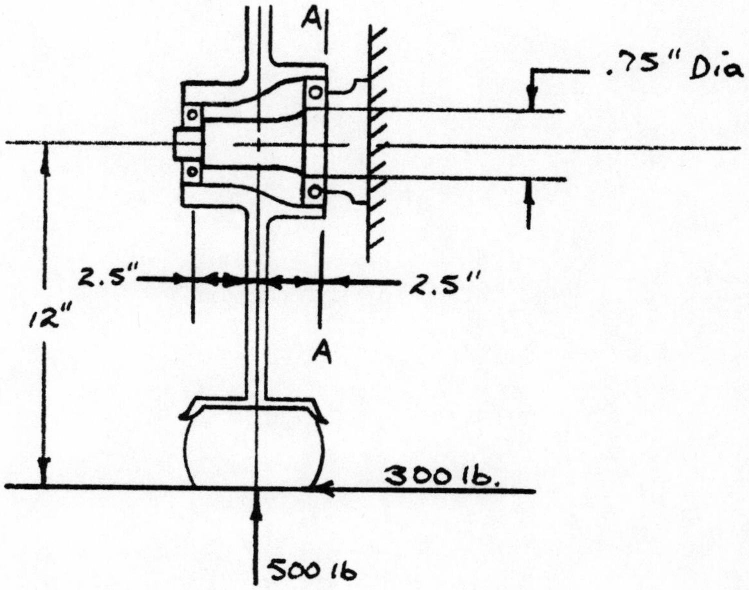

Solution

a) Axial tensile load - 300 lb

Direct Shear load - 500 lb

b) Bending Moment $= 500(2.5) + 300(12)$

$= 1250 + 3600 = 4850$ IN-lb

$$\sigma_{bending} = \frac{Mc}{I} = \frac{32M}{\pi d^3} = \frac{32(4850)}{\pi(.75)^3} = 117{,}000 \text{ psi}$$

$$\sigma = P/A = \frac{300}{\frac{\pi(.75)^2}{4}} = \frac{1200}{\pi(.75)^2} = 679 \text{ psi}$$

Max. Stress at Section A-A

$$\sigma = 117{,}000 + 679 = \underline{\underline{117{,}679 \text{ psi}}}$$

MACHINE DESIGN 7

A spiral bevel gear is mounted as shown. The mean working diameter Dm, tangential force TF, thrust force T_{tf}, and separating force S_{tf} are given. The angular contact bearings are a light press fit on the gear shaft and just slip into the housing. A spacer is matched to the bearings and the housing shoulder to provide a desired preload.

(a) If the preload was zero what are the bearing loads?

(b) What effect does warm-up during operation have on the preload?

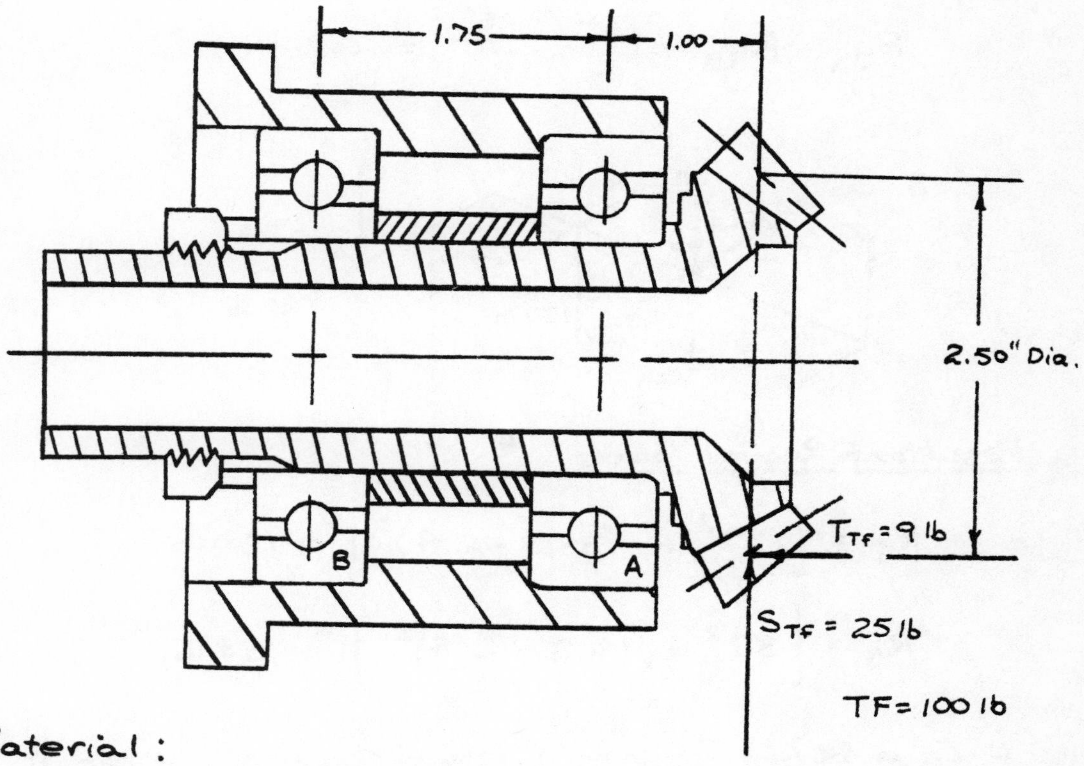

Material:

Gear & Spacer-Steel

Housing - aluminum

Solution

Tangential tooth load

$$\text{Moment} \circlearrowright \text{ bearing } B = 0 = 100(2.75) - R_A\, 1.75$$

$\therefore R_{Z_A} = 157 \text{ lb} \qquad R_{Z_B} = 57 \text{ lb}$

Separating tooth load

$$R_{y_A} = \frac{25(2.75)}{1.75} = 39.3 \text{ lb.}$$

$$R_{y_B} = 14.3 \text{ lb.}$$

Thrust tooth load

$$R_{y_A} = R_{y_B} = \frac{9(1.25)}{1.75} = 6.4 \text{ lb.}$$

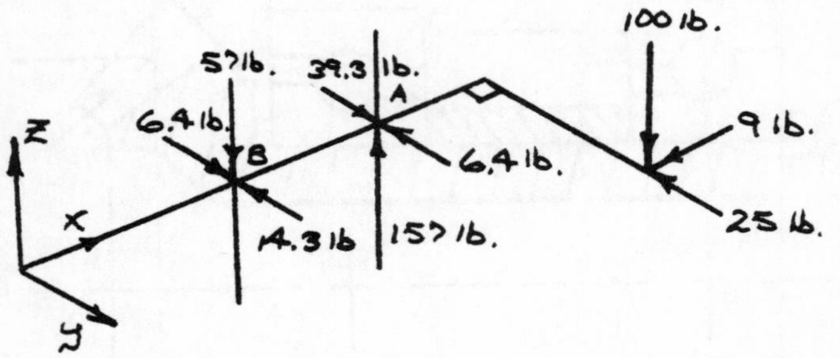

Resultant Radial loads

$$R_A = \left[157^2 + (39.3 - 6.4)^2\right]^{1/2} = 160 \text{ lb}$$

$$R_B = \left[57^2 + (14.3 - 6.4)^2\right]^{1/2} = 58 \text{ lb}$$

* In addition, bearing A would assume the thrust load of 9 lb.

Part b

Warmup increases preload since the aluminum housing will grow more rapidly than the steel gear and spacer.

MACHINE DESIGN 8

The vertical thrust bearing in a large turbine is designed with steel balls operating between thick horizontal steel plates. The sketch shows one of these 0.75 inch diameter balls being loaded by forces P.

Assuming elastic action, what value of P (the force on one ball) corresponds to a maximum contact pressure of 100,000 psi?

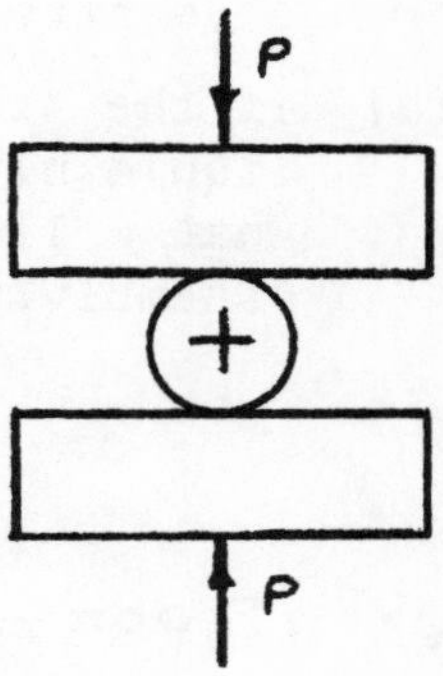

<u>Solution</u>

(Ref.- P 516 #1 Roark)

$$S_c = 0.616\sqrt[3]{\frac{PE^2}{D^2}}$$

<u>Solving for P</u>

$$P = \left(\frac{S_c}{0.616}\right)^3 \frac{D^2}{E^2} = \left(\frac{10\times10^4}{0.616}\right)\frac{(0.75)^2}{(30\times10^6)^2}$$

$$= \underline{\underline{2.69\ lb}}$$

MACHINE DESIGN 9

A hardened steel seat is held in an aluminum valve body with an interference fit. The seat has an outside diameter of 0.6250 inches and the valve body has a wall thickness of 0.125 inches at this location.

REQUIRED: (a) What should be the nominal bore diameter of the valve body, when measured at 70°F, to be given an interference fit that will produce a stress of 15,000 psi in the aluminum? Consider all compressive stresses negligible.

(b) If the steel insert is cooled to -270°F with liquid nitrogen and the valve body heated to 180°F, what will be the diametral clearance during assembly?

Solution

a) $\sigma_{ot} = 15{,}000$ psi

* Employ Eq. # 2-66 P.76 Shigley - Ref.

$$\sigma_{ot} = p \frac{c^2 + b^2}{c^2 - b^2}$$ Solve for contact pressure

$$p = \sigma_{ot} \frac{c^2 - b^2}{c^2 + b^2} = 15{,}000 \frac{(0.437)^2 - (0.312)^2}{(.191) + (.0974)}$$

$$= 4900 \text{ psi}$$

* Employ Eq. b P.77 Shigley

$$\delta_o = \frac{pb}{E_o}\left(\frac{c^2+b^2}{c^2-b^2} + \mu\right)$$

$$= \frac{4900(.312)}{10 \times 10^6}\left[\frac{(.437)^2 + (.312)^2}{(.191) - (0.0974)} + 0.334\right]$$

$$= 1.53 \times 10^{-4}(3.3415) = 0.000523 \text{ in. (radial dimension)}$$

Thus the diameter of the valve body must be,

$$0.625 - 0.000523\,(2) = \underline{\underline{0.624\ in}} \longleftarrow$$

b) Ref. - p.79 Shigley,

$$\alpha_{STL} = 6 \times 10^{-6}\ in/in/^{o}F$$

$$\alpha_{alum.} = 13.3 \times 10^{-6}\ in./in./^{o}F$$

$$\delta_{STL} = 0.625\,(6 \times 10^{-6})(270+70) = 0.0013\ in\ (\text{Diametrical})$$

$$\delta_{alum} = (13.3 \times 10^{-6})(180-70)(0.624) = 0.0009\ in\ (\text{Diametrical})$$

∴ Diametrical Clearance

$$\Sigma\delta = 0.0013 + 0.0009 - 0.001 = \underline{\underline{0.0012\ in}}\ (\text{Dia.})$$

Initial Interference

MACHINE DESIGN 10

The bearings of a steel shaft are to be interference fitted with bronze sleeves as shown so the sleeves will not loosen during operation. Find the minimum and maximum diameters of the bushing.

Maximum operating temperature of bearings will be 160°F, and physical properties of the materials are as follows:

	Steel	Bronze
α	7×10^{-6} in/in°F	0.1×10^{-4} in/in°F
μ	0.30	0.35
E	30×10^{6} psi	16×10^{6} psi
Tensile Strength	80,000 psi	66,000 psi
Yield Strength	36,000 psi	28,000 psi

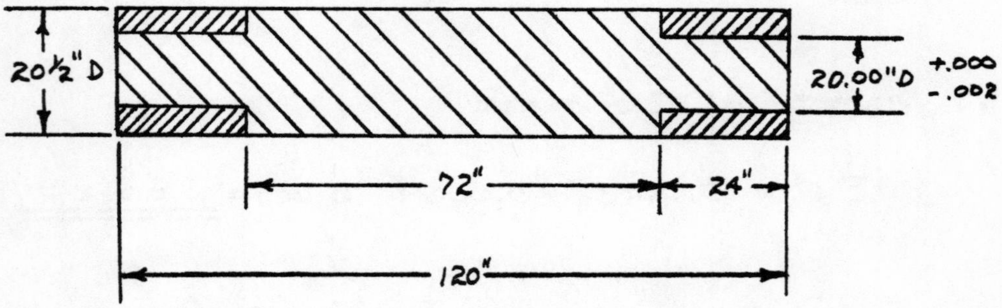

Solution

Ref. p. 76 Shigley

Check growth at 160°F

$$\Delta_{STEEL} = 20 \text{ in } (160-70)(7 \times 10^{-6}) = 0.0126 \text{ in.}$$

$$\Delta_{Bronze} = 20(90)(0.1 \times 10^{-4}) = 0.018 \text{ in}$$

$$\Delta_{Growth} = \Delta_{Bronze} - \Delta_{STEEL} = 0.018 - 0.0126$$

$$= \underline{\underline{0.0054 \text{ in}}}$$

—continued—

Try 0.005 in Interference fit at 160°F. At 70°F the minimum interference fit must be 0.005 + 0.0054 = 0.010 in.

Determine the allowable deflection so that the material will be within the yield stress.

$$\sigma_{ot} = P\left(\frac{c^2+b^2}{c^2-b^2}\right) \quad \& \quad P = \sigma_{ot}\left(\frac{c^2-b^2}{c^2+b^2}\right)$$

$$P = 28{,}000\left(\frac{(10.25)^2-(10)^2}{(10.25)^2+(10)^2}\right) = 28{,}000\left(\frac{5.06}{205}\right)$$

$$= 691 \text{ psi}$$

Now Consider Deformation (Eq. 2-69 - P.77 Shigley)

$$\delta = \delta_o - \delta_i = \frac{bP}{E_o}\left(\frac{c^2+b^2}{c^2-b^2}+\mu_o\right) + \frac{bP}{E_i}\left(\frac{b^2+a^2}{b^2-a^2}-\mu_i\right)$$

$$\delta = \frac{10(691)}{16\times10^6}\left(\frac{205}{5.06}+0.35\right) + \frac{10(691)}{30\times10^6}(1-0.3)$$

$= 0.018$ in. Radial Allowable Bronze Deflection & still be within the yield stress

Thus, the min. interference at 70°F must be 0.010 in. and the max. interference must not be greater than 0.036 in. Diametrical

Size of Bushings

	Max.	Min.
Shaft Diameter	20.000	19.998
Bushing Diameter	19.964	19.988
	0.036	0.010

MACHINE DESIGN 11

The drive shaft on an outboard motor which rotates at speeds up to 5000 rpm is supported by a spline in the crankshaft at the upper end and a bearing at the lower end as shown.

(a) If the engine develops 20 hp at this speed, what diameter solid steel shaft should be used? The allowable tensional stress of 12,000 psi includes the safety factor, endurance limit and service condition considerations.

(b) Is the critical frequency of this shaft below 5,000 rpm?

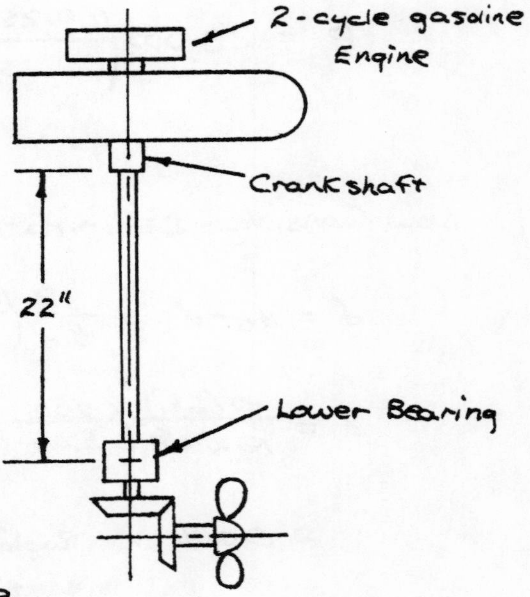

Solution

Assume that the upper end is a fixed support, since it is a splined shaft and assume that the lower end is free to rotate, this will yield a lower critical speed than if considered fixed. Neglect the fact that there will be an overturning moment due to the bevel gear load (can't determine it since no dimensions were given). Also neglect the axial load for the same reason, this load will also decrease the critical speed.

a)

$$T = \frac{63{,}025\,HP}{n} = \frac{63{,}025(20)}{5{,}000} = 252 \text{ in}\cdot\text{lb.}$$

$$\tau = \frac{16 M_T}{\pi d^3} = \frac{16(252)}{\pi d^3} = 12{,}000 \text{ psi.}$$

Solving for Shaft Diameter:

$$d = \sqrt[3]{\frac{4030}{\pi(12 \times 10^3)}} = 0.475 \text{ in}$$

Employ a ½" diameter shaft (Not considering any residual stress or stress concentrations)

b) Employing the Energy method

$$\text{Area} = \pi/4\,(1/2)^2 = 0.197 \text{ in}^2$$

$$\rho = 0.28 \text{ lb/in}^3 \text{ for steel}$$

$$w = A\rho = (0.197)(0.28) = 0.055 \text{ lb/in}$$

$$\delta_{max} = -0.0054\left[\frac{0.055\,(22)^4}{30\times10^6\;\frac{\pi(0.5)^4}{64}}\right] = \underline{\underline{0.000756 \text{ in.}}}$$

<u>Critical Speed</u>

$$\omega_n = \sqrt{g/\delta} = \sqrt{\frac{386 \text{ in/sec}^2}{756\times10^{-6}}}$$

$$= 715 \text{ rad/sec}$$

$$\omega_n = \underline{\underline{6840 \text{ rpm}}}$$

MACHINE DESIGN 12

The drive shaft on an automobile composed of a mild steel tube (3.5" O.D. x 0.80" wall) welded to universal joint, yokes and a splined shaft as shown. If the engine develops 550 hp at 4,000 rpm in high gear, what is the stress in the tube? If the shaft is considered to have uniform properties, end to end, what is the critical speed of the shaft?

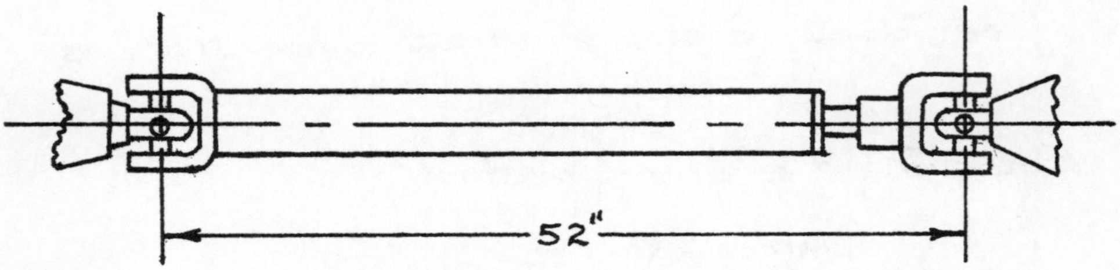

Solution

1) Stress in Shaft due to torsional Shearing Stress.

$$\tau = \frac{Tc}{J} = \frac{16\,T d_o}{\pi(d_o^4 - d_i^4)} \qquad T = \frac{63{,}000\,(HP)}{n}$$

$$= \frac{63{,}000\,(550)}{4000} = 8650 \text{ in}\cdot\text{lb}$$

$$\tau = \frac{16\,(8650)\,3.5}{\pi(3.5^4 - 1.9^4)} = \underline{\underline{1127 \text{ psi}}} \longleftarrow$$

2) $$\text{Weight/in} = \rho \times \text{Area} = 0.28 \text{ lb/in}^3 \ \frac{\pi}{4}\left[(3.5)^2 - (1.9)^2\right]$$

$$= 1.9 \text{ lb/in}$$

consider both ends simply supported

$$\delta_{max} = \frac{5 w \ell^4}{384\,EI} = \frac{5\,(1.9)(52)^4}{384(30\times10^6)\,\frac{\pi}{64}\left[3.5^4 - 1.9^4\right]} = 0.00089 \text{ in.}$$

Critical Speed

$$\omega_n = \sqrt{g/\delta_{static}}$$

$$\omega_n = \sqrt{\frac{386 \text{ in/sec}^2}{8.9\times10^{-4}}} = 630 \text{ rad/sec} = \underline{\underline{6020 \text{ rpm}}}$$

MACHINE DESIGN 13

A scotch yoke mechanism is used to reciprocate a sorting tray in a food processing machine.

If the tray and yoke weigh 50 lbs, what are the peak loads on the slider when running at 100 rpm?

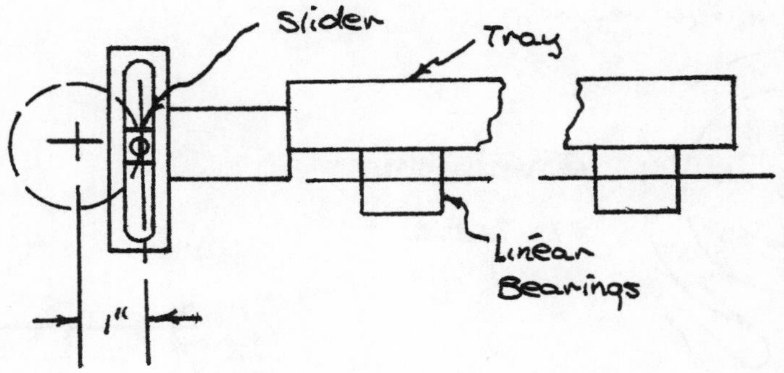

Solution

Assume that the linear bearings are frictionless since no value of μ was given.

$$\text{Acceleration} = \omega^2 R \qquad \omega = \left(\frac{100}{60}\right) 2\pi = 10.5 \text{ rad/sec.}$$

$$A_{max} = (1)(10.5)^2 = 110 \text{ in/sec}^2$$

$$F = MA = \left(\frac{50 \text{ lb}}{386 \text{ in/sec}^2}\right) 110$$

$$= \underline{\underline{14.2 \text{ lb}}} \text{ max. Force}$$

MACHINE DESIGN 14

The Geneva mechanism shown below is part of a binary digital counter. The wheel A rotates at a constant counterclockwise angular velocity of 100 revolutions per minute. Pin P drives the slotted wheel B so that wheel A turns 4 times before wheel B completes a revolution. The pin is located at a distance R = 1.5 inches from the center of wheel A. At the instant shown, wheel B has no angular velocity and the line OP makes a 45° angle with the vertical.

Find the angular velocity of wheel B .075 seconds later.

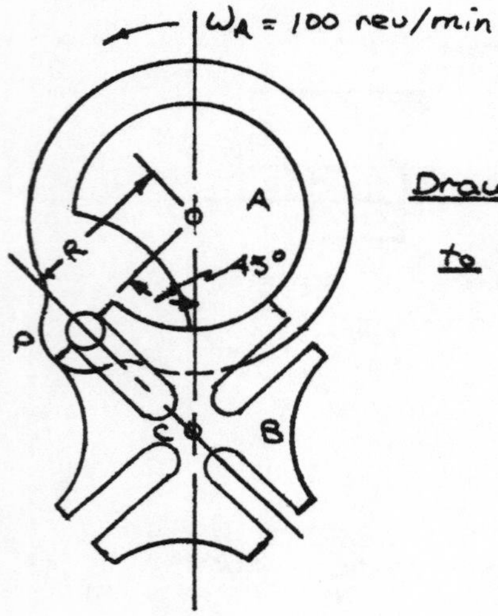

Drawing Not to Scale

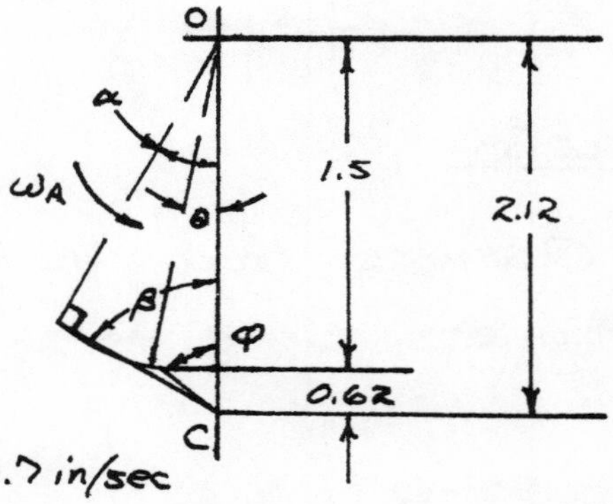

Solution

$$U_p = \omega R = 1.5\left(\frac{100}{60}\right)2\pi = 15.7 \text{ in/sec}$$

One could substitute into the equation of ω_B in terms of ω_A, α & θ where:

$$\omega_B = \omega_a\left(\frac{\cos\alpha(\cos\theta - \cos\alpha)}{1 - 2\cos\alpha\cos\theta + \cos^2\alpha}\right)$$

But Lets try another approach:

$$\omega_A = \frac{100 \text{ rpm}}{60 \text{ sec/min}}(360°) = 600°/\text{sec} \qquad \text{Time} = \frac{45°}{600°} = .075 \text{ sec.}$$

Thus the point P is on the line of centers

$$\omega_B = \frac{U_p}{0.62} = \frac{15.7 \text{ in/sec}}{0.62} = 25.4 \text{ rad/sec} = \underline{\underline{242 \text{ rpm}}}$$

MACHINE DESIGN 15

The lift of a cycloidal cam is given by the equation:

$$Y = L\left(\frac{\theta}{\beta} - \frac{1}{2\pi}\sin\frac{2\pi\theta}{\beta}\right)$$

Where L is the maximum lift, θ is the cam angle, and β is the angle over which the lift occurs.

REQUIRED: If L = 0.75 in., β = 60° and the cam is turning at a constant speed of 1500 rpm, find the maximum acceleration of the follower in ft/sec^2.

Solution

$$y = l\left(\frac{\theta}{\beta} - \frac{1}{2\pi}\sin\frac{2\pi\theta}{\beta}\right)$$

$$\dot{y} = l\left(\frac{\omega}{\beta}\right)\left[1 - \cos\frac{2\pi\theta}{\beta}\right] \quad * \text{ Note: } \frac{d\theta}{dt} = \omega$$

$$\ddot{y} = 2\pi l\left(\frac{\omega}{\beta}\right)^2 \sin\frac{2\pi\theta}{\beta}$$

$$\ddot{y}_{max} = 2\pi l\left(\frac{\omega}{\beta}\right)^2 \text{ since } \sin\frac{2\pi\theta}{\beta} = 1 \text{ at } \ddot{y}_{max}.$$

For $l = 0.75$ in $\beta = 60°$ $\omega = \frac{1500 \text{ rpm}}{60 \text{ sec/min}}(2\pi)$

$= 157$ rad/sec

$$\ddot{y} = 2\pi(0.75)\left(\frac{157 \text{ rad/sec}}{60/57.3 \text{ rad.}}\right)^2 = 106{,}000 \text{ in/sec}^2$$

$$= \underline{\underline{8{,}840 \text{ ft/sec}^2}}$$

MACHINE DESIGN 16

For the planetary gear system shown, arm A is the driver. Gears 2 and 3 are attached together, and gear 4 is held stationary. The numbers of teeth of the various gears are shown in parentheses.

If arm A is turning at 900 rpm clockwise as seen from the left, determine the rpm and direction of rotation of gear 1.

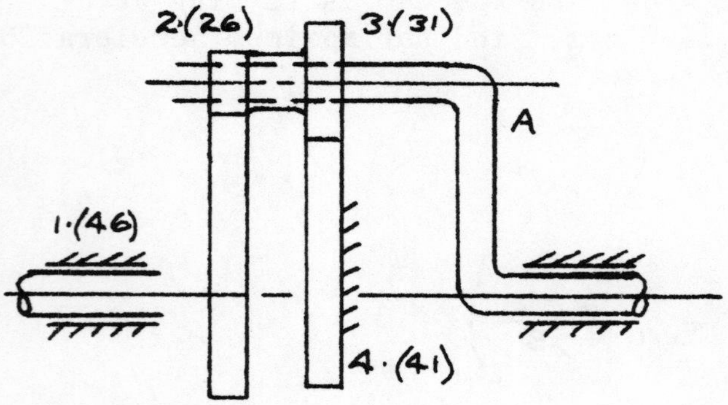

The solution for this type of planetary gear train may be obtained employing the relative velocity equation, the tabular method or looking up this particular arrangement in a handbook like Marks.

The first two methods will be illustrated here.

Reverted Gear Train - Tabular Method

Steps		1	2	3	4	ARM
1	Arm Held Stationary & Fixed Gear Rotate +1 Rev.	$+\frac{26\times 41}{46\times 31}$	$-41/31$	$-41/31$	+1	0
2	Lock all gears & rotate Everything back 1 rev.	−1	−1	−1	−1	−1
3	Total Each Column	$-1+.75$ -0.25			0	−1

Speed of gear 1 = + 0.25 (900) = 225 rpm - same direction as arm - both were negative.

Relative Velocity Equation

$$e = \frac{n_l - n_a}{n_f - n_a}$$

where :

e - gear train value

n_a - Angular vel. of arm

n_f = Angular vel. of 1st gear

n_l = Angular vel. of last gear

Gear Train Value

$$e = \left(\frac{46}{26}\right)\left(\frac{31}{41}\right) = \frac{n_l - n_a}{n_f - n_a} = \frac{0 - 900}{n_f - 900}$$

$$1.34\, n_f - 1200 = -900$$

$$n_f = \frac{300}{1.4} = \underline{\underline{225 \text{ rpm}}}$$

(same direction as the arm)

MACHINE DESIGN 17

A reverted epicyclic gear train is required to have an output speed of approximately 2400 rpm when driven from an electric motor at 1725 rpm. What are the number of teeth required on gears 2, 3 and 4? The motor pinion has 22 teeth, the total number of teeth on each reverted train is 60. Gear 1 has 40 teeth.

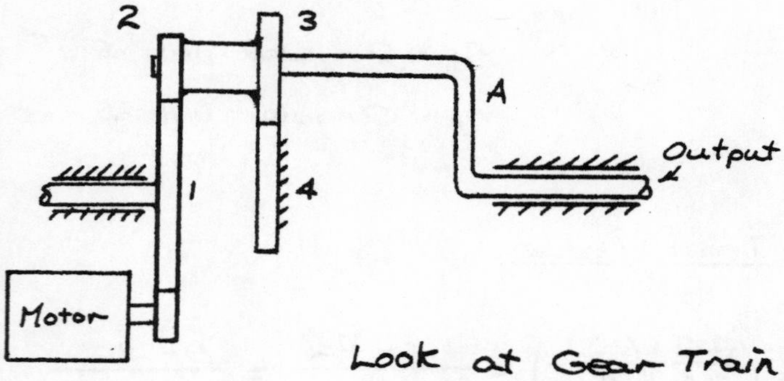

Look at Gear Train From Right side

Assume arm to rotate in a counterclockwise direction. Thus from sketch, the pinion must rotate in the same direction as the arm since gear no. 4 is fixed.

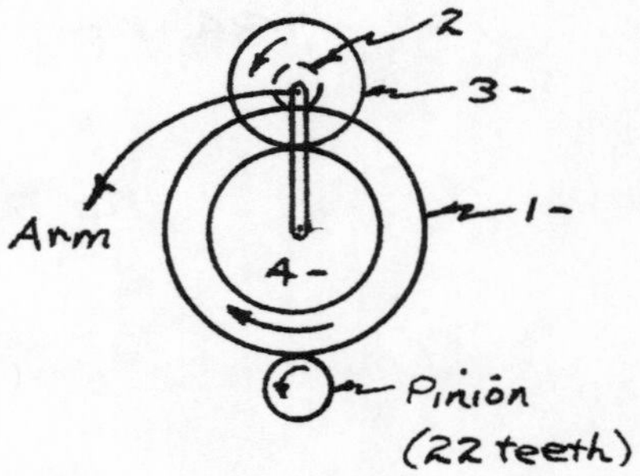

If gear no. 1 has 40 teeth & the total in the gear set is to be 60, then gear no. 2 must have 20 teeth.

Gear 1 rotates @ $1725 \times \frac{22}{40} = 950$ rpm ↻

$$e = \frac{n_\ell - n_a}{n_f - n_a}$$

Assume:

n_f - gear 1

n_ℓ - gear 4

-continued-

$$e = \left(\frac{40}{20} \times {}^{n_3}/_{n_4}\right) = {}^{2n_3}/_{n_4} = \frac{0-(-2400)}{+950-(-2400)}$$

given output speed

Solving for n_3 in terms of n_4

$$n_3 = +\frac{2400}{6700} n_4 = +0.358\, n_4$$

Also, the sum of the teeth of gears 3 + 4 must be 60 or $n_3 + n_4 = 60$

$$n_4 = 60 - n_3$$

$$n_3 = 0.358\,(60 - n_3)$$

Solving, $n_3 = \underline{\underline{16 \text{ teeth}}}$ - OK - no undercut if an involute profile

$$n_4 = 60 - 16 = \underline{\underline{44 \text{ teeth}}}$$

<u>Check This Solution</u>

$$e = ({}^{40}/_{20})({}^{16}/_{44}) = \frac{0 - n_a}{+950 - n_a}$$

Solve for $n_a = -\frac{690}{0.27} = -2550$ rpm opposite in direction to the input member - gear 1

* Note: If n_a were assumed to operate in the opposite direction, the gear 3 has 27 teeth + gear 4 has 33 teeth.

MACHINE DESIGN 18

A hydraulic actuator is needed to provide these forces: maximum force in compression - 4,000 lb, maximum force in tension - 8,000 lb. The rod is made of steel with a tension or compression yield strength of 40,000 psi.

What nominal (nearest 1/16 inch) diameter rod is required for a safety factor of 5 and what nominal bore?

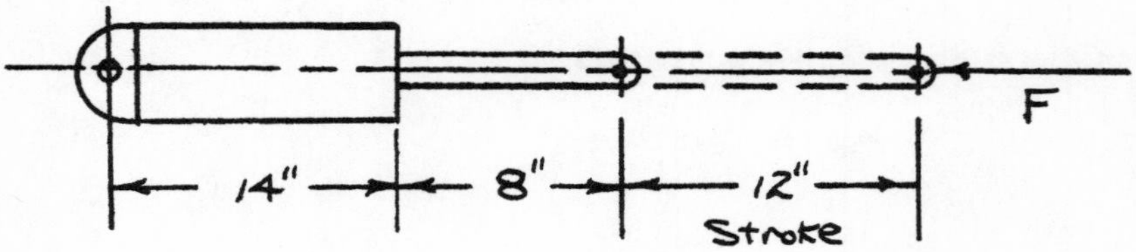

Assume a column - one end fixed and one end free since we know nothing of the linkage design.

$n = 1/4$ (Ref. - p. 145 Shigley)

$$P_{CR} = \frac{n\pi^2 EI}{l^2} = 4000(5) = \frac{\pi^2 (30\times10^6 \text{ psi})\pi D^4}{4(20)^2\, 64}$$

$D^4 = 2.2 \text{ in.}^4$ $\quad D = 1.218$ in Use 1.25"

Design Stress $\quad \sigma_{yield} = P/A = 40{,}000/5 = 8{,}000$ psi

$$A = \frac{\pi}{4}(1.25)^2 = 1.228 \text{ in.}^2$$

$$\sigma_{actual} = 8000/1.228 = 6{,}520 \text{ psi}$$

Design Stress is above this value

$$\text{Factor of Safety} = \frac{8000}{6520} = 1.23$$

—continued—

Net area required to provide the required tension,

$$A = \frac{8000 \text{ lb}}{2000 \text{ psi}} = 4 \text{ in}^2$$

← assumed hydraulic pressure.

∴ Area of piston must be $4 + 1.228 = 5.228 \text{ in}^2$

$$D_{piston} = \left[\frac{4(5.228)}{\pi}\right]^{1/2} = 2.6 \cong 2\tfrac{5}{8} \text{ in Dia.}$$

Extra area helps to overcome seal friction.

MACHINE DESIGN 19

A pulley is keyed to a 2-1/2 inch diameter shaft by a 7/16 x 5/8 x 3 inch flat key. The shaft rotates at 50 rpm. The allowable shear stress for the key is 22 ksi. The allowable compressive stresses for the key, hub and shaft are 66 ksi, 59 ksi, and 72 ksi, respectively.

What is the maximum torque the pulley can safely deliver?

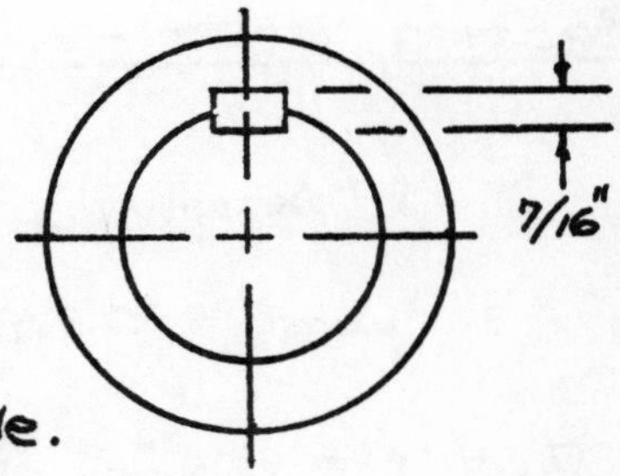

Solution

There are many ways that this key may fail - Calculate the various design stresses allowable.

Since we know nothing about the design loads, employ a safety factor of 2.

$$(\tau_{Design})_{key} = \left(\frac{S_{yield}}{2}\right)\left(\frac{1}{2}\right)$$

← safety factor

↑ Max. Shear Theory of failure

$$(\tau_{Design})_{Key} = \frac{66{,}000}{4} = 16{,}500 \text{ psi}$$

$$(\tau_{Design})_{Hub} = \frac{59{,}000}{4} = 14{,}800 \text{ psi}$$

$$(\tau_{Design})_{Shaft} = \frac{72{,}000}{4} = 18{,}000 \text{ psi}$$

Shear of Key

$A_{Key} = 5/8 \times 3 = 1.87 \text{ in}^2$

Load $= 1.87 \times 16{,}500 = 31{,}000$ lb (Permissible shear load on key)

Bearing Stress - Key

$$(S_{Design})_{Key} = \frac{66{,}000}{2} = 33{,}000 \text{ psi}$$

(2 ← Safety factor)

Area $= (7/16)(1/2)(3) = 0.658$ in.

Permissible Load - Crushing Key $= 0.658(33{,}000)$
$= 21{,}600$ lb.

Bearing Stress - Hub

$$(S_{Design})_{Hub} = \frac{59{,}000}{2} = 29{,}500 \text{ psi}$$

Load $= 0.658(29{,}500) = 19{,}400$ lb

Thus the maximum torque that may be transmitted is:

$$\text{Torque} = \left(\frac{2\tfrac{1}{2} \text{ in. Dia}}{2}\right) 19{,}400 = \underline{\underline{24{,}100 \text{ in}\cdot\text{lb}}}$$

* Note: It is not possible to determine the permissible shear load on the Hub, since the dimensions of the hub were not given.

MACHINE DESIGN 20

An annular flange which is bonded uniformly along the periphery of a round bar is clamped to a fixed base as shown.

REQUIRED: Determine the maximum load P, which can be applied to the end of the bar without exceeding the maximum allowable shearing stress in the bond of 10,000 lbs/in^2. Neglect the weight of the bar.

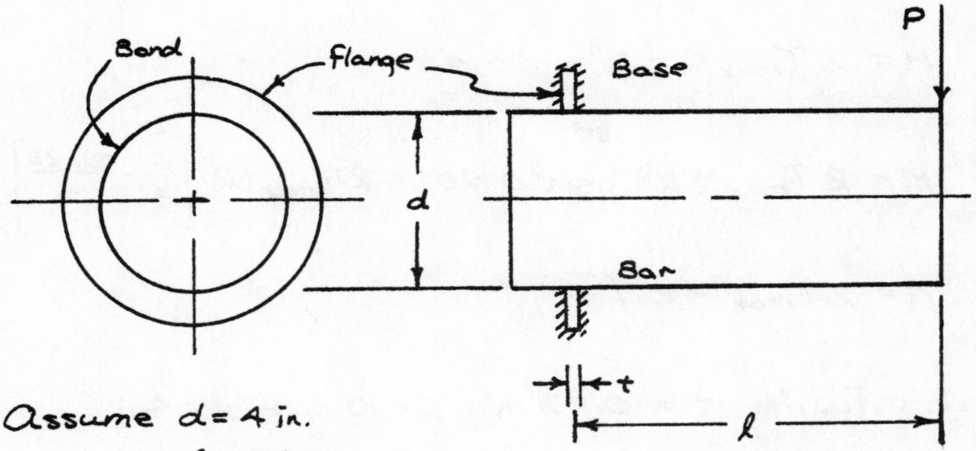

Assume $d = 4$ in.
$\ell = 10$ in.
$t = .02$ in.

Solution

$$\text{Moment} = P\ell = (\tau\, dA)(\rho) \quad \text{(See sketch below)}$$

(Force: $\tau\, dA$; radius: ρ)

End view of Bar

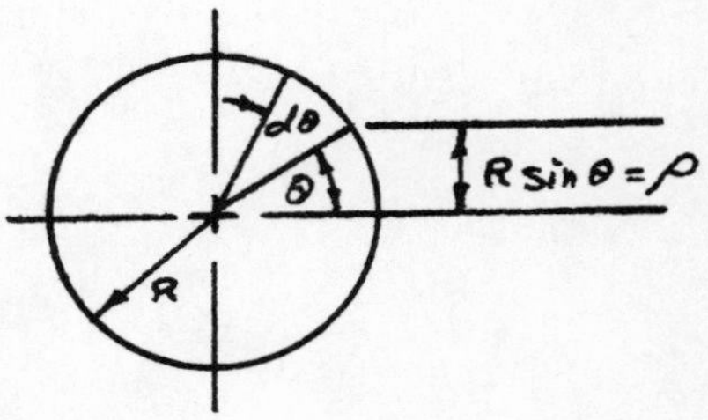

$$\frac{\tau}{\tau_{max}} = \frac{\rho}{R}$$

Assume that the shear varies directly as the distance from the neutral axis.

$$\tau = \tau_{max} \frac{\rho}{R}$$

Permissible Moment (Employing $M = \tau dA(\rho)$)

$$M = 2\int_0^{180°} \tau \rho \, dA = 2\tau_{max} \int_0^{180°} \frac{\rho^2 t R d\theta}{R}$$

Substituting $\tau = \tau_{max} \frac{\rho}{R}$

$$M = 2\tau_{max} \int_0^{180°} t R^2 \sin^2\theta \, d\theta$$

$$M = 2\tau_{max} t R^2 \int_0^{180°} \sin^2\theta \, d\theta = 2\tau_{max} t R^2 \left[\frac{\theta}{2} - \frac{\sin 2\theta}{4}\right]_0^{180°}$$

$$M = 2\tau_{max} t R^2 (\pi/2)$$

Substituting $t = 0.02$ in, $l = 10$ in + $d = 4$ in

$$M = Pl = P(10) = 2(10{,}000)(0.02)(2)^2(\pi/2)$$

$$P = \frac{200(4)(\pi)}{10} = 80\pi = \underline{\underline{251 \text{ lb}}} \longleftarrow$$

MACHINE DESIGN 21

A snap ring is made of 0.177 inch diameter steel wire, bent into an arch of 270° with a mean radius of R = 1.5 in.

Find the tensile forces P required to spread the opening by 3/8 inch.

Use $E = 30 \times 10^6$ psi.

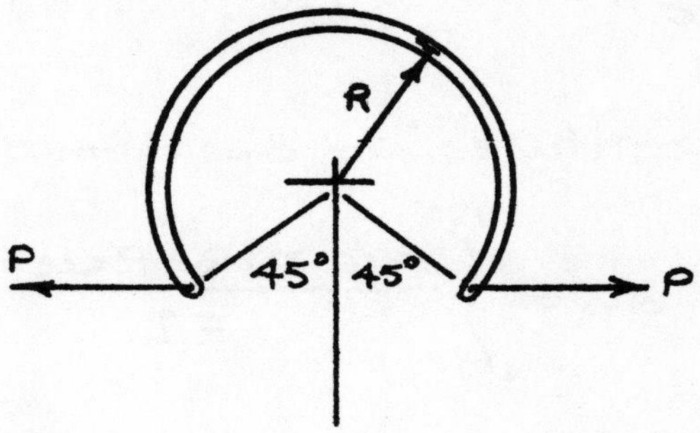

Solution

Employ half of the snap ring since it must be in equilibrium.

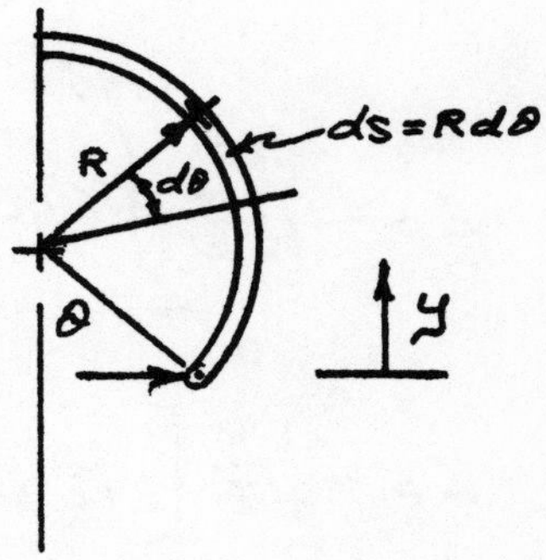

The moment will vary directly as y, so an expression of y as a function of θ & R is:

$$y = 0.707R - R\cos\theta$$

This is only valid between θ = 45° and θ = 180°

Employing Castigliano's Theorem

(Ref. - p.136 Shigley)

Strain Energy in Bending $U = \int \frac{M^2 ds}{2EI}$ & $ds = R d\theta$

$$\delta_P = \int \frac{2M\left(\frac{\partial M}{\partial P}\right) ds}{2EI} = \int_{45^\circ}^{180^\circ} \frac{M\left(\frac{\partial M}{\partial P}\right) R\,d\theta}{EI}$$

Bending Moment

$M = P(\text{Distance})$

$$M = P(0.707R - R\cos\theta)$$

$$\frac{\partial M}{\partial P} = (0.707R - R\cos\theta)$$

Substitute into δ_P formula:

$$\delta_P = \int_{45^\circ}^{180^\circ} \frac{P(0.707R - R\cos\theta)^2 R\,d\theta}{EI}$$

where: $I = \frac{\pi d^4}{64}$

$$I = \frac{\pi(0.177)^4}{64}$$

$$I = \underline{4.93 \times 10^{-5}\ \text{in}^4}$$

$$\delta_P = \frac{PR}{EI}\int_{45^\circ}^{180^\circ} (1.06 - 1.5\cos\theta)^2\,d\theta$$

$$\delta_P = \frac{PR}{EI}\left[1.123\theta - 3.18\sin\theta + 2.25\left(\tfrac{1}{2}\theta + \tfrac{1}{4}\sin 2\theta\right)\right]_{45^\circ}^{180^\circ}$$

$$\delta_P = (3/8)(1/2) = 0.188 = \frac{7RP}{EI}$$

considering only half of the snap ring

$$P = \frac{0.188(30 \times 10^6\ \text{psi})(4.93 \times 10^{-5})}{(7.0)(1.5)} = \underline{\underline{26.5\ \text{lb}}} \leftarrow$$

MACHINE DESIGN 22

A horizontal circular bar, rigidly built in at one end, is to be loaded by a vertical force P = 9.42 kips. The force lies parallel to the X-Z plane shown. The bar's radius is 1.5 inches, its length 9 inches. The horizontal distance "A" is 5.75 inches. Assume a strong bond between lever arm and bar.

REQUIRED: (a) Find an element on the bar that is in pure shear and determine the maximum shear stress present.

(b) Using the maximum-shear stress theory of failure as a basis of design decision (the allowable shear stress is 15 ksi) will the bar fail with this size radius?

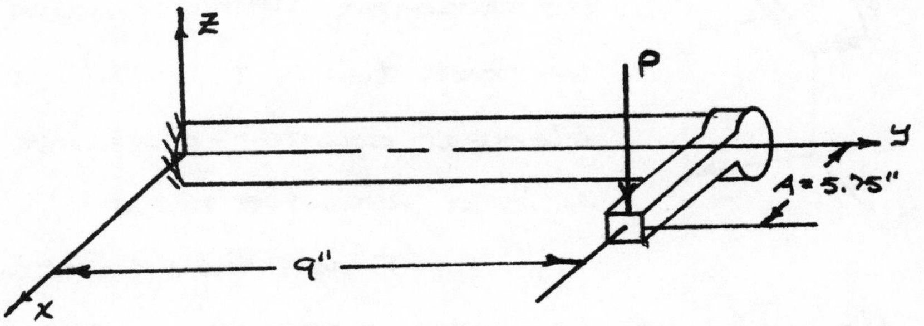

Solution

a) This is a combined stress problem.

$$\sigma_b = \frac{Mc}{I} \leftrightarrow \frac{Tc}{J} = \left(\frac{32M}{\pi d^3} \leftrightarrow \frac{16T}{\pi d^3}\right)_{\text{circular shaft}}$$

where: $M = (9420\text{ lb})(9\text{ in}) = 84{,}780\text{ in}\cdot\text{lb.}$

$T = (9420)(5.75) = 54{,}165\text{ in}\cdot\text{lb.}$

$$\sigma_{\text{Bending-Max at X-Z plane}} = \frac{32(84{,}780)}{\pi(3)^3} \leftrightarrow \frac{16(54{,}165)}{\pi(27)}$$

$$= 31900 \leftrightarrow 10{,}217$$

—continued—

$$\tau_{max} = \left[\left(\frac{31{,}900}{2}\right)^2 + (10{,}217)^2\right]^{1/2} = 18900 \text{ psi}$$

$$\sigma_1 \text{ principal stress} = \frac{31{,}900}{2} + 18{,}900 = 34{,}840 \text{ psi}$$

One could plot Mohr's Circle or simply employ a sketch & determine the location of the plane of pure shear.

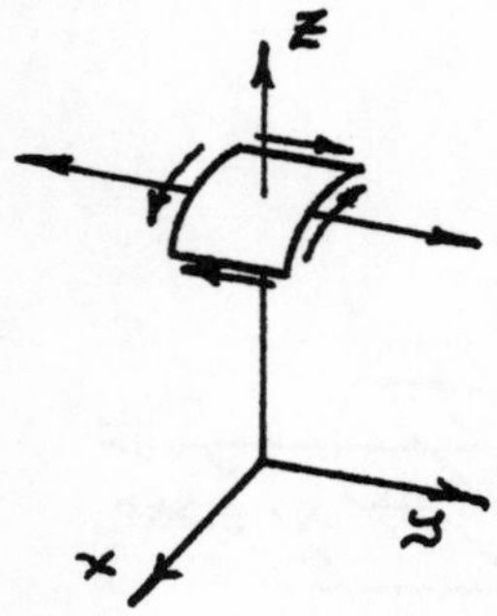

Removing an element from the top of the bar - Notice that the shear on the right hand surface tends to rotate the element counter-clockwise ∴ is a negative shear (Timoshenko Notation).

Therefore, the origin of the circle appears in the 4th quadrant and is $\varphi°$ Below the Horizontal axis.

$$\text{Tan } 2\varphi = \frac{10{,}217 \text{ psi}}{15{,}950 \text{ psi}} = 0.6405 \qquad \varphi = 16° - 19'$$

Pure shear will occur where the bending stress is zero.

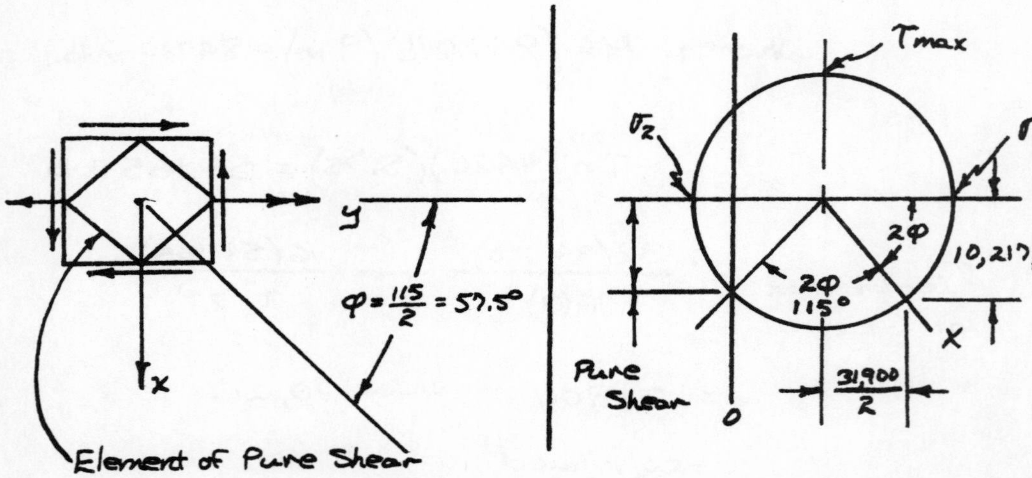

b) This theory asserts that the breakdown of the material depends only on the maximum shearing stresses that are attained in an element.

$$\tau_{max} = \tfrac{1}{2}(\sigma_1 - \sigma_2) = \underline{\underline{18{,}900 \text{ psi}}}$$

∴ this part will fail since this stress is larger than the allowable stress of <u>15,000 psi</u>.

MACHINE DESIGN 23

A punch press is to have a capacity of 8000 lb through a 1/4 inch stroke. For efficiency and cost a small motor with a flywheel was used.

What flywheel effect is required to maintain the flywheel within 20% of its no-load speed of 600 rpm?

<u>Solution</u>

Kinetic Energy of a rotating body

$$K.E. = \tfrac{1}{2} I_o \omega^2 = \tfrac{1}{2} I_o (\omega_1^2 - \omega_2^2)$$

assume that all the energy must be stored in the flywheel or that the frequency of this operation is low.

<u>Energy to be Stored</u>

$$8000 \text{ lb} (\tfrac{1}{4} \text{ in}) = 2000 \text{ in}\cdot\text{lb}$$

$$\omega_1 = \frac{2\pi N}{60} = \frac{2\pi(600)}{60} = 62.8 \text{ rad/sec.}$$

* a 20% drop in speed is permissible

$$A\omega = 20\%(62.8) = 12.56 \text{ rad/sec} \qquad \omega_2 = 62.8 - 12.56 = 50.24 \text{ sec}^{-1}$$

$$K.E. = 2000 \text{ in}\cdot\text{lb} = \tfrac{1}{2} I_o (62.8^2 - 50.24^2)$$

$$\text{Solving for } I_o = \underline{\underline{2.82 \text{ lb}\cdot\text{in}\cdot\text{sec}^2}}$$

MACHINE DESIGN 24

A jack screw was designed as shown. It was well lubricated, and the nut had a sufficient number of threads so that the calculated bearing pressures were well within the allowable limits. The materials were compatible for this usage, and all other calculated stresses were reasonable. Nevertheless, this screw galled on the first usage. Why?

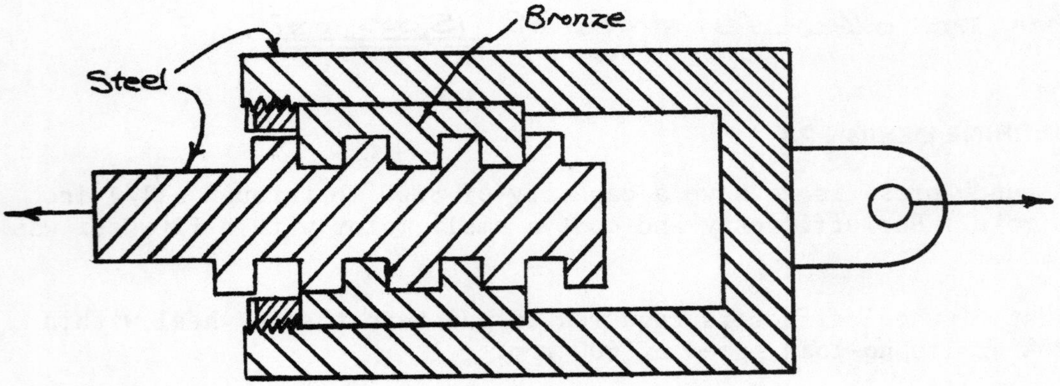

The statement, "That the calculated bearing pressures were well within the allowable limits", bears analysis.

Since we don't have the calculations before us, we can only surmise the assumptions. If the designer neglected to consider the deformation of the threads and considered them all equally sharing the load, this will lead to a high load on the first thread in the nut and a low load on the last thread (Ref.- p382 Shigley). Thus the load on the first thread may be in the range to cause galling.

Another more remote possibility may be operating at a higher temperature environment such that the bronze will "grow" faster than the steel and thus cause galling.

Perhaps the nut was excessively tightened on the bronze threaded insert and this caused the spaces between the bronze threads to deform and increase the bearing pressure above that anticipated.

One assumes that the parts were made to the drawing and that clean lubricant was employed.

MACHINE DESIGN 25

An instrumentation torque limiting device is made as shown.

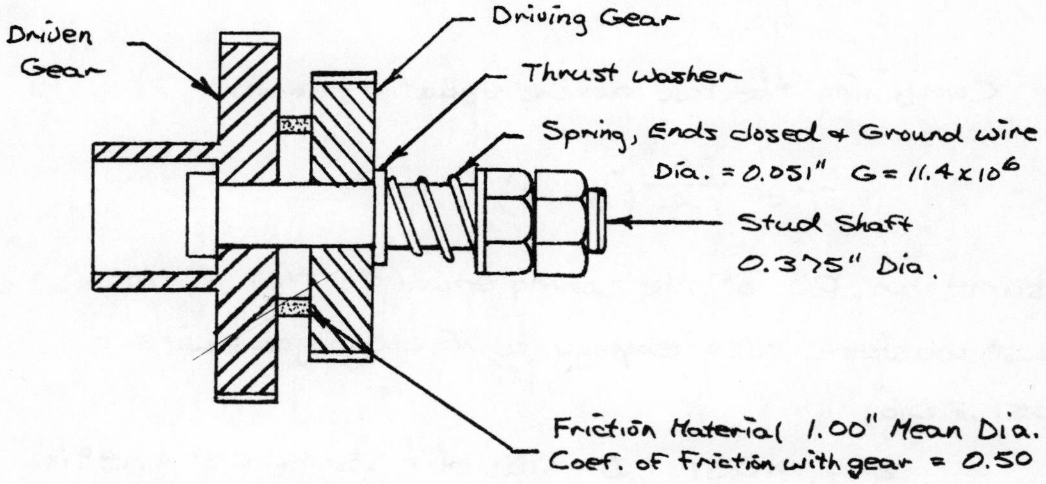

REQUIRED: If the torque is to be adjustable from 0 to 6 in. oz. and the maximum diameter of the thrust washer is 0.50, what are the dimensions of the spring?

Comment: There are numerous springs that will meet the requirements varying in free length, mean diameter and number of active coils. Since this requires a knowledge of spring design and the judgment to make some reasonable assumptions, it is not a problem in which the solution can be obtained by substituting numbers in a handbook equation.

Solution

Equations and Relationships to be used:

$$\delta_{spring} = \frac{8FD^3N}{d^4G} \quad (\text{Ref.- p 446 Shigley})$$

Torque - Integrate the Frictional force times radius (Ref.- p.734 Shigley) uniform wear

$$T = \int_{d/2}^{D/2} 2\pi f p r^2 dr = \frac{\pi f p_a d}{8}(D^2 - d^2) = \frac{Ff}{4}(D+d)$$

Normal Force (Ref-p.734 Shigley)

$$F = \int_{d/2}^{D/2} 2\pi p r dr = \frac{\pi p_a d}{2}(D-d)$$

Combining the two above equations gives,

$$T = \frac{Ff}{4}(D+d)$$

Assume an O.D. of the spring equal to that of the thrust washer, also employ a 16 awg gage wire (0.051 in Dia)

$d = 0.051$ $\quad D = \text{O.D.} - d = .500 - .051 = .449$ in.

I.D. = .500 - .102 = .398 in -ok-

ID Less than 3/8 in. Dia.

$N = 4$ active coils - Thus must add 2 coils for solid height since spring ends are closed and ground.

Solid Height = (4+2) 0.051 = 0.306 in

$$\text{Torque} = 6 \text{ in}\cdot\text{oz.} = 0.375 \text{ in}\cdot\text{lb}$$

Solve for F in the torque Eq. above.

$$F = \frac{4T}{f(D+d)} = \frac{4(0.375)}{0.5\ (2)} \qquad D_{mean} = \frac{D+d}{2}$$

$$D+d = 2D_{mean}$$

$$F = 1.5 \text{ lb.}$$

$$F = 1.5\text{lb} = \frac{\delta d^4 G}{8 D^3 N} = \frac{\delta (0.051)^4 (11.4\times10^6 \text{psi})}{8(.449)^3(4)}$$

$$\delta = \frac{1.5\ (2.88)}{0.0666\ (11.4\times10^6)} = \underline{\underline{0.0568 \text{ in.}}}$$

Thus for zero torque, the spring is free from the thrust washer. For max. torque, the spring must be deflected 0.057 in.

The deflection of the parts has been ignored in the analysis.

5

Fluid Mechanics

R. IAN MURRAY

FLUID STATICS

A fluid continuously deforms when subjected to shear forces. Therefore, if a fluid is at rest or if it moves as a rigid body, there can
be no shear stresses within or on the boundaries of the fluid. The
differential equation of the pressure field is then given by

$$dp = -\gamma \left[\frac{a_x}{g} \, dx + \frac{a_y}{g} \, dy + \left(1 + \frac{a_z}{g}\right) dz \right]$$

p = pressure

$\gamma = \rho g$ = specific weight (i.e., weight density; ρ is mass density)

a_x, a_y, a_z = components of acceleration

g = gravitational force per unit mass

z-axis is parallel to $\vec{g}$ but directed upward

If the density and acceleration are uniform throughout the body of
fluid under consideration, the above equation is easily integrated
to give the pressure distribution. For the important special case
in which the density is uniform and the acceleration is zero, the
pressure distribution is given by

$$\frac{p}{\gamma} + z = C \qquad \text{density uniform} \\ \text{accel} = 0$$

C is a constant that can be evaluated if the density is known and
the pressure is known at some reference elevation. From the pressure distribution the resultant force on a submerged surface can
readily be determined by applying the principles of statics for
distributed loads. The difference in pressure between discrete
points in the fluid can also be determined and is important in
instrumentation problems.

$$p_1 - p_2 = \gamma (z_2 - z_1)$$

IDEAL FLUID FLOW

When a fluid flows, shear stresses are produced within the fluid and at the boundaries. However, there are many important problems in which the effect of the shear stresses can be neglected. This tends to be true for low viscosity fluids when the distance along the flow path is short and the streamlines under consideration are not close to a fixed boundary. If, in addition, the flow is steady and the variation of density is negligible*, Bernoulli's equation is obtained:

$$H_1 = H_2 \quad \text{where} \quad H = z + \frac{p}{\gamma} + \frac{V^2}{2g} = \text{total head}$$

1 and 2 refer to two points on the same streamline.

$z + \frac{p}{\gamma}$ is the piezometric head; it decreases toward the center of curvature when streamlines are curved.

z is the potential head.

$\frac{p}{\gamma}$ is the pressure head.

$\frac{V^2}{2g}$ is the velocity head.

When changes of elevation are negligible Bernoulli's equation may be written:

$$p_1 + \rho \frac{V_1^2}{2} = p_2 + \rho \frac{V_2^2}{2} = p_o$$

Here p is the static pressure, $\rho \frac{V^2}{2}$ is the dynamic (or velocity) pressure, and p_o is the stagnation (or total) pressure.

REAL FLUID FLOW

When the effect of shear stress cannot be neglected, several alternative procedures are available:

1. Solution of the general equations of motion. This is beyond the scope of this review.

2. Use of empirical coefficients to modify ideal fluid flow solutions. Appropriate coefficients will be defined as they are encountered in the review problems. These coefficients may be used to cope with difficult geometric conditions as well as to correct for the effect of assuming ideal fluid flow conditions.

---

*For problems in which the variation of density is not negligible, see Chapter 8.

3. Augmentation of Bernoulli's equation with a head loss term for flow in pipes and ducts,

$$H_1 = H_2 + H_L$$

$$H_L = f \frac{\ell}{D} \frac{V^2}{2g} + \Sigma K_L \frac{V^2}{2g} = f(\frac{\ell}{D} + \Sigma \frac{L}{D}) \frac{V^2}{2g} = \text{head loss}$$

f is the Darcy friction factor and is a function of Reynolds' number and relative roughness. $f = f(N_R, e/D)$ can be obtained from a Moody diagram.

ℓ is the total length of pipe of diameter D along a single flow path connecting points 1 and 2.

K_L is a loos coefficient for a valve or fitting.

L is an equivalent length of pipe for a valve or fitting

$$N_R = \frac{VD\rho}{\mu} = \frac{VD}{\nu}$$

ρ = mass density

μ = dynamic viscosity

ν = kinematic viscosity

e = equivalent sand grain roughness

Good numerical data can be obtained from "Technical Paper 410", Crane Co., from "Standard Handbook for Mechanical Engineers" by Baumeister & Marks, or from most textbooks on fluid mechanics.

For noncircular ducts, D can be interpreted as the hydraulic diameter defined as 4A/P where A is the flow cross-sectional area and P is the wetted perimeter.

4. Use of experimental coefficients based on a dimensional analysis of the problem. In particular, for the problem of drag,

$C_D = f(N_R, N_F, N_M)$ for geometrically similar systems.

$$C_D = \frac{F_D}{A\rho \frac{V^2}{2}}$$ = drag coefficient; A is a characteristic area. F_D is drag force.

Letting L represent a characteristic dimension and a represent the velocity of sound,

$$N_R = \frac{VL}{\nu} = \text{Reynolds' number.}$$

$$N_F = \frac{V}{\sqrt{Lg}} = \text{Froude number}$$

$$N_M = \frac{V}{a} = \text{Mach number}$$

Note: Froude number enters into problems involving free surfaces; Mach number enters into problems of compressible flow. It is difficult to imagine a problem in which both would be involved. Frequently, the drag coefficient may be considered constant (limited range of independent variables).

CONSERVATION PRINCIPLES

A control volume is an open system whose boundaries are fixed with respect to an inertial coordinate system. An open system is a system across whose boundaries mass, energy, and momentum can flow. When the principles of mass, energy and momentum conservation are employed to obtain equations applicable to a control volume, the resulting equations relate changes within the control volume to conditions at the boundaries. For steady state the rate of change of mass, energy, and momentum <u>within</u> the control volume is zero.

In the following equations summations (Σ) are to be interpreted as addition for inflow quantities, subtraction for outflow quantities. In expressions for evaluating terms, one-dimensional flow is assumed. t is time.

1. Mass (continuity)

$$\Sigma\dot{m} = \frac{dm}{dt} \qquad \dot{m} = \rho AV = \rho Q = \text{mass flow rate}$$

A is area normal to V, Q is volumetric flow rate, $\frac{dm}{dt}$ is the rate of change of mass <u>within</u> the control volume.

2. Energy (1st law of thermodynamics)

$$\dot{Q} + P + \Sigma(\dot{m}e) = \frac{dE}{dt}$$

$\dot{Q}$ is net heat transfer rate (positive for net input). Avoid confusion with volumetric flow rate.

P is power (mechanical or electrical) (positive for net input).

Note: For a pump $P = \frac{Q\gamma H}{\eta}$ where η = pump efficiency and H is the <u>increase</u> in total head produced by the pump.

$\frac{dE}{dt}$ is rate of change of energy <u>within</u> the control volume.

e = specific energy associated with a flow

$= u + \frac{p}{\rho} + gz + \frac{V^2}{2} = u + gH$ (H = total head at section of boundary under consideration)

u = specific internal energy: $u + \frac{p}{\rho} = h$ = specific enthalpy

3. Momentum (Newton's 2nd law of motion)

a.) $\Sigma\vec{F} + \Sigma(\dot{m}\vec{V}) = \frac{d\vec{M}}{dt}$ (Linear Momentum Equation)

$\vec{F}$ = external force acting on the control volume.

$\vec{M}$ = momentum within the control volume.

Note: This is a vector equation. Summations must be made vectorially or else the equation must be resolved into three algebraic component equations.

b.) $\Sigma T + \Sigma\dot{m}\vec{r} \times \vec{V} = \frac{d\vec{\mathcal{H}}}{dt}$ (Angular Momentum Equation)

$\vec{T}$ = external torque or moment of force with respect to a chosen moment center.

$\vec{r}$ = position vector (relative to the moment center) of the centroid of the area over which the velocity is V.

$\vec{\mathcal{H}}$ = angular momentum within the control volume.

FLUID MECHANICS 1

A cylindrical tank in a rocket contains 10 gallons of water. The internal tank dimensions are 8" in diameter and 100" long. The tank is 3000 feet above sea level and is being accelerated parallel to its length at 20 feet per second at an angle 20 degrees above horizontal. The space not occupied by water is air at 100 psia.

Determine the angle of the liquid surface exposed to the air.

Solution

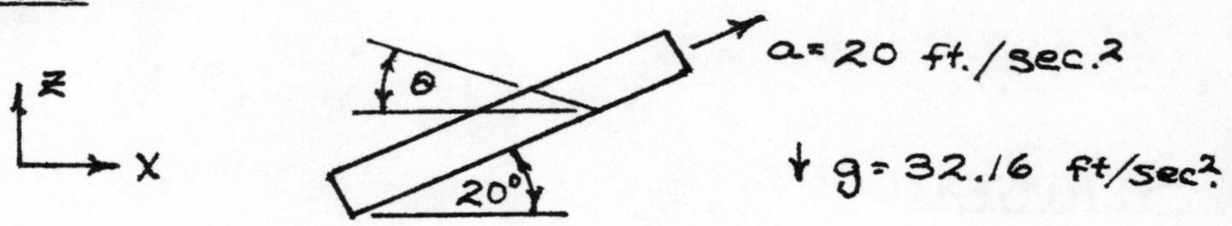

$$dp = -\gamma\left[\frac{a_x}{g}\,dx + \left(1+\frac{a_z}{g}\right)dz\right] = 0 \quad \text{(on interface)}$$

$$\theta = \tan^{-1}\left(-\frac{dz}{dx}\right) = \tan^{-1}\frac{a_x}{g+a_z}$$

$$a_x = a\cos 20^\circ = 18.80 \text{ ft./sec.}^2$$

$$a_z = a\sin 20^\circ = 6.85 \text{ ft./sec.}^2$$

$$\theta = \tan^{-1}\frac{18.80}{39.01} = \underline{25.7^\circ}$$

FLUID MECHANICS 2

Water flows through a perfect nozzle in the side of a water tank. The water level in the tank is held constant 20 feet above the ground level.

What height y should the nozzle be to make the stream from the orifice travel a maximum horizontal distance before it strikes the ground?

Ignore any friction.

Show all calculating and reasoning.

Solution

at nozzle exit: $v^2/2g = H = 20\text{ ft} - y$

$$\therefore v = \sqrt{2g(20\text{ ft.} - y)}$$

$x = vt$ and $y = \frac{1}{2}gt^2 \rightarrow t = \sqrt{2y/g}$

$$x = \sqrt{2g(20\text{ft.} - y)(2y/g)} = 2\sqrt{20\text{ ft.}\ y - y^2}$$

$\frac{dx}{dy} = 0$ when $20\text{ ft.} - 2y = 0$

$$y = \underline{10\text{ ft.}}$$

FLUID MECHANICS 3

Six measurements are made on a ventilating duct system as shown in the sketch below. The measurements are all made within two feet of the fan. The air velocity and duct resistance on the suction side of the fan are 2,000 feet per minute and 0.80 inches of water respectively. The air velocity and duct resistance on the discharge side of the fan are 4,005 feet per minute and 2.5 inches of water respectively. The inlet of the suction duct is in the weather and the discharged air is in spaces maintained at atmospheric pressure.

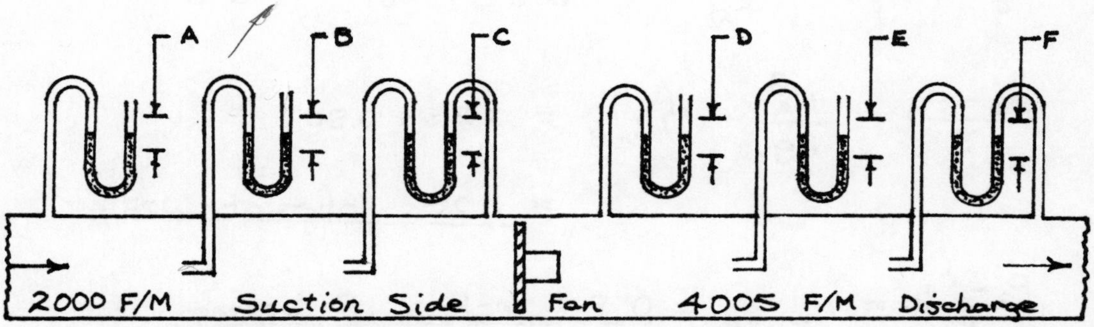

To find:

(a) With the fan in operation, what will be the manometer readings in inches of water at A, B, C, D, E, and F?

(b) Which of the manometers in the sketch indicates velocity pressure, total pressure, and static pressure for both the suction and discharge side of the fan?

Solution

Let subscripts refer to conditions at the point to which the left leg of the manometer of the subscript is attached. Let the subscript zero refer to the atmosphere.

a. $$\frac{V_A^2}{2g} = \frac{\left(\frac{2000}{60}\frac{ft.}{sec.}\right)^2}{64.4\ ft/sec^2} = 17.25\ ft.\ of\ air \times \frac{.075}{62.4} \times \frac{12\ in}{ft.}$$

$$= 0.25 \text{ inches of water}$$

$$\frac{U_D^2}{2g} = 4\frac{V_A^2}{2g} = 1.00 \text{ inch. of water}$$

$$\frac{P_O}{\gamma} = \frac{P_A}{\gamma} + \frac{V_A^2}{2g} + H_{L_{O-B}} = \frac{P_B}{\gamma} + H_{L_{O-B}}$$

$$\frac{P_O - P_A}{\gamma} = \frac{U_A^2}{2g} + H_{L_{O-B}} = 0.25 + 0.80$$

$$= \underline{1.05 \text{ inches of water}}$$

$$\frac{P_O - P_B}{\gamma} = H_{L_{O-B}} = \underline{0.80 \text{ inches of water}}$$

$$\frac{P_B - P_A}{\gamma} = \frac{V_A^2}{2g} = \underline{0.25 \text{ inches of water}}$$

* Note: Assume that the size of the duct on the discharge side of the fan is the same as on the suction side. The higher velocity on the discharge side is due to the blockage effect of the fan's wake. FUTHER DOWNSTREAM THE VELOCITY WILL AGAIN BE EQUAL TO V_A

$$\frac{P_E}{\gamma} = \frac{P_D}{\gamma} + \frac{U_D^2}{2g} = \frac{P_O}{\gamma} + \frac{U_A^2}{2g} + H_{L_{E-O}}$$

AN ALTERNATE ASSUMPTION WOULD BE THAT THE DUCT AREA ON THE DISCHARGE SIDE IS HALF THAT ON THE SUCTION SIDE. THIS WOULD YIELD A DIFFRENT SOLUTION

$$\frac{P_D - P_O}{\gamma} = H_{L_{E-O}} - \frac{V_D^2 - V_A^2}{2g} = 2.50 - (1.00 - 0.25)$$

$$= \underline{1.75 \text{ inches of water}}$$

$$\frac{P_E - P_O}{\gamma} = H_{L_{E-O}} + \frac{V_A^2}{2g} = 2.50 + 0.25$$

$$= \underline{2.75 \text{ inches of water}}$$

$$\frac{P_E - P_D}{\gamma} = \frac{V_D^2}{2g} = \underline{1.00 \text{ inches of water}}$$

b. Velocity Pressure : C & F

Total Pressure : B & E

Static Pressure : A & D

FLUID MECHANICS 4

Your plant operates a sheet plastic extrusion press with a nozzle .005" thick x 96" wide. You wish to increase production by installing a larger motor, speed reducer and pump combination. At 10 ft per second extrusion velocity, the pressure at the nozzle will be 1250 psi.

What size electric motor and what pump capacity are required to obtain the 10 ft per second rate?

<u>Solution</u>

$$Q = AV = 0.005 \text{ in} \times 96 \text{ in.} \times 10 \text{ ft/sec} = 4.80 \text{ in}^2 \frac{\text{ft.}}{\text{sec.}}$$

$$= 4.80 \text{ in}^2 \frac{\text{ft}}{\text{sec}} \times \frac{\text{ft}^2}{144 \text{ in}^2} \times \frac{7.48 \text{ gal.}}{\text{ft.}^3} \times 60 \frac{\text{sec.}}{\text{min.}}$$

$$= \underline{15 \text{ gpm}}$$

$$P_{ideal} = \dot{m}\frac{\Delta p}{\rho} = Q\Delta p$$

$$= \frac{4.80\ in^2\ \frac{ft.}{sec.} \times 1250\ \frac{lb}{in.^2}}{550\ \frac{ft \cdot lb}{hp \cdot sec}} = \underline{10.9\ Hp}$$

* Assuming reasonable efficiencies for pump, and speed reducer, a 20 Hp motor will be required.

FLUID MECHANICS 5

A 16" x 16" x 24" steel piping tee is to carry 10,000 gpm water (5,000 gpm from each branch) at 50 psi pressure. The joints are to be made with Dresser-type couplings that will not transmit axial thrust.

Find the reaction at its anchor.

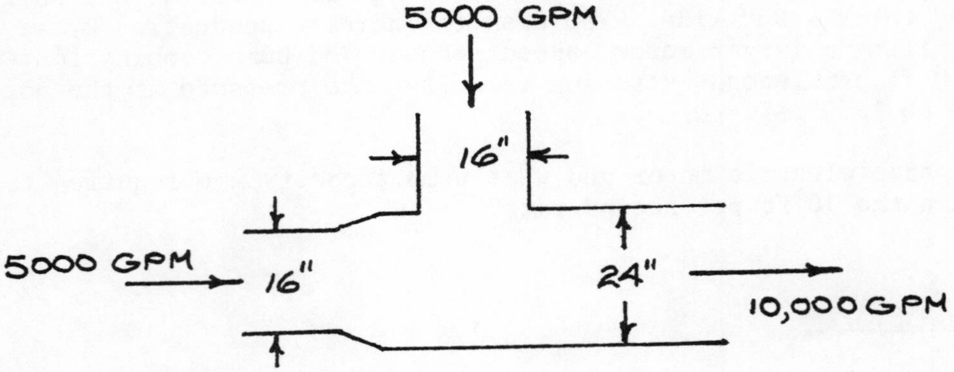

Solution

$$V_1 = V_2 = \frac{Q_1}{A_1} = 7.97\ ft./sec.$$

$$V_3 = \frac{Q_3}{A_3} = 7.09\ ft./sec.$$

* Note: Since changes in velocity are small, assume uniform pressure.

$$F_1 = F_2 = pA_1 = 50\ \text{lb/in}^2 \times \frac{\pi}{4}(16)^2\ \text{in}^2 = 10050\ \text{lb}$$

$$F_3 = pA_3 = 50 \times \pi/4\,(24)^2 \qquad = 22650\ \text{lb}$$

$$(\dot{m}V)_3 = \rho A_3 U_3^2 = \frac{\gamma A_3 U_3^2}{g} = \frac{62.4 \times \pi/4\,(2)^2(7.09)^2}{32.2}$$

$$= 307\ \text{lb.}$$

* Momentum changes are negligible

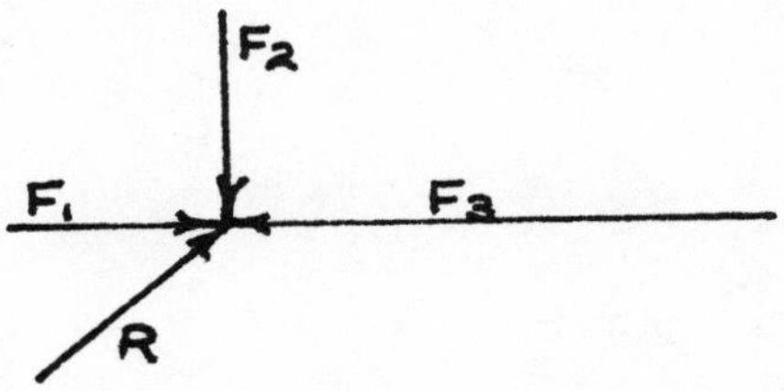

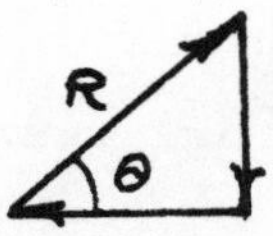

$F_2 = 10050$ lb

$F_3 - F_1 = 12600$ lb

$$\theta = \tan^{-1}\frac{10050}{12600} = \underline{38.6^\circ}$$

$$R = F_2/\sin\theta = \underline{16100\ \text{lb}}$$

FLUID MECHANICS 6

A recent book describing World War II submarines is "Iron Coffins", 1st edition, Holt, Rinehart & Winston, New York, 1969, by H. A. Werner. On page 15 of the book it is stated that "Two electric motors, operating on gigantic storage batteries, ran the ship when she was submerged; they would propel the boat for one hour at the top speed of nine knots, or for three days at a cruising speed of one or two knots".

Give a technical explanation of the variation of underwater duration with speed noted by Captain Werner.

Solution

$$\text{Power output} = -\eta \frac{dE}{dt} \qquad \text{where } \eta = \text{efficiency}$$

$$= F_D V = \tfrac{1}{2} C_D A \rho V^3$$

$$-\frac{dE}{dt} = \frac{C_D A \rho}{2\eta} V^3$$

$$E_{max} - E_{min} = \frac{C_D A \rho}{2\eta} \int_0^t V^3 \, dt$$

* if $\dfrac{C_D A \rho}{2\eta}$ is constant (an assumption)

$$\text{then} \int_0^{t_1} V_1^3 \, dt = \int_0^{t_2} V_2^3 \, dt$$

If V_1 and V_2 are constant (given),

$$\text{then } V_1^3 t_1 = V_2^3 t_2$$

$$V_2 = \sqrt[3]{\frac{t_1}{t_2}} \, V_1 = \frac{9 \text{ Knots}}{\sqrt[3]{72}} = 2.16 \text{ Knots}$$

Therefore, 1 to 2 Knots for 3 days is feasible.

FLUID MECHANICS 7

A high-rise 54-story office building in San Francisco has 2 water tanks located in the penthouse. The centrifugal pumps for supplying water to these tanks are located on the basement floor. Each pump discharges 400 gallons per minute through a 6-inch schedule 40 pipe. The pressure of the water entering the pump is 60 pounds per square inch.

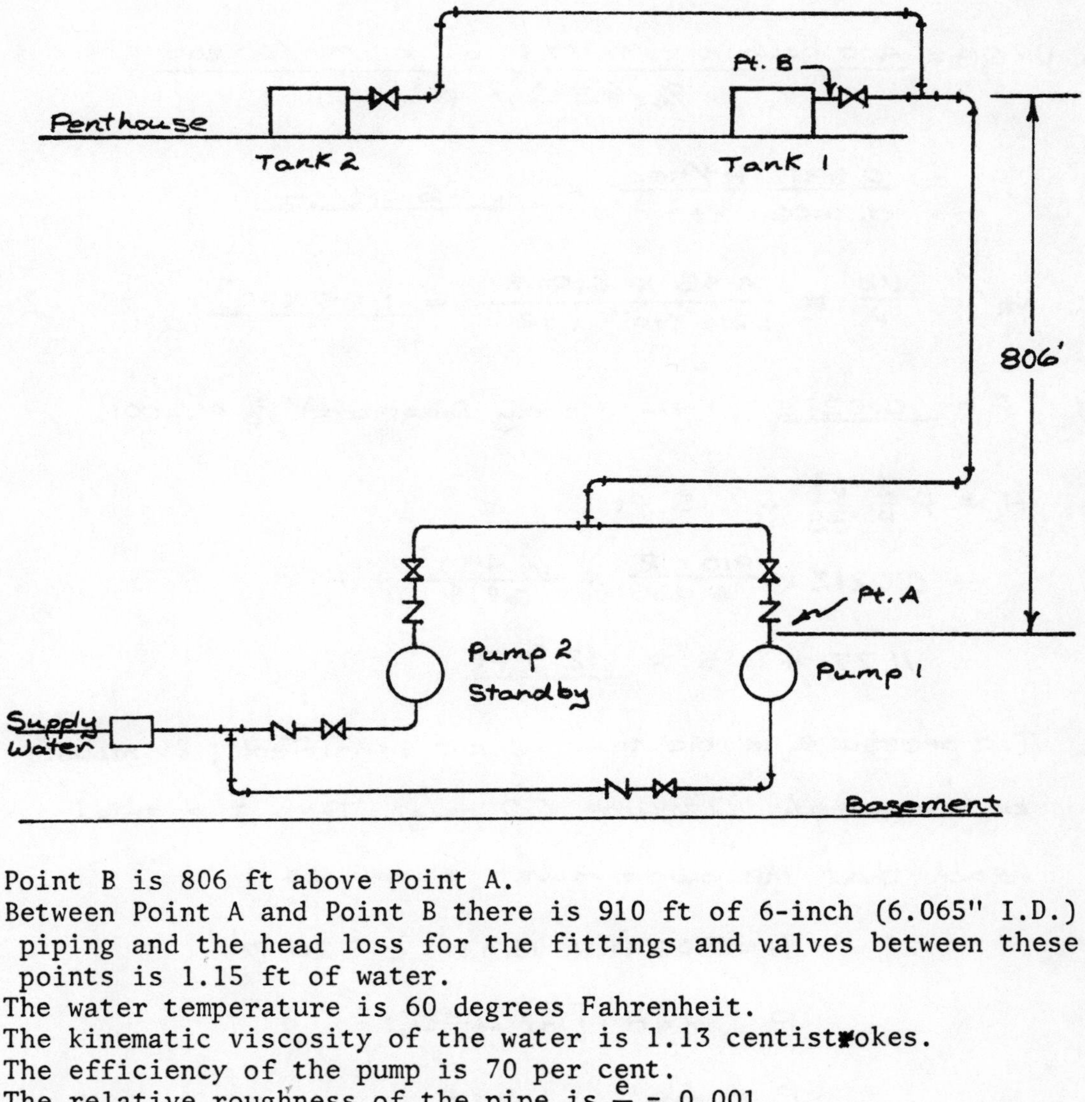

Point B is 806 ft above Point A.
Between Point A and Point B there is 910 ft of 6-inch (6.065" I.D.) piping and the head loss for the fittings and valves between these two points is 1.15 ft of water.
The water temperature is 60 degrees Fahrenheit.
The kinematic viscosity of the water is 1.13 centistokes.
The efficiency of the pump is 70 per cent.
The relative roughness of the pipe is $\frac{e}{D} = 0.001$.

Determine:
a. Kinematic viscosity of the water in ft^2/sec.
b. Velocity of the water in the pipe in feet per second.
c. Reynolds number.
d. Friction factor (f).
e. Total head drop between Point A and Point B.
f. The pump horsepower required.

Solution

a. $1.13 \text{ centistokes} \times \dfrac{cm^2}{100 \text{ centistoke sec}} \times \dfrac{in^2}{(2.54 cm)^2} \times \dfrac{ft^2}{144 in^2}$

$= \underline{1.216 \times 10^{-5} \quad ft^2/sec.}$

b. $V = Q/A = \dfrac{400 \text{ gal/min} \times ft^3/7.48 \text{ gal} \times \text{min}/60 \text{ sec}}{\frac{\pi}{4}\left(\frac{6.065}{12}\right)^2 ft^2}$

$= \dfrac{0.891 \quad ft^3/sec}{0.2006 \quad ft^2} = \underline{4.45 \text{ ft/sec}}$

c. $N_R = \dfrac{VD}{\nu} = \dfrac{4.45 \times 6.065}{1.216 \times 10^{-5} \times 12} = \underline{1.85 \times 10^5}$

d. $f = \underline{0.0212}$ (from Moody Diagram)($\frac{e}{D} = 0.001$)

e. $H_L = f \dfrac{\ell}{D} \dfrac{V^2}{2g} + 1.15 \text{ ft}$

$= 0.0212 \times \dfrac{910 \times 12}{6.065} \times \dfrac{(4.45)^2}{64.4} + 1.15$

$= 11.77 + 1.15 = \underline{12.9 \text{ ft}}$

f. The pressure in the tank is not specified; it must be assumed. Assume 60 psig. Then the total head that the pump must supply is:

$$H = 806 \text{ ft.} + 13.3 \text{ ft.} \approx 820 \text{ ft}$$

$$P = \dot{m} g H + [\dot{m}(u_2 - u_1) - \dot{Q}]$$

$$P = \frac{\dot{m} g H}{\eta} = \frac{Q \gamma H}{\eta}$$

$$= \frac{0.891 \text{ ft}^3/\text{sec} \times 62.4 \text{ lb/ft}^3 \times 820 \text{ ft.}}{0.70 \times 550 \quad \text{ft} \cdot \text{lb/hp} \cdot \text{sec.}}$$

$= \underline{118 \text{ hp}}$

FLUID MECHANICS 8

A ranch complex requires a maximum water supply rate of 300 cubic feet per minute. The supply is a reservoir with its surface 250 ft above the point of delivery, and one-quarter mile distant. Minimum pressure required at point of delivery is 50 psig.

What minimum size of steel pipe would be needed for this pipe line from the reservoir to the delivery point?

Solution

$$H_1 = H_2 + H_L$$

$$250 \text{ ft} = \frac{50 \times 144}{62.4} + \frac{V^2}{2g} + \left(f\frac{\ell}{D} + \Sigma K_L\right)\frac{V^2}{2g}$$

$$\left(f\frac{\ell}{D} + 1.5\right)\frac{V^2}{2g} = 250 - 115 = 135 \text{ ft.}$$

(assuming a sharp-edged entrance)

$$V = \frac{Q}{A} = \frac{300 \text{ ft}^3/\text{min} \times \frac{1 \text{ min}}{60 \text{ sec}}}{\frac{\pi}{4} \times \frac{D^2}{144}} \qquad \left\{ \begin{array}{l} V - \text{ ft/sec} \\ D - \text{ inches} \end{array} \right.$$

* Try D = 10 inches : $V = 9.16$ ft/sec , $\frac{V^2}{2g} = 1.30$ ft

$$VD = 91.6$$

$$\frac{e}{D} = 0.00018 \qquad f = 0.0149$$

$$f\frac{\ell}{D} = \frac{0.0149 \times 5280 \times 12}{4 \times 10} = 23.6$$

$$(28.6 + 1.5) \times 1.30 = 32.6 < 135$$

* Try D = <u>8 inches</u> : $100 < 135$

FLUID MECHANICS 9

Determine a standard size pipe for the drain line for the tank system shown on the sketch so that it will drain a full tank of water in as close to 9 hours as possible.

Show your calculations.

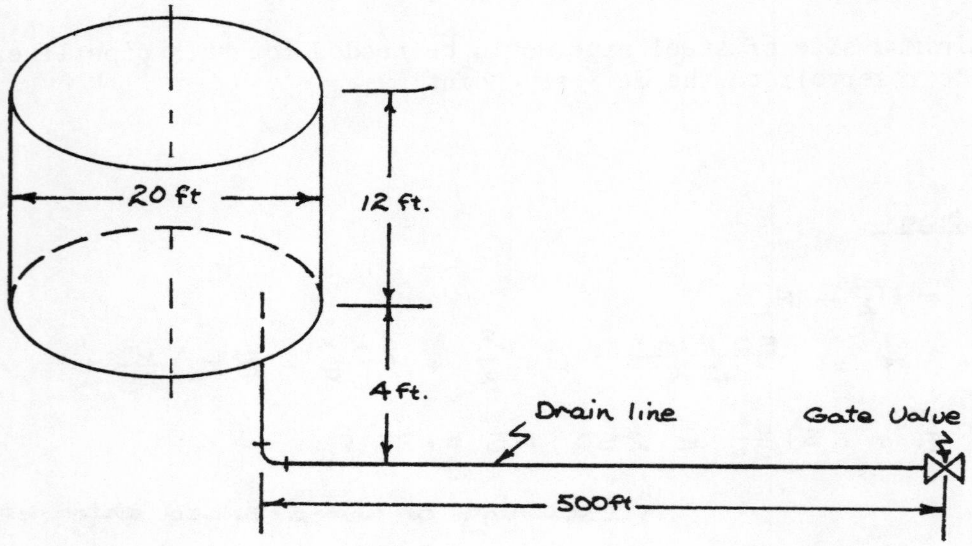

Solution

* Assume quasi-steady flow ; i.e., Bernoulli's augmented equation holds at any time t even though H changes with time.

$$H_1 = H, \quad H_2 = \frac{U^2}{2g} = \frac{Q^2}{2gA^2}, \quad H_L = f\left(\frac{\ell}{D} + \Sigma \frac{L}{D}\right)\frac{U^2}{2g}$$

Thus,

$$H = \left[1 + f\left(\frac{\ell}{D} + \Sigma \frac{L}{D}\right)\right]\frac{Q^2}{2gA^2} = KQ^2, \text{ assuming } f \text{ is constant}$$

By Continuity $Q = \dfrac{\pi (20\text{ ft})^2}{4}\left(-\dfrac{dH}{dt}\right) = \sqrt{\dfrac{H}{K}}$

$$\int_{16\text{ ft}}^{4\text{ ft}} H^{-1/2}\, dH = \int_0^{9\text{ hrs.}} \frac{-dt}{100\pi\sqrt{K}\ \text{ft}^2}$$

$$2H^{1/2}\Big]_{16\text{ ft.}}^{4\text{ ft.}} = \frac{-9\text{ hrs}}{100\,\pi\sqrt{K}\ \text{ft}^2} = 2(2-4)\ \text{ft}^{1/2} = -4\ \text{ft}^{1/2}$$

$$\sqrt{K} = \frac{9\text{ hrs.} \times 3600\text{ sec/hr}}{400\,\pi\ \text{ft}^{5/2}} = 25.8\ \text{sec. ft}^{-5/2}$$

$$K = 665\ \text{sec}^2\ \text{ft}^{-5}$$

$$Q_{min} = \sqrt{\frac{H_{min}}{K}} = \frac{\sqrt{4\text{ ft}}}{25.8\text{ sec ft.}^{-5/2}} = 0.0775\ \text{ft}^3/\text{sec}$$

$$Q_{max} = \sqrt{\frac{H_{max}}{K}} = \frac{4}{25.8} = 2\,Q_{min}$$

* Assume: wide open gate valve $L/D = 13$

Standard 90° Elbow $L/D = 30$

Sharp-Edged Entrance $K_L = 0.5$

* Crane : "Flow of Fluids" Tech. paper 410

Then,

$$K = \frac{1.5 + f\left(\frac{504\text{ ft}}{D} + 43\right)}{2g\left(\frac{\pi}{4}\right)^2 D^4}$$

$$\frac{1.5 + f\left(\frac{504\text{ ft}}{D} + 43\right)}{D^4} = 665 \times 64.4\left(\frac{\pi^2}{16}\right)$$

$$= 26{,}500\ \text{ft}^{-4}$$

* Assume $D = 3'' = .25$ ft and steel

$$V_{min} = Q/A = \frac{.0775}{\frac{\pi}{4}\left(\frac{1}{4}\right)^2} = 1.58\ \text{ft/sec.}$$

$$VD_{min} = 4.74\ \frac{\text{ft}}{\text{sec}}\ \text{in} \qquad VD_{max} = 9.48\ \frac{\text{ft}}{\text{sec}}\ \text{in}$$

$$\frac{e}{D} = .0006 \qquad f_{mean} = .023$$

$$[1.5 + .023(2016 + 43)] \times 256\ \text{ft}^{-4} = 12{,}500 < 26{,}500\ \text{ft}^{-4}$$

*Assume $D = 2\frac{1}{2}''$

$$U_{min} = 1.58 \times \left(\frac{3}{2.5}\right)^2 = 2.28 \text{ ft/sec}$$

$$UD_{min} = 5.7 \qquad UD_{max} = 11.4$$

$$e/D = .0007 \qquad f_{mean} = 0.023$$

$$\left[1.5 + .023(2420 + 43)\right]\left(\frac{12}{2.5}\right)^4 = 30{,}900 \text{ ft}^{-4} > 26{,}500 \text{ ft}^{-4}$$

Therefore, use <u>2½" steel pipe</u>

FLUID MECHANICS 10

A pipe line with diameter 36 in and length 2.5 miles, connects two fresh-water open reservoirs which have their water surfaces at elevations 240 ft and 165 ft, respectively. In order to increase the rate of flow between the reservoirs by exactly 30 per cent, it is decided to lay an additional 36 in diameter pipe line from the upper reservoir. The second pipe line is to lie parallel to the original pipe line and is to be connected to the latter at the point which produces this increase in flow.

Determine this point of connection, assuming that the friction factor, f, has the value 0.01 for each pipe line.

Minor losses may be considered to be negligible.

Solution

$$H_L = f \frac{\ell}{D} \frac{U^2}{2g} = \left(\frac{8f}{\pi^2 g D^5}\right) \ell Q^2$$

By assumption $\frac{8f}{\pi^2 g D^5}$ is the same throughout old and new systems.

Let x = length of new pipe and note that total head loss is the same for both systems.

Thus, $\ell Q^2 = x\left(\frac{1.3}{2} Q\right)^2 + (\ell - x)(1.3Q)^2$ and $x = 0.544\,\ell$

$= \underline{1.36 \text{ mi.}}$

FLUID MECHANICS 11

Fuel oil is to be removed from a barge by a portable pump at a rate of 200 gallons per minute. The pump is to discharge the oil into a 6-inch schedule 40 steel pipe oil transfer line that runs from the dock to a storage tank, as shown schematically. Point A and Point B are at the same elevation

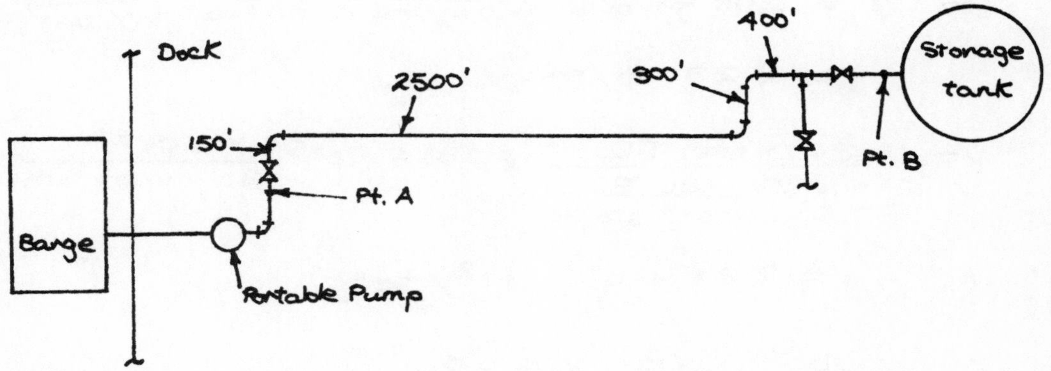

Required fittings and valves:

4 long-radius flanged 90-degree ells
1 tee, flanged
2 gate valves, flanged, wide open

Oil tempeature is 50°F, and its viscosity is 3400 Saybolt Seconds Universal. Specific gravity of the oil at 50°F is 0.934.

Determine the pump power and head required to perform this task.

Solution

* Assume entrance and exit losses are negligible and that there is no change in elevation.

$$H_L = f\left(\frac{\ell}{D} + \Sigma \frac{L}{D}\right)\frac{V^2}{2g} = H \text{ (required from pump)}$$

$$\ell/D = \frac{150 + 2500 + 300 + 400}{1/2} = 6700$$

$$L/D = 4 \times 20 = 80 \quad 90^\circ \text{ ells (long radius)}$$

$$1 \times 20 = 20 \quad \text{Tee (flow through run)}$$

$$2 \times 13 = \underline{26} \quad \text{Gate values (wide open)}$$

$$\Sigma L/D = 126 \quad \text{(turbulent flow)}$$

$$U = Q/A = 200 \text{ gal/min.} \times \frac{ft^3}{7.48 \text{ gal.}} \times \frac{min}{60 \text{ sec}} \times \frac{4 \times 144}{\pi (6.06)^2 \, ft^2}$$

$$= 2.23 \text{ ft/sec}$$

$$\nu = \frac{3400 \text{ sec. S.U.}}{4.635 \ \frac{\text{sec. S.U.}}{\text{centistoke}}} \times 1.076 \times 10^{-5} \ \frac{ft^2}{\text{sec centistoke}}$$

$$= 7.88 \times 10^{-3} \ ft^2/sec$$

$$N_R = \frac{UD}{\nu} = \frac{2.23 \times 6.06}{7.88 \times 10^{-3} \times 12} = 143 \quad (\therefore \text{ Laminar flow})$$

$$f = \frac{64}{N_R} = \frac{64}{143} = 0.448$$

$$(L/D)_{\text{Laminar}} = \frac{N_R}{1000} \left(\frac{L}{D}\right)_{\text{Turbulent}} \quad \text{for } N_R < 1000$$

(from Crane, technical paper 410)

$$\Sigma L/D = 0.143 \times 126 = 18$$

$$H_L = 0.448 \times 6720 \times \frac{(2.23)^2}{64.4} = 233 \text{ ft.} = H$$

$$P = \frac{Q \gamma H}{\eta} \quad \text{assume } \eta = 60\%$$

$$P = \frac{200 \text{ gal/min}}{7.48 \text{ gal/ft}^3} \times \frac{.934 \times 62.4 \text{ lb/ft}^3 \times 233 \text{ ft.}}{.6 \times 33000 \ \frac{\text{ft} \cdot \text{lb}}{\text{hp} \cdot \text{min}}}$$

$$= \underline{18.3 \text{ hp}}$$

FLUID MECHANICS 12

A hydro-pneumatic tank is to be used as an accumulator in a five-story building water system, as shown.

The pressures are a maximum of 90 psig on the first floor and a minimum of 50 psig on the fifth floor. First floor elevation is 120 feet, fifth floor, 160 feet, and tank, 170 feet. The pump is started and stopped by a pressure switch on the tank. The pump has a constant flow of 100 gpm when it runs. The system will use water at anywhere from zero to 100 gpm.

What capacity tank would you specify to allow a maximum of 6 pump starts per hour? Neglect friction loss.

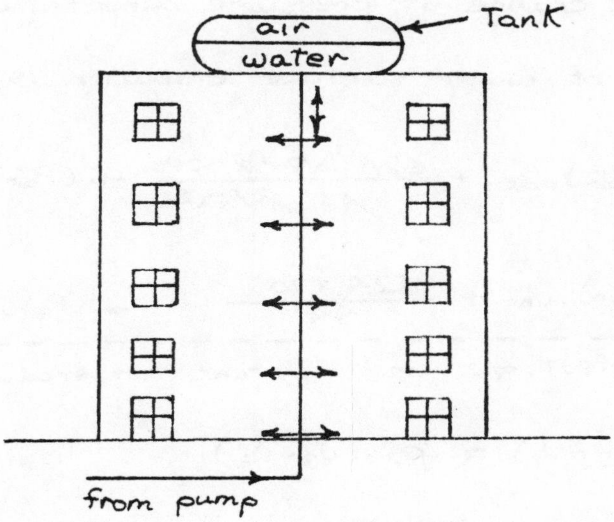

Solution

Hydro-pneumatic Tank

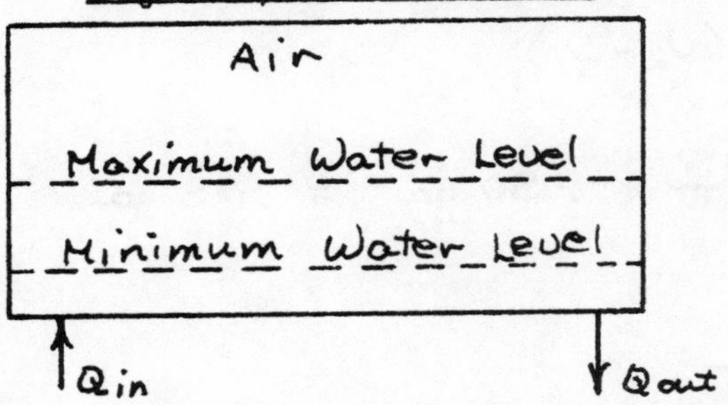

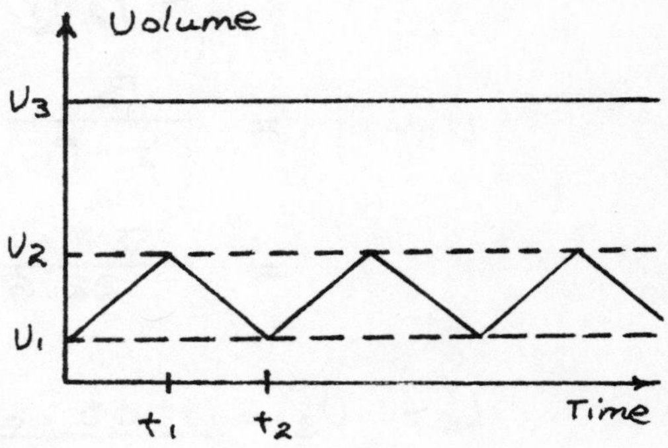

$\Sigma \dot{m} = \frac{dm}{dt}$; for ρ = constant $\Sigma Q = \frac{dV}{dt}$; V = volume

$$(Q_{in} - Q_{out})t_1 = Q_{out}(t_2 - t_1) = U_2 - U_1$$

Let $x = \dfrac{Q_{out}}{Q_{in}}$ Then $(1-x)t_1 = x(t_2 - t_1)$ and $t_1 = x t_2$

$$U_2 - U_1 = Q_{in}(1-x)x t_2 \ ; \quad \frac{d(U_2 - U_1)}{dx} = Q_{in} t_2 (1-2x)$$

$(1-2x) = 0 \rightarrow x = 1/2$ for maximum value of $U_2 - U_1$

$$U_2 - U_1 = 100 \text{ gpm} \times 1/2 \times 1/2 \times 10 \text{ min} = 250 \text{ gal.}$$

---

* Assume the change of pressure directly due to variation of water surface elevation is negligible.

$$90 \text{ psig} = (P_{air})_{max} + \frac{62.4 \text{ lb/ft}^3 \times 50 \text{ ft}}{144 \text{ in}^2/\text{ft}^2} \rightarrow (P_{air})_{max} = 68.3 \text{ psig}$$

$$50 \text{ psig} = (P_{air})_{min} + \frac{62.4 \times 10}{144} \rightarrow (P_{air})_{min} = 45.7 \text{ psig}$$

---

* Assume perfect gas and constant temperature

Then $p_1(U_3 - U_1) = p_2(U_3 - U_2)$

$$= p_2[(V_3 - V_1) - (V_2 - V_1)]$$

$$U_3 - U_1 = \frac{p_2}{p_2 - p_1}(U_2 - U_1)$$

$$= \frac{68.3 + 14.7}{22.6} \times 250 \text{ gal} = 918 \text{ gal.}$$

Let $U_3 = \underline{1000 \text{ gal}}$

FLUID MECHANICS 13

Two identical pumps, each pump 800 gallons per minute of water against a total delivery head of 475 feet, as shown.

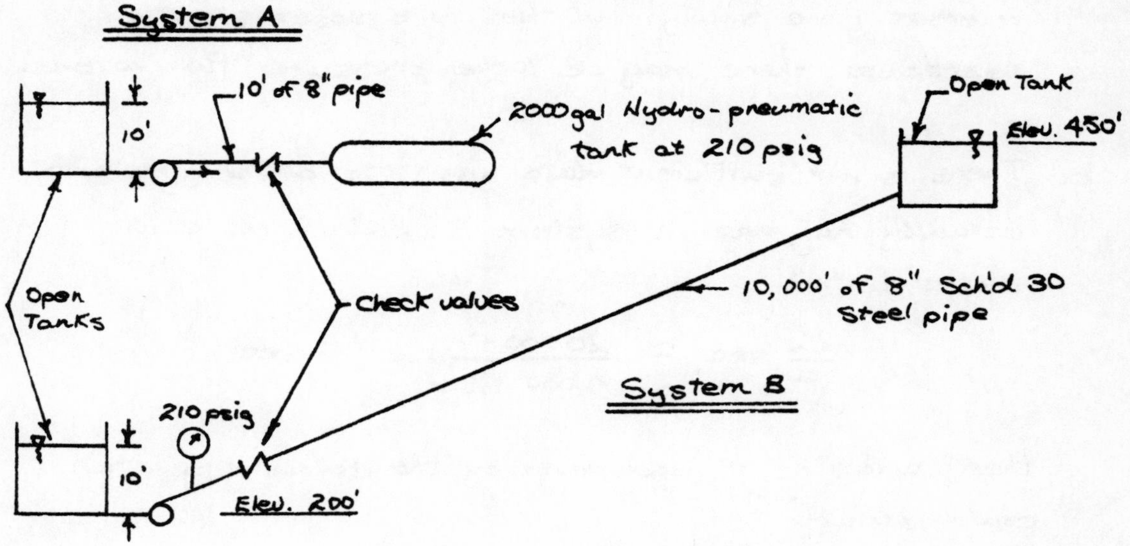

An observer is puzzled that the swing check valve on A slams noisily when the pump stops, while an identical valve on B closes quite acceptably. He feels that this is contrary to his understanding that water hammer is a problem only on long pipe lines.

(a) Explain the phenomenon.

(b) The valve flanges are rated for 400 psi. Is this adequate?

(c) Is the interior of the pipe line of system B in good, fair, or poor condition?

Solution

a. The noisy closing of a valve does not necessarily imply water hammer, nor does the quiet closing of a valve imply that there is no water hammer problem. Water hammer is associated with large pressure fluctuations due to the inertial effects in a transcient flow. In system A there is very little inertia, the flow quickly reverses and the valve slams shut; but the pressure is maintained at approximately 210 psi. In system B there is a great deal of inertia; the flow reverses more slowly and the valve closes quietly. Nevertheless, there may be large pressure fluctuations.

b. There is not sufficient data given to determine what actually happens in system B, but if the valve closes within

$$\frac{2L}{a}\text{ sec} \simeq \frac{20{,}000\text{ ft}}{4\,000\text{ ft/sec}} = 5\text{ sec}$$

then variations of head given by the following equation could occur.

$$\Delta H = \frac{aU}{g} \approx \frac{4000 \times 5\ \text{ft}^2/\text{sec}^2}{32.2\ \text{ft/sec}^2} = 620\text{ ft.}$$

* See below for calculation of U.

This corresponds to a pressure variation of 269 psi. This much pressure reduction would not occur, for the pressure can not fall below the vapor pressure. Column separation occurs and the pipe would probably collapse. In any event, on rejoining a large pressure rise occurs. The flanges will be subject to pressures greater than their rating for short periods of time. They probably will not fail, however, because of the safety factor in the rating.

c. $$Q = 800\text{ gpm} \times \frac{\text{ft}^3}{7.48\text{ gal}} \times \frac{\text{min}}{60\text{ sec.}} = 1.78\ \text{ft}^3/\text{sec}$$

$$U = Q/A = \frac{1.78}{\frac{\pi}{4}\left(\frac{8.071}{12}\right)^2} = 5.02\ \text{ft/sec}$$

$$\frac{V^2}{2g} = 0.391 \text{ ft.}$$

$$H_L = 475 - (450 - 210) = 235 \text{ ft.}$$

* Assume negligible minor losses

$$H_L = f \frac{l}{D} \frac{V^2}{2g} \qquad f = \frac{235}{15000 \times 0.391} = 0.04$$

* For new steel pipe :

$$e/D = \frac{0.00015}{2/3} = 0.000225$$

$$VD = 5 \text{ ft/sec} \times 8 = 40$$

$f = 0.017$ (Moody Diagram)

$$0.04 >> 0.017$$

pipe is in poor condition

FLUID MECHANICS 14

You are selecting the pipe diameter for an 800-foot pipe transmission line with one outlet to irrigate a field of yet undetermined size. The required flow is 8 gallons per minute per acre and the consumption is 4.1 acre-feet per acre, per year. The water is pumped with a pump of 75 per cent overall efficiency at a cost of 0.7¢ per kilowatt-hour. At 325 gallons per minute, friction is 101 feet per 1000 feet in 4 in pipe, and 12 feet per 1000 feet in 6 in pipe. Economic life is 40 years and money is worth 7%. The 4 in pipe costs \$1.25 and the 6 in pipe costs \$1.50 per foot installed.

Find the economic break-even point between the two pipe sizes.

Solution

Let A be the number of acres to be irrigated

Let T be the pumping time per year

Then $Q = 8A$ gpm ; $QT = 4.1A$ acre-ft./year

$4.1/8$ acre·ft/gpm = 2783 hrs./year = T

Note : 1 acre·ft = 43560 ft^3 ; 7.48 gal. = 1 ft^3

$$\frac{H_4}{101\text{ ft}} = \frac{H_6}{12\text{ ft.}} = \frac{800}{1000}\left(\frac{8A}{325}\right)^2 = 4.847\times10^{-4}A^2$$

Note: the above is from $H_L = f\frac{L}{D}\frac{v^2}{2g}$ and $v = \frac{4Q}{\pi D^2}$

assumes f is independent of N_R

$$P = \frac{Q\gamma H}{\eta} \quad \therefore \frac{P_4}{101\text{ ft}} = \frac{P_6}{12\text{ ft.}} = \frac{62.4\text{ lb/ft}^3 \times 8 \times 4.847\times10^{-4}A^3\text{ gpm}}{0.75}$$

$$\frac{62.4\frac{\text{lb}}{\text{ft}^3}\times 8\times 4.847\times10^{-4}A^3\frac{\text{gal}}{\text{min}}}{0.75}\times\frac{\text{ft}^3}{7.48\text{ gal}}\times\frac{\text{hp}\cdot\text{min}}{33\times10^3\text{ ft}\cdot\text{lb}}\times\frac{0.746\text{ kw}}{\text{hp}}$$

$$\frac{P_4}{101} = \frac{P_6}{12} = 9.75\times10^{-7}A^3\text{ kw}$$

Annual pumping cost due to pipe friction = $PT\times\$.007 = \$19.48\,P$

4" pipe : $(\$19.48\times9.75\times10^{-7}A^3)\times101 = \$1.918\times10^{-3}A^3$

6" pipe : $(\$1.899\times10^{-5}A^3)\times12 = \$0.228\times10^{-3}A^3$

Annual investment cost : (capital recovery factor = .0750)

4" pipe : 800 ft × \$1.25/ft × .0750 = \$75

6" pipe : $\$75\times\frac{1.50}{1.25}$ = \$90

For break-even point, increment of annual investment cost equals increment of savings in operating costs. Thus

$\$15 = \$1.690\times10^{-3}A^3 \rightarrow$ <u>A = 20.7 acres</u>

6

Thermofluid Mechanics

RICHARD K. PEFLEY

The more elaborate problems associated with this topic may require a complex mix of determinations involving system behavior, physical principles and property relationships. A brief summary of the major features relating to such problems follows:

A. SOLUTION PROCEDURE CHECK LIST

1. Establish whether the system is an open (control volume) or closed system--recall that open and closed relates to matter (working substance) crossing the system's boundaries.

2. Specify whether the event being examined constitutes a process or cycle.

3. Determine the phase region in which the working substance is situated.

4. Identify and apply the physical principles (commonly Newton's second law of motion, the first and second laws of thermodynamics, and conservation matter) that give insight to the problem.

5. Introduce working substance property relationships and process or cycle specifications.

6. Achieve a solution and reflect to see if the answer makes sense.

B. POINTS TO REMEMBER RELATIVE TO PURE SUBSTANCE MOLECULAR STATE BEHAVIOR

1. In the single phase region the substance has two degrees of freedom, i.e., two properties such as p and T may be independently varied.

2. If two phases are in equilibrium, each phase component loses one degree of freedom, i.e., pressure and temperature become dependent and specifying one establishes all other properties such as density and viscosity.

3. If three phases coexist, the molecular state of each phase component is fixed, i.e., each phase has zero degrees of freedom and only the amount of each phase can be varied.

4. For most processes involving pure solids or liquids, the density and specific volume can usually be treated as constants. As a second approximation the coefficient of compressibility (κ) and cubical expansion (β) can be treated as constants which allows the density and volume change to be determined by

$$\ell n \frac{\nu_2}{\nu_1} = \ell n \frac{\rho_1}{\rho_2} = \kappa(p_2 - p_1) + \beta(T_2 - T_1)$$

5. For the perfect gas region the density is related to pressure and temperature by

$$\frac{p}{\rho} = RT$$

C. PROCESS DESCRIPTIONS — PROCESSES ARE DESCRIBED BY PROPERTIES THAT REMAIN CONSTANT

isothermal - constant temperature
isobaric - constant pressure
isometric - constant volume
isentropic - constant entropy
throttling - constant enthalpy

Note that processes produce different effects depending upon where they take place in the phase regions. For example, a throttling process results in a slight increase in temperature for liquids, a drop in temperature for a mixture of liquid and vapor and a constant temperature for a perfect gas.

D. PROCESS EFFICIENCIES ARE PROCESS RATINGS

For examples consider volumetric and isentropic efficiencies of a reciprocating compressor.

$$\eta_{vol} = \frac{\text{actual volume of fluid displaced}}{\text{ideal volume of fluid displaced}}$$

$$\eta_{isen} = \frac{\text{reversible adiabatic work of compression}}{\text{actual work of compression}}$$

One efficiency has to do with the breathing quality of the process while the other has to do with energy use efficiency.

E. CYCLE AND CYCLE RATINGS

The Carnot cycle is the basic cycle for considering conversion of heat to work--a heat engine, or for pumping heat--a heat pump.

Cycle rating definitions and their particular values for the Carnot cycle are as follows:

$$\eta_{t_{\text{heat eng.}}} = \frac{W_{net}}{Q_{added}} = 1 - \frac{T_c}{T_h}$$

$$\text{C.O.P.}_{\text{heat pump}} = \frac{Q_{delivered}}{W_{net}} = \frac{1}{\eta_t} = \frac{1}{1 - \frac{T_c}{T_h}}$$

$$\text{C.O.P.}_{\text{refrigerator}} = \frac{Q_{picked\ up}}{W_{net}} = \text{C.O.P.}_{\text{heat pump}} - 1$$

Note that all quantities here are cyclic quantities and are easily identified on the T-s plane since the Carnot cycle is a rectangle on this plane.

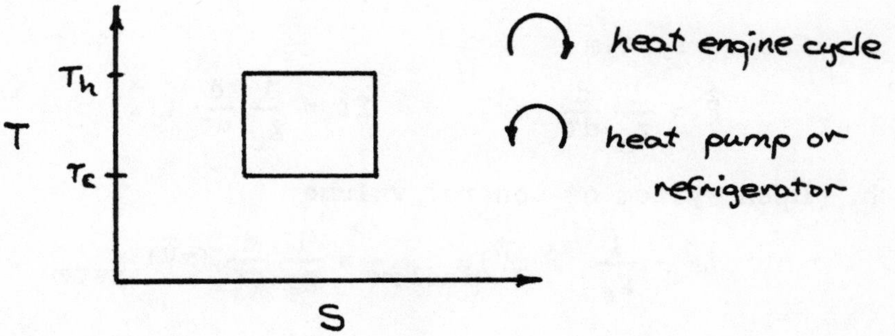

F. SUMMARY OF PHYSICAL PRINCIPLES AND PROPERTY RELATIONSHIPS

1. Property Relationships for a Pure Substance

   a. p-v-T surface of state

   b. Incompressible fluid - ρ is a constant:

   $$du = c\,dT \qquad ds = \frac{c\,dT}{T}$$

   $$dh = du + v\,dp$$

   c. Mixed phase:

   $$x = \frac{m_g}{m_{f+g}} \qquad v_x = v_f + x v_{fg} \qquad \text{same for } h_x,\ u_x,\ s_x$$

   d. Perfect gas:

   $$pv = \frac{R_o}{M}T \quad du = c_v dT,\ ds = \frac{c_v dT}{T} + R\frac{dv}{v},\ k = c_p/c_v$$

   $$c_p - c_v = R, \quad dh = c_p dT,\ ds = \frac{c_p dT}{T} - R\frac{dp}{p}$$

2. Continuity Principle

$$\Sigma \dot{m} = \frac{dm}{d\tau}$$

3. First Law of Thermodynamics

a. Closed system

$$Q + W = \Delta E \begin{matrix}\text{final} \\ \text{initial}\end{matrix}$$

$$\Delta E = \Delta U + \Delta KE + \Delta PE$$

b. Open system or control volume

$$\frac{dQ}{d\tau} + P + \Sigma \dot{m}e)_{\text{inflow}} = \frac{dE}{d\tau}$$

$$e = u + pv + z\frac{g}{g_c} + \frac{V^2}{2g_c} = h + z\frac{g}{g_c} + \frac{V^2}{2g_c}$$

4. Momentum Principle

a. Closed system

$$\vec{F} = \frac{1}{g_c}\frac{d}{d\tau}(m\vec{V}) \qquad \Sigma\vec{T} = \frac{1}{g_c}\frac{d}{d\tau}(I\vec{\omega})$$

b. Open system or control volume

$$\Sigma\vec{F} + \frac{1}{g_c}\ \Sigma\dot{m}\vec{V})_{\text{inflow}} = \frac{1}{g_c}\frac{d}{d\tau}(m\vec{V})_{\text{system}}$$

5. Second Law of Thermodynamics

a. Closed system

$$\frac{dQ}{T}_{\text{surroundings}} \leq dS_{\text{system}}$$

b. Open system or control volume

$$\frac{1}{T_{\text{sur}}}\frac{dQ}{d\tau} + \Sigma\dot{m}s \leq \frac{dS}{d\tau})_{\text{system}}$$

G. TERMINOLOGY FOR THERMOFLUID MECHANICS REVIEW

c, c_p, c_v	specific heats
E	total energy
e	specific energy
F	force
f	Darcy-Stanton friction factor
g	acceleration of gravity
g_c	proportionality in Newton's second law of motion
h	enthalpy

H	head
H_L	head loss
I	mass moment of inertia
k	specific heat ratio
m	mass
$\dot{m}$	mass flow rate
M	mass equivalent of molecular weight
p	pressure
P	power
Q	heat
R	gas constant
R_o	universal gas constant
r	radius
T	temperature
s	specific entropy
v	specific volume
V	velocity
W	work
x	quality
z	elevation
u	internal thermal energy
U	total internal thermal energy
ρ	density
γ	specific weight
ω	angular velocity
τ	time
η	efficiency

THERMOFLUID MECHANICS 1

Air is compressed reversibly and adiabatically to 3 atmospheres. It is then cooled back to room temperature and stored in a reservoir until used. If it expands reversibly and adiabatically through a positive displacement motor:

(a) What is the ratio of work input to work output?

(b) Explain whether or not it would be advantageous to prevent the air from cooling back to room temperature before entering the air motor.

Solution

a) Assume: negligible change in K.E., perfect gas behavior, $\bar{C}_p$ = constant.

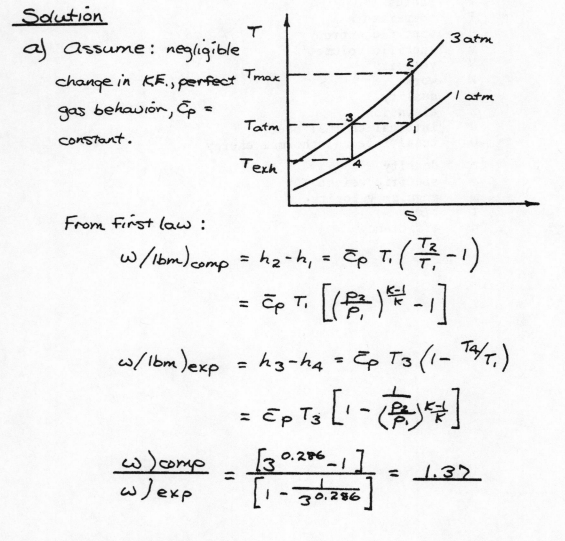

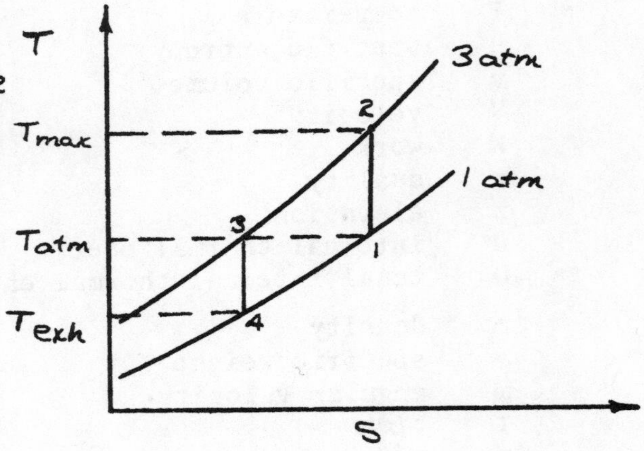

From First law:

$$w/lbm)_{comp} = h_2 - h_1 = \bar{C}_p\, T_1\left(\frac{T_2}{T_1} - 1\right)$$

$$= \bar{C}_p\, T_1\left[\left(\frac{P_2}{P_1}\right)^{\frac{k-1}{k}} - 1\right]$$

$$w/lbm)_{exp} = h_3 - h_4 = \bar{C}_p\, T_3\left(1 - T_4/T_1\right)$$

$$= \bar{C}_p\, T_3\left[1 - \frac{1}{\left(\frac{P_2}{P_1}\right)^{\frac{k-1}{k}}}\right]$$

$$\frac{w)_{comp}}{w)_{exp}} = \frac{\left[3^{0.286} - 1\right]}{\left[1 - \frac{1}{3^{0.286}}\right]} = \underline{1.37}$$

b) Process will yield more work per pound by

-continued-

not cooling. Condensation and possibly icing in lines and expander may be avoided. The expander will run hotter which if handled may be a disadvantage.

THERMOFLUID MECHANICS 2

A hilsch tube is a device whereby compressed air at room temperature is caused to spin by discharging through a nozzle into a cylinder. The core air is bled off through one exhaust line and is found to be cold and the peripheral air is bled off through another exhaust pipe and is found to be warm. During this process there is no heat transfer or work exchange with the surroundings. The steady state data taken from this system follows:

	Inlet Air	Hot Outlet Air	Cold Outlet Air
Velocity (ft/sec)	10.0	30.0	20.0
Pressure (psia)	40.0	21.8	14.7
Temperature (F)	70.0	112.0	- 60.0
Flow Area (in^2)	2.00	1.00	0.50

(a) What physical laws are available for checking the authenticity of this data?

(b) Use these laws to check the data to confirm its authenticity.

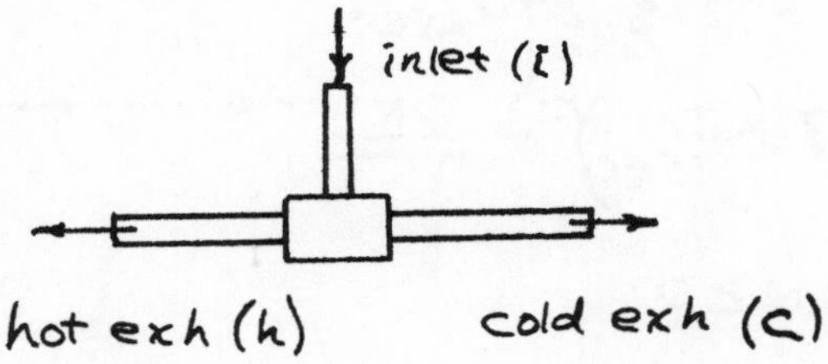

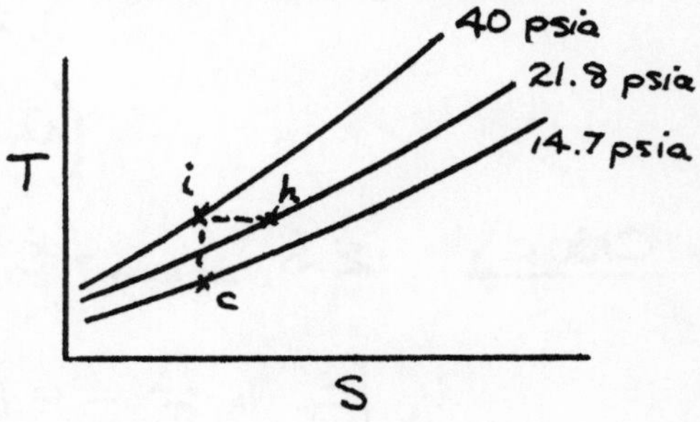

Solution

Principles: a) conservation of matter

b) 1st Law of thermodynamics

c) 2nd Law of thermodynamics

Assume : steady, 1-dimensional, adiabatic flow of a perfect gas.

Chk a) $\dot{m}_i = U\rho A = \frac{Up A}{RT} = \frac{10(40)2}{53.3\,(530)} = 0.0283\,\frac{lbm}{sec}$

$\dot{m}_c = \frac{20(14.7)\,0.5}{53.3\,(400)} = 0.0069\,\frac{lbm}{sec}$

$\dot{m}_h = \frac{30\,(21.8)\,1.0}{53.3\,(572)} = 0.0214\,\frac{lbm}{sec}$

$\Sigma\, \dot{m}_c + \dot{m}_h = 0.0283\,\frac{lbm}{sec}$

(0.0128 Kg/SEC)

a) OK ←

Chk b) $\dot{m}_h \left[c_p (T_h - T_i) + \frac{(U_h^2 - U_i^2)}{2g_c} \right]$ (neg.)

$= \dot{m}_c \left[c_p (T_i - T_c) + \frac{(U_i^2 - U_c^2)}{2g_c} \right]$ (neg.)

$0.0214 \left[0.24\,(112-70)\right] \overset{?}{=} 0.0069 \left[0.24\,(70+60)\right]$

$0.215 = .215$ BTU/SEC

(0.227 KJ/SEC) b) OK ←

Chk c) $\Sigma\, \dot{S}_{outflow} - \dot{S}_{inflow} \geq 0$

$\dot{m}_h \left[c_p \ln \frac{T_h}{T_i} - R \ln\left(\frac{p_h}{p_i}\right) \right] + \dot{m}_c \left[c_p \ln\left(\frac{T_c}{T_i}\right) - R \ln\left(\frac{p_c}{p_i}\right) \right] \geq 0$

$0.214 \left[0.0185 + .0418\right] + 0.0069 \left[-0.067 + 0.069\right] > 0$

Note: $\frac{BTU}{SEC\,°R} \times 1.90$ YIELDS $\frac{KJ}{SEC\,°K}$ c) OK ←

THERMOFLUID MECHANICS 3

A two stage compressor develops a 1.27:1 pressure ratio in the first stage with a total-to-total efficiency of 90% and a 1.31:1 pressure ratio in the second stage with a total-to-total efficiency of 88%. If the specific heat ratio was 1.400, what is the compressor efficiency?

Solution

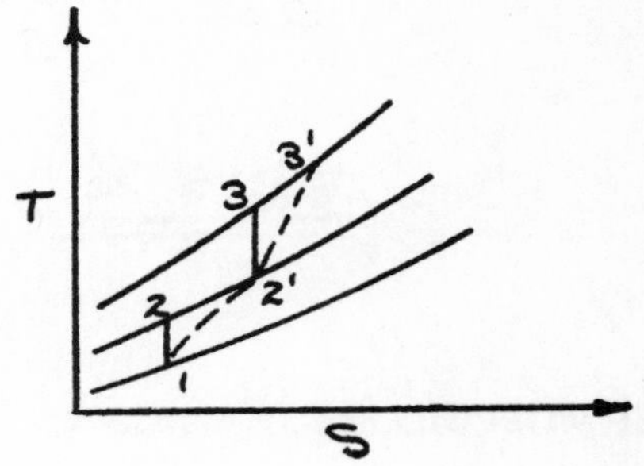

Assume: Total to total means stagnation enthalpy to stagnation enthalpy, perfect gas, constant c_p

η is isentropic efficiency

$$\eta_{overall} = \frac{w_{(S)}}{w_{actual}} = \frac{w_{(S)}}{w_{1^{st}stage} + w_{2^{nd}stage}}$$

Let $p_2/p_1 = r_{p_1}$, $p_3/p_2 = r_{p_2}$, $\frac{k-1}{k} = \alpha$

$$w_{1^{st}stage} = h_{2'} - h_1 = \frac{c_p T_1 (T_2/T_1 - 1)}{\eta_1} = \frac{c_p T_1 (r_{p_1}^{\alpha} - 1)}{\eta_1}$$

$$w_{2^{nd}stage} = \frac{c_p T_{2'} (r_{p_2}^{\alpha} - 1)}{\eta_2} \quad \text{but } T_{2'} - T_1 = \frac{T_2 - T_1}{\eta_1}$$

$$\therefore w_{2^{nd}stage} = \frac{c_p (r_{p_2}^{\alpha} - 1)}{\eta_2} (T_1)\left(1 + \frac{r_{p_1}^{\alpha} - 1}{\eta_1}\right)$$

$$= \frac{c_p T_1}{\eta_1 \eta_2} (r_{p_2}^{\alpha} - 1)(\eta_1 + r_{p_1}^{\alpha} - 1)$$

Then
$$\eta_{overall} = \frac{c_p T_1 \left[(r_{p_1} r_{p_2})^{\alpha} - 1\right]}{\frac{c_p T_1}{\eta_1}(r_{p_1}^{\alpha} - 1) + \frac{c_p T_1}{\eta_1 \eta_2}(r_{p_2}^{\alpha} - 1)(\eta_1 + r_{p_1}^{\alpha} - 1)}$$

$$\eta_{overall} = \frac{\left[(r_{p_1} r_{p_2})^{\alpha} - 1\right] \eta_1 \eta_2}{\eta_2 (r_{p_1}^{\alpha} - 1) + \eta_1 (r_{p_2}^{\alpha} - 1) + (r_{p_2}^{\alpha} - 1)(r_{p_1}^{\alpha} - 1)}$$

$$= \frac{\left[(1.27 \times 1.31)^{0.286} - 1\right] 0.90 \times 0.88}{0.88 (1.27^{0.286} - 1) + 0.90 (1.31^{0.286} - 1) + (1.27^{.286} - 1)(1.31^{.286} - 1)}$$

$$= \underline{\underline{88.5\ \%}}$$

THERMOFLUID MECHANICS 4

A single-state air compressor having 7.0% clearance is driven by a constant speed electric motor. It has been tested in San Francisco (at sea level) at an air temperature of 60°F, at atmospheric pressure, and found to deliver 100 cubic feet per minute of dry air when the discharge pressure is 100 psig (7.64 pounds/minute of free air).

What will be the capacity of this air compressor, in pounds of dry air per minute, if it is operated in a mountain location at an altitude of 10,000 feet, where the pressure is 10.10 psia and the temperature is 60°F?

Assume pressure drop at suction is negligible, polytropic exponent of compression is n = 1.3, and discharge pressure remains 100 psig.

<u>Solution</u>

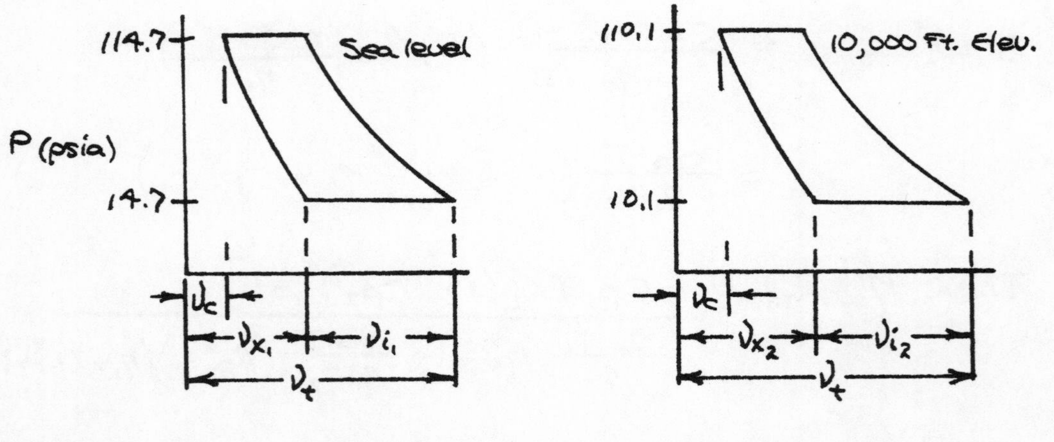

assume: 1. $V_c = 0.07\, V_t$

2. $pV^{1.3} = C$ is ok for all comp. + exp. processes

$$\therefore V_{x_1} = V_c \left(\frac{114.7}{14.7}\right)^{\frac{1}{1.3}} = 0.336\, V_t$$

$$V_{x_2} = V_c \left(\frac{110.1}{10.1}\right)^{\frac{1}{1.3}} = 0.442\, V_t$$

$$V_{i_1} = 0.664\, V_t \qquad V_{i_2} = 0.558\, V_t$$

$$\text{new intake rate} = \frac{V_{i_2}}{V_{i_1}} \times 100 = 84.5 \text{ cfm}$$

$$\dot{m} = \text{cfm}\, \rho)_{\text{intake}} = \text{cfm} \frac{p}{RT}$$

$$= 4.38 \frac{\text{lbm.}}{\text{min.}}$$

(1.99 Kg/min)

THERMOFLUID MECHANICS 5

A large hotel has a boiler room where steam is generated for heating, laundry, and culinary purposes. The two boilers in this boiler room were formerly oil-fired but are being converted to natural gas. It is necessary to run a gas pipe line a distance of 265 feet from the meter to the gas burners in the boilers.

What size gas line would you specify, based on the recommendation of the Uniform Plumbing Code? Allow for a steaming rate of 200 per cent rated.

Specifications:

Boilers	-	two in number, each rated at 1500 lbs steam per hour
Steam	-	125 psia and 95 per cent quality
Feedwater	-	100°F
Natural Gas	-	1100 Btu per cu ft; 0.65 specific gravity
Gas pressure leaving the meter	-	5 psig
Gas pressure required at the burners	-	1.5 psig
Gas line length	-	265. feet

Solution

$$\dot{E}_{steam} = \dot{m}_s (h_{out} - h_{in})$$

$$= 6000 \frac{lbm}{hr} (h_f + 0.95\, h_{fg} - h_{in})$$

$$= 6000(315.6 + 0.95 \times 875.4 - 68.0)$$

$$= 6.46 \times 10^6 \text{ Btu/hr.}$$

Assume $\eta_{boiler} = 85\%$

$$\dot{E}_{gas} = \frac{6.46 \times 10^6}{0.85} = 7.59 \times 10^6 \frac{Btu}{hr.}$$

$$Q_{gas} = \frac{7.59 \times 10^6}{1100} = 6900 \left.\frac{ft^3}{hr}\right)_{s.t.p.}$$

$$= 5{,}150 \left. ft^3/hr \right)_{5\, psig}$$

$$= 6{,}250 \left. ft^3/hr \right)_{1.5\, psig}$$

Code is not specific for these conditions (Suggests 4"–6" d)

Estimate $U = 20$ ft/sec @ 1.5 psig & check Δp

$$VA = Q = \frac{6250}{3{,}600}\ \text{ft}^3/\text{sec}$$

$$\therefore d = 4.0''$$

$$H_L = f\,\frac{l}{d}\,\frac{U^2}{2g} \qquad f = .03 \text{ assuming fully turbulent}$$

$$= .03 \times \frac{265}{1/3} \times \frac{400}{64.4} = 148\ \text{ft.}$$

Since $\rho \cong .05\ \frac{\text{lbm}}{\text{ft}^3}$ $\quad \Delta p = .05$ psi

4" d ok

3" d might be ok

THERMOFLUID MECHANICS 6

It is proposed to use energy radiated from the sun as an energy supply for a heat engine on the surface of the earth. The best engine cycle will approximate the Carnot cycle. It will receive energy from a solar energy collector and reject waste energy to the surrounding atmosphere. Assuming that the temperature of the sun is 10,000 R, the temperature of the surroundings is 60°F and the solar collector is perfectly insulated from the surroundings, determine the desired operating temperature of the collector for a maximum power output from the prime mover.

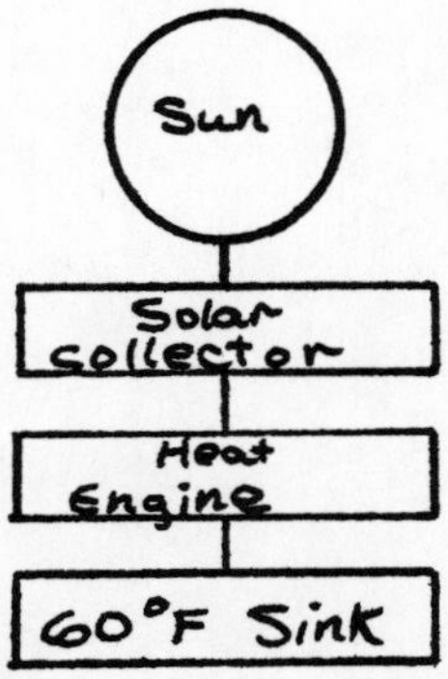

Solution

$$q_{input} = c_1\left(T_S^4 - T_h^4\right)$$

$$Power = \eta_t\, q_{input} = \left(1 - \frac{T_S}{T_h}\right)c_1\left(T_S^4 - T_h^4\right)$$

want P to be max. as T_h is varied

$$\text{Let } T' = \frac{T}{1000}$$

$$\text{Then } P/c_1 = \left(1 - \frac{0.520}{T_h'}\right)\left(10^4 - T_h'^4\right) \quad T_h' \text{ is } \frac{T_{collector}}{1000}$$

Substitute values for T_h' and plot three or four points.

P/c is found to be a maximum when $T_h \simeq$ **4200°R** (2333°K)

THERMOFLUID MECHANICS 7

An evaporator operating at a shell pressure of 38 psia is producing 25,000 lbs/hr of water vapor with a purity of 25 ppm solids. Make up water to the evaporator is softened city water at 60°F with 774 ppm solids. Saturated water from a boiler operating at 2100 psia is being added to the shell of the evaporator at a rate of 1000 lbs/hr and 30 ppm solids. Terminal difference at the blow down heat exchanger is 40°F. The heating steam to the evaporator is supplied at 103.5 psia and 653° and is discharged at a saturation pressure of 99.4 psia.

Below is a flow diagram of the evaporator.

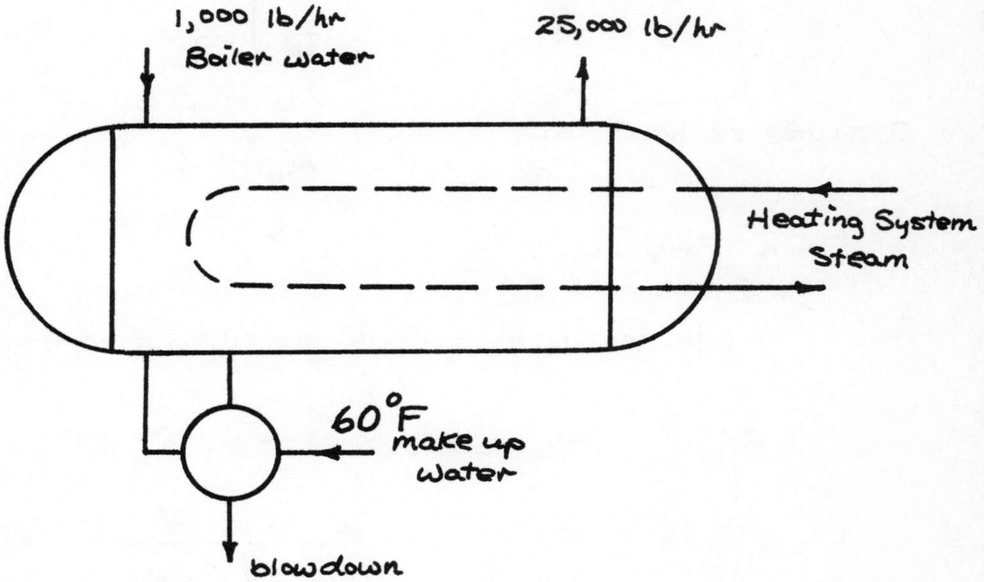

(a) What rate of evaporator blow down (in lbs/hrs) is required to maintain a 3000 ppm solids concentration in the evaporator shell?

(b) What rate of heating steam flow is necessary to produce the 25,000 lbs/hr of water vapor under the above conditions?

Solution

Assume steady state

$$c_w \dot{m}_w + c_s \dot{m}_s = c_b \dot{m}_b + c_v \dot{m}_v$$

$$\dot{m}_w = \dot{m}_v + \dot{m}_b - \dot{m}_s$$

$$\therefore C_w(\dot{m}_v + \dot{m}_b - \dot{m}_s) + C_s\dot{m}_s = C_b\dot{m}_b + C_v\dot{m}_v$$

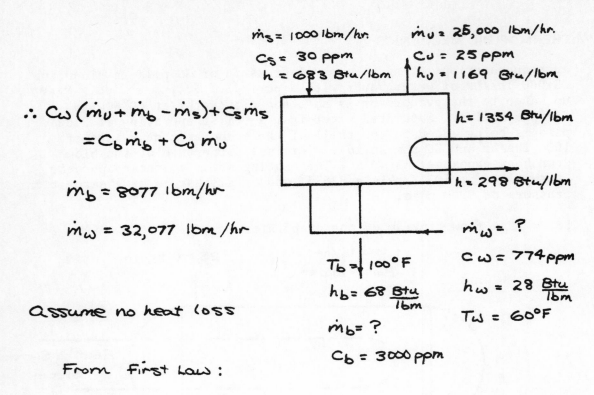

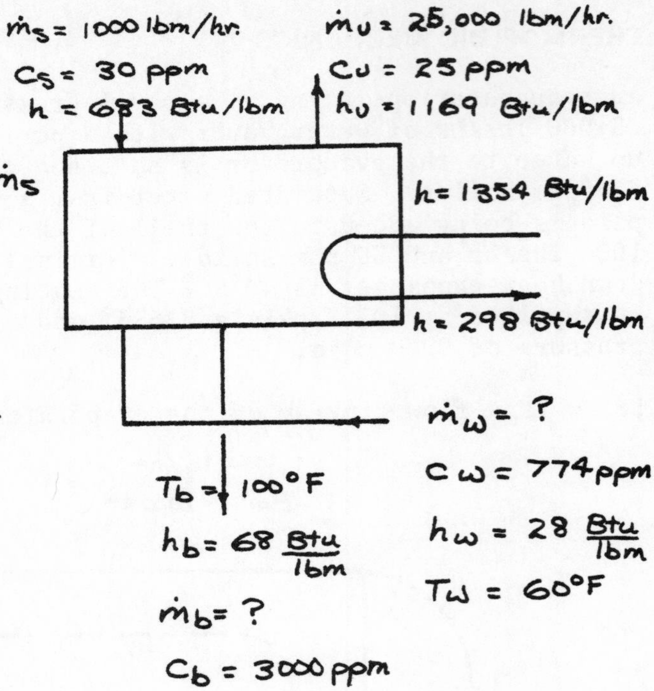

$$\dot{m}_b = 8077 \text{ lbm/hr}$$

$$\dot{m}_w = 32{,}077 \text{ lbm/hr}$$

assume no heat loss

From First Law:

$$\dot{m}_v h_v + \dot{m}_b h_b - \dot{m}_s h_s - \dot{m}_w h_w = \dot{m}\,\Delta h)_{\text{steam heat source}}$$

$$\dot{m}_v h_v = 25{,}000 \times 1169 = 29.2 \times 10^6 \text{ Btu/hr}$$

$$\dot{m}_b h_b = 8{,}100 \times 68 = \underline{0.6 \times 10^6 \text{ Btu/hr}}$$

$$29.8 \times 10^6 \text{ Btu/hr}$$

$$\dot{m}_s h_s = 1{,}000 \times 683 = 0.7 \times 10^6 \text{ Btu/hr}$$

$$\dot{m}_w h_w = 32{,}100 \times 28 = \underline{0.9 \times 10^6 \text{ Btu/hr}}$$

$$1.6 \times 10^6 \text{ Btu/hr}$$

$$\Sigma = 28.2 \times 10^6 \text{ Btu/hr}$$

$$m_{\text{steam}} = \frac{2820 \times 10^4}{1354 - 298} = \underline{26{,}600 \text{ lbm/hr}}$$

$$(12{,}065 \text{ Kg/hr})$$

THERMOFLUID MECHANICS 8

The figure illustrates the general arrangement of an experimental vapor compression still.

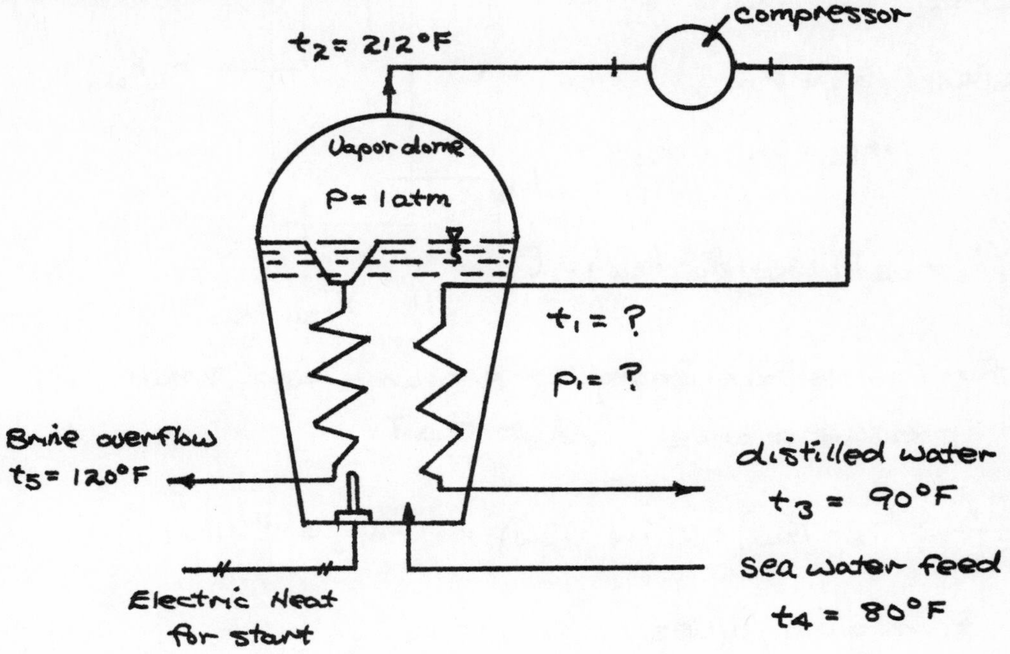

(a) Analyze the process to determine:

(i) The work expended by the compressor per pound of water distilled.

(ii) The pressure and the temperature of the compressed vapor.

(b) Enumerate the advantages and disadvantages of the process.

(c) If you were given the project for improving the system for commercial operation, what would be your recommendations?

Assume that the temperature of the inlet sea water feed is 80°F, the brine overflow is 120°F and the vapor chamber is 212°F. The brine overflow is 200% of the distillate (i.e., each 3 lbs of feed yields 1 lb of distillate). The specific heat of sea water is 0.97.

Solution

Assume: Steady State, adiabatic operation, isentropic compression, vapor is a perfect gas.

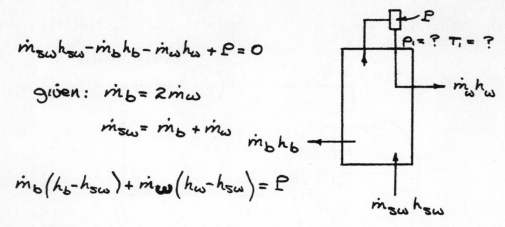

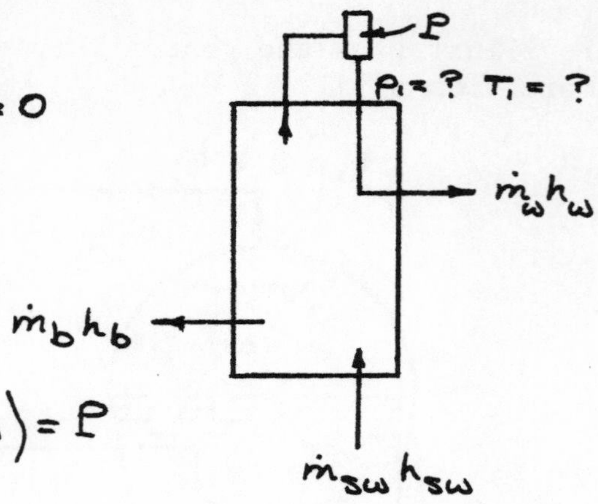

$$\dot m_{sw} h_{sw} - \dot m_b h_b - \dot m_w h_w + P = 0$$

given: $\dot m_b = 2\dot m_w$

$$\dot m_{sw} = \dot m_b + \dot m_w$$

$$\dot m_b(h_b - h_{sw}) + \dot m_w(h_w - h_{sw}) = P$$

* for liquids experiencing temp. change and small pressure change, $\Delta h \simeq c\Delta T$

$$\frac{\dot m_b}{\dot m_w} c(T_b - T_{sw}) + c(T_w - T_{sw}) = P/\dot m_w = W/\text{lbm}$$

From given values

$$2(0.97)(120-80) + (0.97)(90-80) = w = 87.2\ \frac{\text{Btu}}{\text{lbm}}$$

$$(202.8\ \text{KJ/Kg})$$

for compressor

$$w = h_1 - h_{g,212} = c_p(T_1 - 212°F)$$

$$T_1 = \frac{87.2}{0.45} + 212°F = 406°F$$

for isentropic compression

$$\frac{p_1}{p_{14.7}} = \left(\frac{T_2}{T_1}\right)^{k/k-1} = \left(\frac{866}{672}\right)^{\frac{1.33}{0.33}}$$

$$p_1 = 39.6\ \text{psia}$$

$$(273\ \text{KN/m}^2)$$

b) advantage – no high pressure, temperature or vacuum.
disadvantage – large mechanical work investment and compressor size

c) Water cool compressors, and recover some of waste energy in exhaust streams.

THERMOFLUID MECHANICS 9

Water is forced out of Tank No. 2 by gas under pressure in Tank No. 1. The desired water pressure at the outlet valve is 500 psig when Tank No. 2 is empty. Both tanks are insulated.

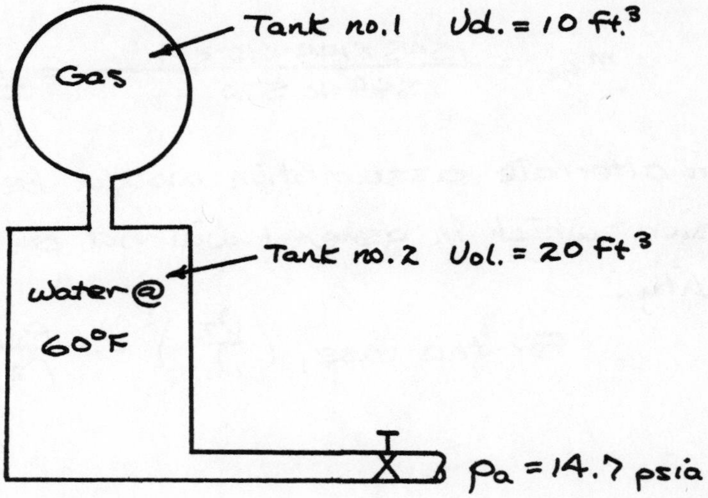

(a) If the pressurizing gas is N_2, starting at 60°F, what should the starting gas pressure be?

(b) If the gas is helium under the same conditions, what would be the starting pressure?

(c) What will be the weight of gas used in each case described above?

<u>Solution</u>

Assume isothermal expansion, as mass of gas is small compared to mass of tank and water, and blowdown will not be abrupt so there is time for heat transfer.

a) $\dfrac{V_2}{V_1} = \dfrac{p_1}{p_2}$ $\qquad p_1 = \dfrac{515 \times 30}{10} = \underline{\underline{1545 \text{ psia}}}$

b) same as a)

c) $p_1 V_1 = m \frac{R_0}{M} T_1$

$$m_{N_2} = \frac{1545 \times 144 \times 10 \times 28}{1544 \times 520} = \underline{\underline{77 \text{ lbm}}} \; (34.9 \text{ Kg})$$

$$m_{He} = \frac{1545 \times 144 \times 10 \times 4}{1544 \times 520} = \underline{\underline{11 \text{ lbm}}} \; (4.99 \text{ Kg})$$

An alternate assumption would be adiabatic expansion which in general will not be as close to reality.

For this case, $\left(\frac{V_2}{V_1}\right)^K = \frac{p_1}{p_2}$.

THERMOFLUID MECHANICS 10

In a particular steam power plant air is believed to leak into the condenser. To check whether or not this is so, the plant is run until conditions are steady, whereupon the steam supply from the engine is shut off. Simultaneously, the air and condensate extraction pumps are closed down so that the condenser is isolated. At shutdown the temperature and vacuum in the condenser are observed to be 101.7°F and 27.3 in of Hg. After 5 minutes the values are 82.5°F and 19.25 in of Hg. The barometer reading is 29.7 in of Hg. The effective volume of the condenser is 12 cu ft.

Determine from this data:

(a) The weight of air leakage into the condenser during the observed period.

(b) The weight of water vapor condensed during this same time period.

Assume (the gas constant) R for air is 53.3 (Ft $\text{lb}_f/\text{lb}_m\,^\circ\text{R}$)

Solution

Assume that at any instant T is constant throughout the condenser.

$$\boxed{V = 12\ ft^3}$$

At Shutdown,

$T = 101.7°F \quad p = 29.7 - 27.3 = 2.4$ in Hg abs.

$= 1.18$ psia

For $T = 101.7°F$, $p_{sat} = 1$ psia

* Therefore, some air in condenser at shutdown.

$$m_{i,a} = \frac{pV}{RT} = \frac{.18 \times 144 \times 12}{53.3 \times 561.7} = 0.0104\ lbm$$

After 5 minutes,

$T = 82.5°F \quad p = 29.7 - 19.3 = 10.4$ in Hg abs

$= 5.11$ psia

$p_{sat} = .55$ psia

$$m_{f,a} = \frac{4.56 \times 144 \times 12}{53.3 \times 542.5} = 0.0272\ lbm$$

a) $\Delta m_a = 0.0272 - 0.0104 = \underline{\underline{0.0168\ lbm}}$

b) $$\Delta m_w = \frac{V}{R}\left(\frac{p_{i,s}}{T_i} - \frac{p_{f,s}}{T_f}\right) = \frac{12 \times 144}{85.6}\left(\frac{1}{461.7} - \frac{.55}{542.5}\right)$$

$$= 20.2\,(.00217 - .00101)$$

$$= \underline{\underline{0.0232\ lbm}}$$

(0.0105 Kg)

7

Heat Transfer

RICHARD K. PEFLEY

BASIC HEAT TRANSFER RELATIONSHIPS

1. Conductive Heat Transfer

   a. Defining equation

$$dq = - k\ dA \frac{\partial t}{\partial n}$$

   b. Temperature field equation in rectangular and cylindrical coordinates

$$\frac{\partial^2 t}{\partial x^2} + \frac{\partial^2 t}{\partial y^2} + \frac{\partial^2 t}{\partial z^2} = \frac{1}{\alpha}\frac{\partial}{\partial \tau} - \frac{p}{k}$$

$$\frac{\partial^2 t}{\partial r^2} + \frac{1}{r}\frac{\partial t}{\partial r} + \frac{1}{r^2}\frac{\partial^2 t}{\partial \theta^2} + \frac{\partial^2 t}{\partial z^2} = \frac{1}{\alpha}\frac{\partial t}{\partial \tau} - \frac{p}{k}$$

   c. Thermal conductive resistance for one-dimensional heat flow in a rectangular slab, cylinder and a sphere.

$$\frac{w}{kA}, \quad \frac{\ln(r_o/r_i)}{2\pi k \ell}, \quad \frac{r_o - r_i}{4\pi k\ r_o r_i}$$

2. Thermal Radiation

   a. Black body emission behavior

      i. Hemispherical radiation intensity--Stefan Boltzmann law

$$W_B = \sigma T^4 \quad \sigma = 0.171 \times 10^{-8} \frac{\text{Btu}}{\text{hr ft}^2 {}^\circ\text{R}^4}$$

ii. Wave length intensity--Planck's distribution law

$$W_{B,\lambda} = \frac{c_1/\lambda^5}{e^{\frac{c_2}{\lambda T}} - 1} \qquad c_1 = 1.187 \times 10^{-8} \frac{\text{Btu } \mu^4}{\text{ft}^2\text{hr}}$$

$$c_2 = 2.5896 \times 10^4 \,°\text{R } \mu$$

iii. Wave length of most intense radiation--Wein's displacement law

$$\lambda_{max} T = c_3 \qquad c_3 = 5215.6 \ \mu°\text{R}$$

iv. Directional radiation intensity--Lambert's cosine law

$$\frac{dE}{d\tau} = \frac{\sigma}{\pi} T_1^4 \cos\phi \, dA$$

v. Radiation exchange between black, isothermal surfaces separated by a non-absorbing medium

$$q_{1\to 2} = \frac{\sigma}{\pi} (T_1^4 - T_2^4) \int_{A_1} \int_{A_2} \frac{\cos\phi_1 \cos\phi_2}{r^2} dA_1 \, dA_2$$

b. Grey body behavior

i. Definition

$$\alpha = \alpha\,(T)$$

ii. Fractional distribution of incident radiant energy

$$\alpha + r + tr = 1$$

iii. Circuit network for grey body enclosure composed of isothermal surfaces

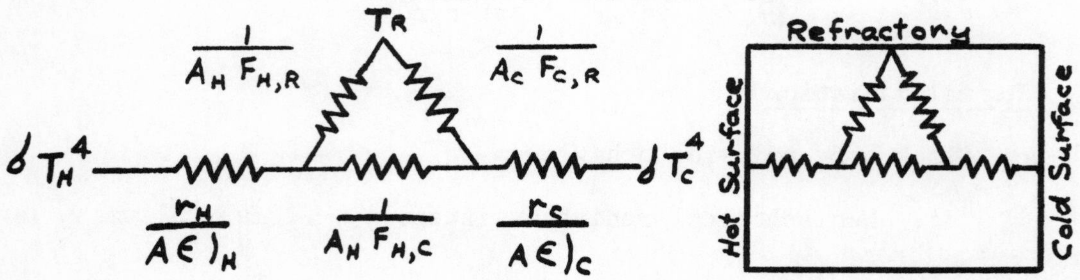

3. Convective Heat Transfer

a. Defining equation

$$dq = h\, dA\, (t_{surface} - t_{fluid})$$

b. Natural convection, t_{fluid} is defined as $t_{surrounding\ bath}$

$$h = h\,(L,\ \beta,\ \Delta t,\ \rho,\ \mu,\ g,\ k,\ c_p,\ g_c)$$

Resulting groups are

$$\frac{hL}{k} = f\left(\frac{c_p\ \mu\ g_c}{k},\ \frac{L^3 \rho^2 g^2 \beta\ \Delta t}{u^2\ g_c^2}\right)$$

c. Unestablished, forced convective heat transfer, t_{fluid} is defined as $t_{free\ stream}$

$$h = h\,(V,\ x,\ \rho,\ \mu,\ c_p,\ k,\ g_c)$$

$$\frac{hx}{k} = f\left(\frac{Vx}{\mu g_c},\ \frac{c_p \mu}{k g_c}\right)$$

d. Established forced convection in closed conduits, t_{fluid} is defined as

$$\bar{t}_f = \frac{\int_A t\ V\ dA}{Q}$$

$$h = h\,(V,\ d,\ \rho,\ \mu,\ c_p,\ k,\ g_c) \qquad \frac{hd}{k} = f\left(\frac{V\ d\rho}{g_c},\ \frac{c_p\ \mu\ g_c}{k}\right)$$

e. Convective resistance for uniform convective coefficient

$$R = \frac{1}{hA}$$

4. Heat flow rate for parallel or counter flow heat exchangers assuming uniform fluid properties and heat transfer coefficients and transverse heat transfer only

$$q = \frac{\Delta t_m}{\Sigma R} = UA\ \Delta t_m \qquad \Delta t_m = \frac{\Delta t_a - \Delta t_b}{\ln \dfrac{\Delta t_a}{\Delta t_b}}$$

where Δt_a and Δt_b are temperature differences at opposite ends of the exchanger.

If Δt_a, Δt_b are not known for exchanger, use heat exchanger effectiveness concept

$$\xi = \frac{q_{actual}}{C_{min}(T_{h,in} - T_{c,in})} = f\left(\frac{C_{min}}{C_{max}}, \frac{AU}{C_{min}} = \frac{1}{\Sigma RC_{min}}\right)$$

(See reference for graphs of ξ versus $\frac{AU}{C_{min}}$)

Definition of Terms:

A	area
$C_{min,max}$	thermal capacitance rate
d,L,ℓ	length
f	function
F	radiation geometry factor
g	acceleration of gravity
g_c	proportionality factor in Newton's 2nd law
h	connective heat transfer coefficient
k	thermal conductivity
n	length in direction of heat flow
p	heat generation per unit volume
Q	volumetric flow rate
q	heat transfer rate per unit area
R,ΣR	thermal resistance, composite resistance
r	reflectivity or radius
t,T	temperature, absolute
tr	transmissivity
U	
V	velocity
w	thickness in direction of heat flow
x,y,z,r,θ	coordinate lengths
τ	time
α	absorbtivity or diffusivity
φ	radiation angle relative to surface normal
ρ	density
μ	viscosity
β	coefficient of cubical expansion
λ	wave length
ξ	heat exchange effectiveness

Reference: Kreith, F., Principles of Heat Transfer, International Textbook Company.

HEAT TRANSFER 1

The steady state heat leakage through a large composite wall is desired. The wall and accompanying heat transfer data are shown in the sketch below. You are asked to find two approximate solutions which lie on opposite sides of the true solution and hence predict the true value by averaging the results. Report the leakage in Btu/hr-ft^2 of the face area.

mat'l	k $\frac{Btu}{hr\ ft\ °F}$
Brick	0.40
Doug. Fir	0.062
Fibreglass	0.020
Plasterbd.	0.300

Outside

t = 90°F

$h_{cr} = 3.5 \frac{Btu}{hr\ ft^2\ °F}$

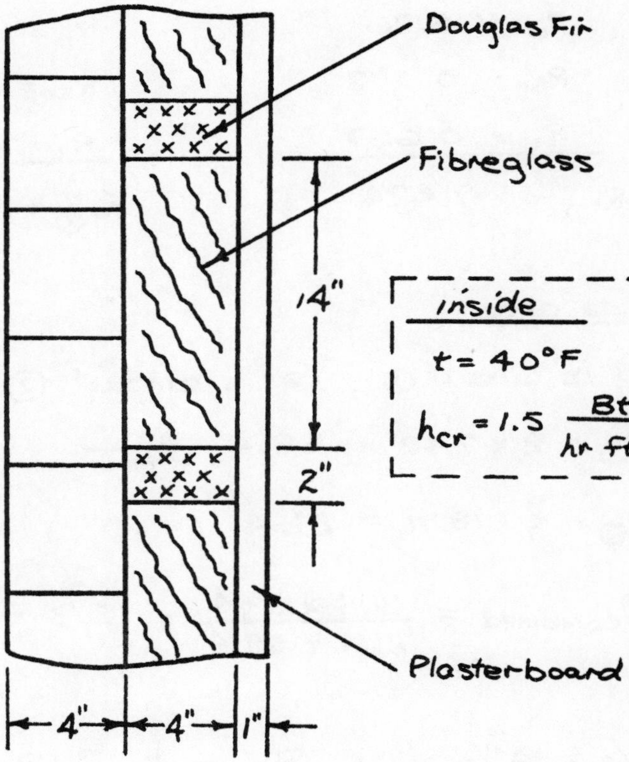

inside

t = 40°F

$h_{cr} = 1.5 \frac{Btu}{hr\ ft^2\ °F}$

Solution

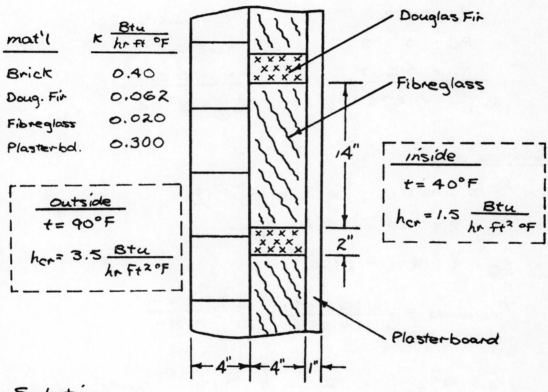

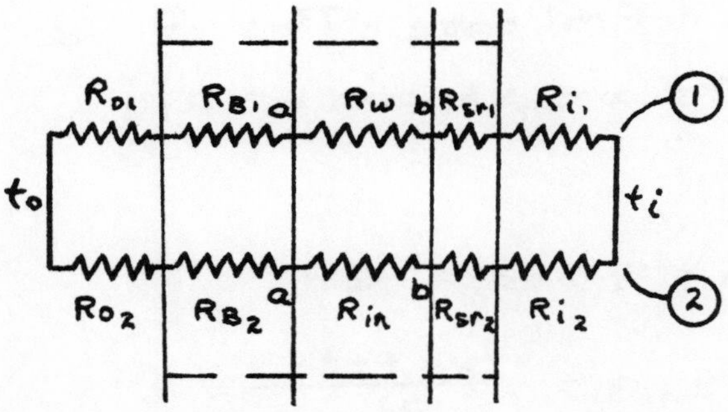

Resistances/unit area $\qquad \frac{ft^2\ hr\ °F}{Btu}$

$$R_o = \frac{1}{h} = 0.286$$

$$R_B = t/k = \frac{0.333}{0.40} = 0.833 \quad ''$$

$$R_w = \frac{0.333}{0.062} = 5.36 \quad ''$$

$$R_{in} = \frac{0.333}{0.020} = 16.65 \quad ''$$

$$R_{sr} = \frac{1}{12 \times 0.30} = 0.278 \quad ''$$

$$R_i = \frac{1}{1.5} = 0.667 \quad ''$$

Consider circuits ① + ② as separate, i.e., infinite transverse resistance

1 ft² face

①	②
$R_o = 0.286$	$R_o = 0.286$
$R_B = 0.833$	$R_B = 0.833$
$R_w = 5.36$	$R_{in} = 16.65$
$R_{sr} = 0.278$	$R_{sr} = 0.278$
$R_i = 0.667$	$R_i = 0.667$
$R_{t①} = 7.424$	$R_{t②} = 18.714$

Composite area

1/8 is circuit ①, 7/8 is circuit ②

$$R_{①} = 8 \times 7.42 = 59.4 \ \frac{hr\ ^{o}F}{Btu}$$

$$R_{②} = \frac{8}{7} \times 18.71 = 21.4 \quad ''$$

$$R_{combined} = \frac{21.4 \times 59.4}{21.4 + 59.4} = 15.7 \ \frac{hr\ ^{o}F}{Btu}$$

$$q/A = \Delta t / R_{comb.} = 50/15.7 = 3.18 \ Btu/hr\ ft^2$$

Consider transverse resistance as zero --- opposite side of reality from first case. This is achieved by imagining circuits are shorted at a-a and b-b.

$$R_{in} = 8/7\ 16.65 = 19.0 \qquad R_w = 8(5.36) = 42.9$$

$$R_{combined} = \frac{19.0 \times 42.9}{19.0 + 42.9} = 13.2 \ \frac{hr\ ft^2\ ^{o}F}{Btu}$$

$R_o = 0.286$

$R_B = 0.833$

$R_{comb} = 13.2$

$R_{sr} = 0.278$

$R_i = 0.667$

$R_{t\,comb} = 15.3$

$$q/A = \frac{50}{15.3} = 3.27\ \frac{\text{Btu}}{\text{hr ft}^2}$$

$$(q/A)_{actual} = \frac{3.18 + 3.27}{2}$$

$$= 3.23\ \frac{\text{Btu}}{\text{hr ft}^2}$$

$$= (10.2\ \text{WATTS}/\text{M}^2)$$

HEAT TRANSFER 2

A properly designed steam-heated, tubular preheater, with gases passing through the tubes, is heating 45,000 pounds per hour of air from 70°F to 170°F when using saturated steam at 5 psig. It is proposed to double the air flow rate through the heater and yet heat the air from 70°F to 170°F; this is to be accomplished by increasing the steam pressure. State reasonable assumptions, and determine the new steam pressure required to meet the changed conditions. Express answer in psig.

Solution

Assume:

1) air side heat transfer resistance dominates — turbulent flow

2) steady state

3) no heat leakage

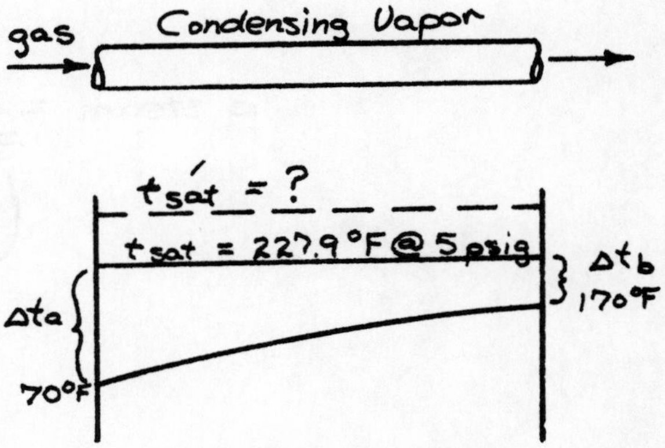

$$q = \frac{\Delta t_m}{R_t} = \frac{\Delta t_m}{\frac{1}{h_a \pi d_a l} + \frac{\ln r_o/r_i}{2\pi k l} + \frac{1}{h_s \pi d_s l}} \simeq \frac{\Delta t_m}{\frac{1}{h_a \pi d_a l}}$$

Let q' = new heat transfer rate

$$\frac{q'}{q} = \frac{h_a'}{h_a}\,\frac{\Delta t_m'}{\Delta t_m} \qquad \Delta t_m = \frac{\Delta t_a - \Delta t_b}{\ln \Delta t_a/\Delta t_b} = 100°F$$

for turbulent flow heat transfer in pipes

$$\frac{hd}{k} = 0.023\left(\frac{vd\rho}{\mu g_c}\right)^{0.80}\left(\frac{c_p \mu g_c}{k}\right)_f^{1/3}$$

$$\therefore \frac{h_a'}{h_a} \simeq \frac{(v\rho)'^{0.80}}{(v\rho)^{0.80}} \simeq (2)^{0.80}$$

Substituting $q'/q = 2$ and $h_a'/h_a = (2)^{0.80}$

$$\Delta t_m' = \Delta t_m \frac{(2)}{(2)^{0.80}} = 100(2)^{0.20} = 114.8°F$$

Guess $t'_{sat} = 241°F$

$$\Delta t_m' = \frac{(241-70)-(241-170)}{\ln \frac{241-70}{241-170}} = 114.8°F$$

$$p'_{steam} = \underline{\underline{25 \text{ psia}}}$$

$$\left(170 \frac{KN}{M^2}\right)$$

HEAT TRANSFER 3

From the data given in the figure, calculate the temperature of the earth. Assume earth to be a perfect absorber.

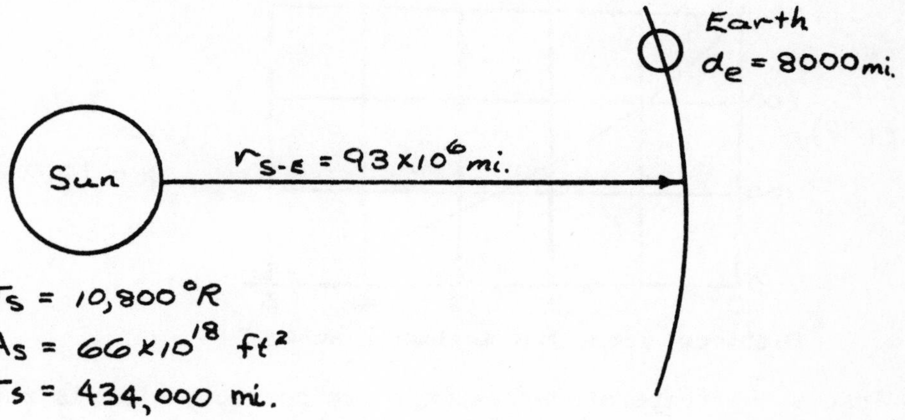

Solution

Assume sun is black radiator, and emits as much as it absorbs.

Radiation rate from sun $\frac{dE}{d\tau} = \sigma A_S T_S^4$

Fraction of this absorbed by Earth $= \left(\frac{\pi d_E^2}{16 \pi r_{S-E}^2}\right) \sigma A_S T_S^4$

Radiation rate from earth $= \sigma \pi d_e^2 T_e^4$

$$\therefore \quad \sigma \pi d_e^2 T_e^4 = \left(\frac{\pi d_e^2}{16 \pi r_{S-E}^2}\right) \sigma A_S T_S^4$$

$$T_e = \sqrt[4]{\frac{A_S}{16 \pi r_{S-E}^2}}\ T_S$$

$$T_e = \underline{\underline{520°R = 60°F}}$$

$$(288.9°K)$$

HEAT TRANSFER 4

Temperature readings taken at the same time at various points in a slab of homogeneous material were plotted as shown below.

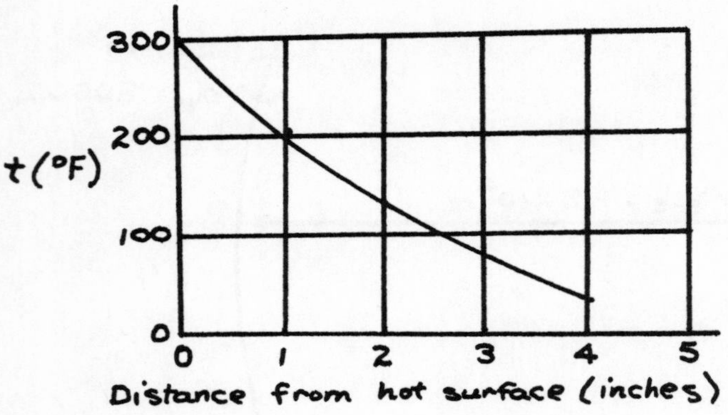

$$c = 0.40 \ \frac{\text{Btu}}{\text{lbm °F}}$$

$$k = 10.0 \ \frac{\text{Btu in}}{\text{hr ft}^2 \text{ °F}}$$

Assume the temperature readings are correct, what is the exact rate of heat flow per square foot past the plane two inches from the hot surface? Is temperature rising or falling?

Solution

a) Determine gradient at 2" position

$$q/A = -k\frac{dt}{dn} = 530 \ \frac{\text{Btu}}{\text{hr ft}^2}$$

$$(1671 \text{ WATTS/m}^2)$$

b) Temperature is rising because inflow of heat (steeper gradient) on left of the 2" plane is greater than outflow on the right.

HEAT TRANSFER 5

Find the area in square feet required for a heating coil to heat 5000 pounds of air per hour from 70°F to 200°F using steam at 250°F.

Material of the coil shall be aluminum with wall thickness of 0.049 inches and having a thermal conductivity of 1570 Btu/hr/ft^2/°F/inch.

Assume the coil to be clean and clear. The conductance of the air surface film is 30 Btu/hr/ft^2/°F. Specific heat of air is assumed constant at 0.24 Btu/pound/°F.

Assume d_i is approximately equal to d_o for the coil.

h_{steam} is approximately equal to 1000 Btu/hr ft^2 °F

The logarithmic mean temperature difference shall be used.

Solution

$$q = \dot{m} c_p \Delta T_{air} = 5\times10^3 \times 0.24 \times 1.3\times10^2 = 1.56\times10^5 \ \frac{Btu}{hr}$$

$$q = \frac{\Delta T_{mean}}{R_{air} + R_{wall} + R_{steam}}$$

$$\Delta T_{mean} = \frac{(250-70)-(250-200)}{\ln\dfrac{(250-70)}{(250-200)}} = 101\ °F$$

$$R_{air} = \frac{1}{hA} = \frac{1}{30\times A} = 3.33\times10^{-2}\ \frac{1}{A}$$

$$R_{wall} = \frac{w}{kA} = \frac{.049}{1.57\times10^3 A} = .003\times10^{-2}\ \frac{1}{A}\ \text{(neglect)}$$

$$R_{steam} = \frac{1}{hA} = \frac{1}{1\times10^3 A} = .10\times10^{-2}\ \frac{1}{A}$$

$$\Sigma R = 3.43\times10^{-2}\ \frac{1}{A(ft^2)}$$

$$A = \frac{1.56\times10^5\ Btu/hr}{101°F} \times 3.43\times10^{-2}\ \frac{hr\ °F\ ft^2}{Btu}$$

$$= \frac{1.56\times3.43\times10}{1.01} = \frac{52.5\ ft^2}{(4.87\ m^2)}$$

HEAT TRANSFER 6

A cylindrical fuel element in a nuclear reactor is shown below:

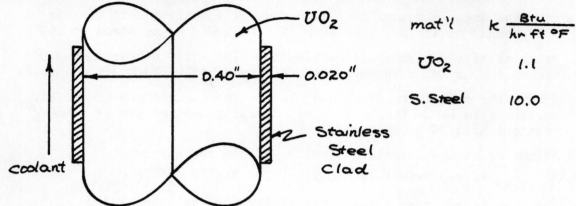

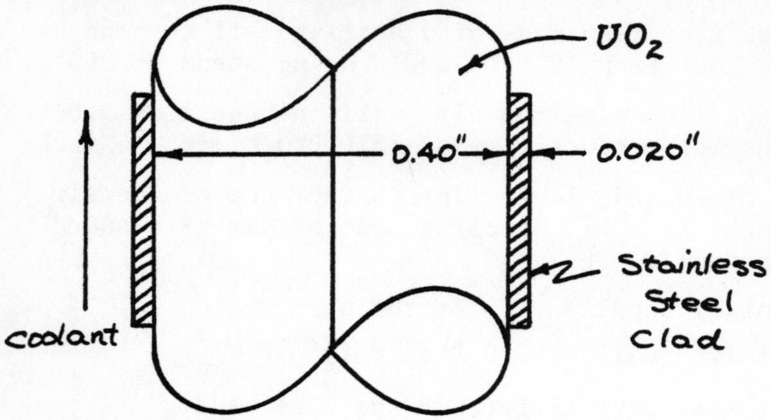

mat'l	k $\frac{Btu}{hr\ ft\ °F}$
UO_2	1.1
S. Steel	10.0

What is the center-line temperature in the UO_2 when the fuel element is producing 40×10^6 Btu/ft^{3}hr in the UO_2. Assume the temperature of the coolant is 500°F; and that the heat transfer coefficient for the coolant (h) = 10,000 Btu/ft^{2}hr°F.

Solution

For radial heat flow with heat generation

$$\frac{d^2t}{dr^2} + \frac{1}{r}\frac{dt}{dr} = -\frac{P}{k}$$

$$\frac{d}{dr}\left(r\frac{dt}{dr}\right) = -\frac{P}{k} r$$

$$r\frac{dt}{dr} = -\frac{P}{2k} r^2 + C_1 \qquad @\ r=0\ \frac{dt}{dr} = 0 \quad \therefore C_1 = 0$$

$$t = -\frac{P}{4k} r^2 + C_2 \qquad @\ r=0 \quad t = t_{max}$$

$$t_{max} - t_r = \frac{Pr^2}{4k}$$

at $r_o = 0.20''$

$$t_{max} - t_{r_o} = \frac{Pr_o^2}{4k} = \frac{40 \times 10^6 \times 0.04}{4 \times 1.1 \times 144} = 2520°F$$

(1382.2°K)

-continued-

$$q/ft = p \pi r_o^2 = \frac{40 \times 10^6 \times \pi \times 0.04}{144} = 3.49 \times 10^4 \frac{Btu}{hr\ ft.}$$

$$q/ft = \frac{t_{r_o} - t_f}{R_{clad} + R_{film}} = \frac{t_{r_o} - t_f}{\frac{\omega}{k A_m} + \frac{1}{hA}}$$

$$t_{r_o} - t_f = 3.49 \times 10^4 \left(\frac{0.02}{10 \times \pi \times 0.42} + \frac{12}{10{,}000 \times \pi \times 0.44} \right)$$

$$= 83°F$$

$$t_{max} = t_f + (t_{r_o} - t_f) + (t_{max} - t_{r_o})$$

$$= 500°F + 83°F + 2520°F = \underline{\underline{3103°F}}$$

HEAT TRANSFER 7

A two-pass surface condenser is to be designed using an overall heat transfer coefficient of 480 Btu/hr-sq ft-°F with reference to the square feet of outside tube surface. The tubes are to be 1 inch O.D. with 1/16 inch walls.

Entering circulating water velocity is to be 6 fps.

Steam enters the condensers at a rate of 1,000,000 lb/hr at a pressure of 1 psia and an enthalpy of 1090 Btu/lb.

Condensate leaves as saturated liquid at 1 psia.

Circulating water enters the condenser at 85°F and leaves at 95°F.

Calculate the required number and length of condenser tubes.

Solution

$$\dot{m}(h_i - h_o)_{s\,(steam)} = \dot{m}c(t_o - t_i)_{w\,(water)}$$

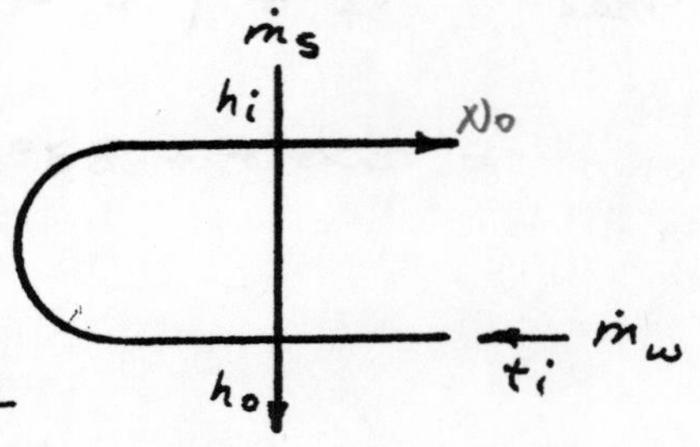

$$\dot{m}_w = \frac{1\times10^6(1090-69.7)}{1\times10} = 1.02\times10^8\ \frac{lbm}{hr}$$

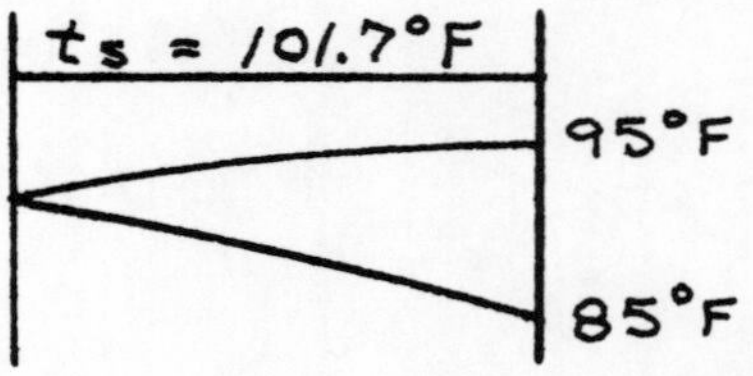

Let n = no. of tubes/pass

$$n\,U\rho A = m_w$$

$$n = \frac{1.02\times10^8\times4\times144}{6\times3.6\times0.624\times\pi\times(0.875)^2\times10^5}$$

$$= \underline{\underline{18{,}100\ \text{tubes/pass}}}$$

$$q = UA\,\Delta t_m \qquad \Delta t_m = \frac{(101.7-85)-(101.7-95)}{\ln\frac{16.7}{6.7}}$$

$$A = \frac{q}{U\,\Delta t_m} = \frac{\dot{m}_w c\,(t_o - t_i)}{U\,\Delta t_m}$$

$$= \frac{1.02\times10^9}{480\times10.9} = 1.94\times10^5\ ft^2 \text{ of outside surface}$$

– continued –

$$2n\pi d_o l = 1.94 \times 10^5$$

$$l = \frac{1.94 \times 10^5}{2\pi n d_o}$$

$$= \frac{1.94 \times 10^5 \times 12}{1.81 \times 10^4 \times \pi \times 1 \times 2} = 20.5 \text{ ft.}$$

(6.25 m)

HEAT TRANSFER 8

Assume you are the designer of a line of heat exchangers in which warm water flowing inside of copper tubes is used to heat air outside the tubes. Because of size limitations you find it necessary to use finned surface to achieve a rating of 50,000 Btu/hr for the exchanger. An inventor claims to have a new technique for inserting longitudinal fins inside the tubes at a cost only one quarter that of placing fins outside the tubes for the same net increase in capacity over bare tubes. The water velocity in the tubes is 6 ft per second and the water temperatures are 190°F inlet and 160°F outlet. The air is warmed from 50°F to 110°F. Is the inventor's claim worth pursuing? Justify your answer.

Convective coefficients for air commonly are an order of magnitude smaller than water. Since the copper wall will offer negligable resistance, 90% of the resistance is on the air side. If the water side resistance were completely eliminated, only a 10% improvement in heat exchange rate would be possible.

∴ Adding fins to water side is not worth it.

HEAT TRANSFER 9

The heat capacity of the interior structure and contents of a building is 100,000 Btu/°F. The conductance (net effect of walls, roof, windows, etc.) is 6500 Btu/hr°F. At 5:00 P.M., the heat supply is stopped and the interior temperature is 70°F. The outdoor temperature is constant at 40°F. Estimate, by appropriate calculations, the interior temperature at 1:00 A.M.

Solution

Let c = capacitance of bldg.

Let K = overall conductance

$$-c\frac{dt}{d\tau} = K(t-t_o)$$

$$\int_{t_i}^{t_F}\frac{dt}{t-t_o} = -\frac{K}{C}\int_0^{\tau} d\tau$$

$$\ln\frac{t_F-t_o}{t_i-t_o} = \frac{K\tau}{C} \quad \text{or} \quad t_F = t_o + (t_i-t_o)e^{-\frac{K\tau}{c}}$$

$$t_F = 40°F + 30°F\left(e^{-\frac{6500\times 8}{1\times 10^5}}\right)$$

$$= 57.9°F$$

$$(14.4°C)$$

HEAT TRANSFER 10

A 5 inch diameter polished copper sphere is heated electrically to maintain a surface temperature of 300°F. If the sphere is suspended in a large room where the air and wall temperatures are 75°F, how many Btu per hour must be supplied to the sphere to compensate for the radiation loss?

Solution

$$q_{rad.} = \delta A \epsilon \left(T_s^4 - T_r^4 \right)$$

$$= 0.171 \times \pi D^2 \times 0.05 \left(\overline{7.6}^4 - \overline{5.35}^4 \right)$$

$$= 0.171 \times \pi \left(\frac{5}{12} \right)^2 \times 0.05 \left(\overline{7.6}^4 - \overline{5.35}^4 \right)$$

$$= 12.2 \ \text{Btu/hr}$$

$$(3.57 \ \text{WATTS})$$

HEAT TRANSFER 11

The figure below describes the core geometry of a gas turbine regenerator for a 5,500 shp open cycle gas turbine plant. Operating conditions and the heat transfer surfaces are specified as follows:

Operating Conditions

Air flow rate, humid	200,000 lbs/hr
Air humidity	0.015 lbs/H_2O/lb dry air
Fuel-air ratio	0.015 lbs/lb
Fuel: fuel oil	H/C = 0.15, lbs/lb

Air Side

Louvered plate-fin surface	3/8 - 6.06
Entering pressure	130 lbs/in^2 abs
Entering temperature	350°F = 810°R

Gas Side

Plain plate-fin surface	11.1
Entering pressure	15 lbs/in^2 abs
Entering temperature	800°F = 1260°R

The objective of this problem is to determine for the specified conditions and the basic heat transfer and flow friction characteristics of the surface, (1) the regenerator effectiveness, and (2) the pressure drops for both the air and gas sides.

The steps in the general analysis require the determination of eleven specific factors. Name at least 8 of the factors.

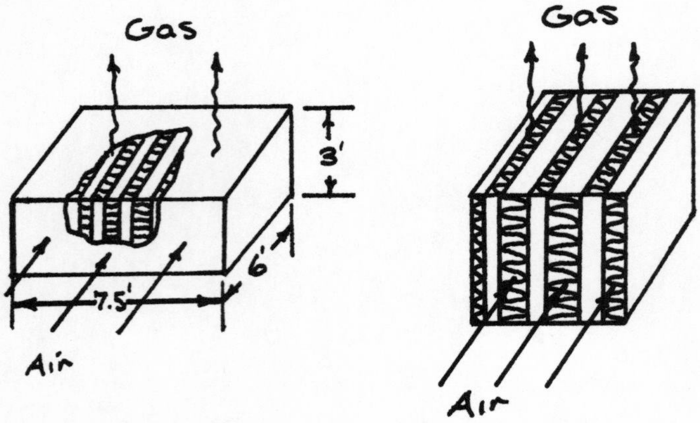

Air Side - Surface Louvered - Plate - Fin 3/8-6.06
Air Side - Surface Plain - Plate - Fin 11.1

Solution

Given:

1) Heat exchanger size & form

2) Inlet gas states

3) Mass flow rates

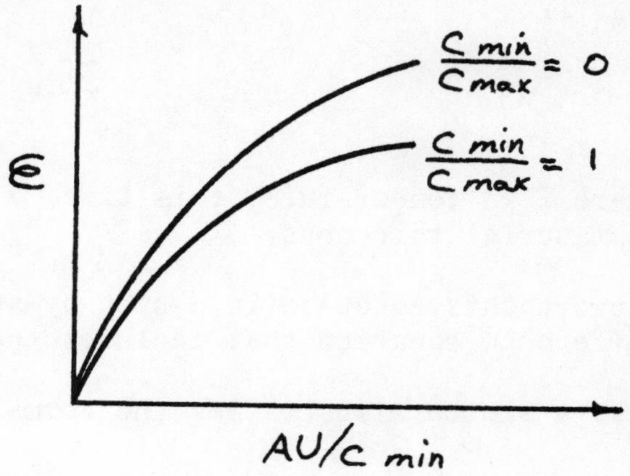

The following must be determined to find ϵ:

1. C_{min}
2. C_{max}
3. C_{min}/C_{max}
4. h_{hot}
5. h_{cold}
6. $\eta_{fin,cold}$
7. $\eta_{fin,hot}$
8. U
9. AU/C_{min}

The following must be determined to find Δp:

10. $\bar{\rho}_{hot}$
11. $\bar{\rho}_{cold}$
12. f_{hot}
13. f_{cold}

HEAT TRANSFER 12

The partial differential equation for one-dimensional heat conduction is:

$$\frac{\partial T}{\partial t} = \alpha \frac{\partial^2 T}{\partial x^2}$$

where T is temperature, t is time, α is thermal diffusivity and x is material thickness.

Convert this equation in a step-by-step procedure into a finite-difference equation that includes the heat balance.

Make a sketch illustrating the terms of the final equation.

Solution

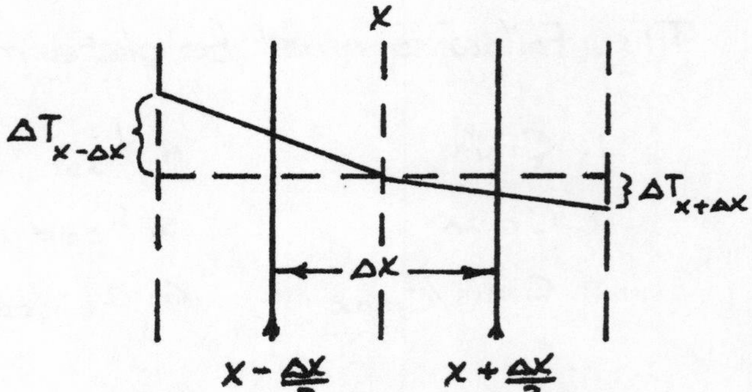

Consider application of equation to plane-x

$$\frac{\partial T}{\partial \tau} = \frac{k}{\rho c} \frac{\partial^2 T}{\partial x^2}$$

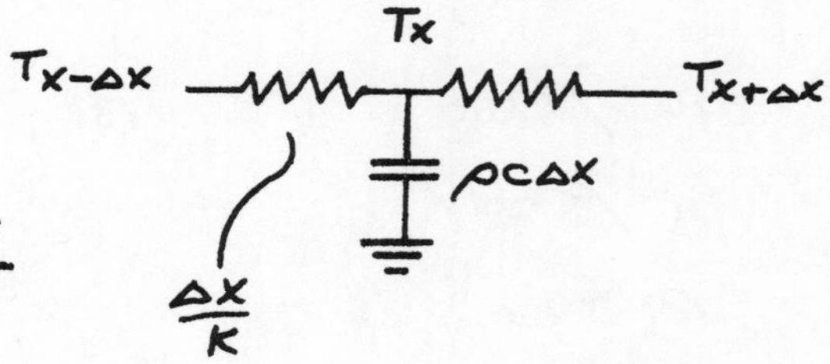

in difference form,

$$\frac{\Delta T}{\Delta \tau} = \frac{k}{\rho c} \frac{\Delta}{\Delta x} \frac{\Delta T}{\Delta x}$$

by rearrangement,

$$\underbrace{\rho c \Delta x \frac{\Delta T}{\Delta \tau}}_{\text{rate of energy storage}} = \Delta\left(k \frac{\Delta T}{\Delta x}\right)$$

$\Delta\left(k \frac{\Delta T}{\Delta x}\right)$ = inflow heat − outflow heat

inflow heat: $-k \frac{\Delta T}{\Delta x}\Big)_{x-\frac{\Delta x}{2}}$ outflow heat: $-k \frac{\Delta T}{\Delta x}\Big)_{x+\frac{\Delta x}{2}}$

8

Gas Dynamics and Combustion

MICHEL A. SAAD

GAS DYNAMICS

While problems can be solved by use of the Mach function relationships which follow, it is expeditious to use Mach function tables and the problem solutions which are presented utilize the tables. For reference see "The Dynamics and Thermodynamics of Compressible Flow" by A. H. Shapiro.

Fundamental Relations:

Speed of sound $c = \sqrt{g_c \left(\frac{\partial p}{\partial \rho}\right)_s}$

For Perfect Gas $c = \sqrt{k g_c RT}$

For Air ($k = 1.4$, $R = 53.34$ ft-lbf/lbm°R)
$c = 49\sqrt{T}$ ft/sec (T°R)

Mach Number $M = \frac{V}{c}$

Energy Equation $h_o = h + \frac{V^2}{2g_c J}$

$$T_o = T + \frac{V^2}{2g_c J c_p} \quad \text{(perfect gas)}$$

Note: subscript o refers to stagnation state; superscript * refers to Mach 1 condition in the relations that follow.

Isentropic Relations for Perfect Gas:

$$\frac{p}{p_o} = \left(\frac{\rho}{\rho_o}\right)^k = \left(\frac{T}{T_o}\right)^{\frac{k}{k-1}} = \left(\frac{c}{c_o}\right)^{\frac{2k}{k-1}}$$

(Adiabatic Perfect Gas)

$$\frac{T_o}{T} = 1 + \frac{k-1}{2} M^2$$

(Isentropic Perfect Gas)

$$\frac{p_o}{p} = \left(1 + \frac{k-1}{2} M^2\right)^{\frac{k}{k-1}}$$

(Isentropic Perfect Gas)

$$\frac{\rho_o}{\rho} = \left(1 + \frac{k-1}{2} M^2\right)^{\frac{1}{k-1}}$$

(Adiabatic Perfect Gas)

$$\frac{c_o}{c} = \left(1 + \frac{k-1}{2} M^2\right)^{1/2}$$

At M = 1 (and k = 1.4):

$$\frac{T^*}{T_o} = \frac{c^{*2}}{c_o^2} = \frac{2}{k+1} = 0.8333$$

$$\frac{\rho^*}{\rho_o} = \left(\frac{2}{k+1}\right)^{\frac{1}{k-1}} = 0.6339$$

$$\frac{p^*}{p_o} = \left(\frac{2}{k+1}\right)^{\frac{k}{k-1}} = 0.5283$$

Mass rate of flow per unit area:

$$\frac{\dot{m}}{A} = \sqrt{\frac{kg_c}{R}} \frac{p_o}{\sqrt{T_o}} \frac{M}{\left(1 + \frac{k-1}{2} M^2\right)^{\frac{k+1}{2(k-1)}}}$$

$$\left(\frac{\dot{m}}{A}\right)_{max} = \frac{\dot{m}}{A^*} = \sqrt{\frac{kg_c}{R} \left(\frac{2}{k+1}\right)^{\frac{k+1}{k-1}}} \frac{p_o}{\sqrt{T_o}}$$

For air:

$$\frac{\dot{m}}{A^*} = 0.532 \frac{p_o}{\sqrt{T_o}}$$

where $\dot{m}$ is in lbm/sec, A^* is ft^2, p_o in lbf/ft^2 and T_o in °R.

Isentropic Flow through a Nozzle:

$$\frac{dA}{A} = \frac{dV}{V} (M^2-1) = g_c \frac{(1-M^2)}{\rho V^2} dp = \frac{1-M^2}{M^2} \frac{d\rho}{\rho}$$

	M < 1			M > 1		
	dV	dp	dρ	dV	dp	dρ
dA +	-	+	+	+	-	-
dA -	+	-	-	-	+	+

$$\frac{A}{A^*} = \frac{\dot{m}/A^*}{\dot{m}/A} = \frac{1}{M}\left[\frac{2 + (k-1)\,M^2}{k+1}\right]^{\frac{k+1}{2(k-1)}}$$

Frictional Flow Relations (Constant Area, Adiabatic Flow):

$$\frac{V}{V^*} = M\sqrt{\frac{k+1}{2 + (k-1)M^2}}$$

$$\frac{p}{p^*} = \frac{1}{M}\sqrt{\frac{k+1}{2 + (k-1)M^2}}$$

$$\frac{T}{T^*} = \frac{k+1}{2 + (k-1)M^2}$$

$$\frac{\rho}{\rho^*} = \frac{V^*}{V} = \frac{1}{M}\sqrt{\frac{2 + (k-1)M^2}{k+1}}$$

$$\frac{p_o}{p_o^*} = \frac{1}{M}\sqrt{\left(\frac{2 + (k-1)M^2}{k+1}\right)^{\frac{k+1}{(k-1)}}}$$

$$\frac{4\bar{f}L^*}{D} = \frac{1-M^2}{kM^2} + \frac{k+1}{2k}\ln\left[\frac{(k+1)M^2}{2 + (k-1)M^2}\right]$$

$$\frac{4\bar{f}L_{1-2}}{D} = \left(\frac{4\bar{f}L^*}{D}\right)_{M_1} - \left(\frac{4\bar{f}L^*}{D}\right)_{M_2}$$

Simple Heating - Rayleigh Line:

$$\frac{T}{T^*} = \frac{(1+k)^2M^2}{(1+kM^2)^2}$$

$$\frac{p}{p^*} = \frac{1+k}{1+kM^2}$$

$$\frac{V}{V^*} = \frac{\rho^*}{\rho} = \frac{(k+1)M^2}{1+kM^2}$$

$$\frac{T_o}{T_o^*} = \frac{(k+1)M^2(2+(k-1)M^2}{(1+kM^2)^2}$$

$$\frac{p_o}{p_o^*} = \frac{k+1}{1+kM^2}\left[\frac{(2+(k-1)M^2}{k+1}\right]^{\frac{k}{k-1}}$$

COMBUSTION

Fundamental Concepts:

Air-Fuel Ratio (A/F) -
Mass or molal ratio of air to fuel in a combustion process.

Stoichiometric -
A/F such that sufficient oxygen is present to fully oxidize the fuel with no excess oxygen.

Heating Value (HV) or Heat of Reaction -
The amount of energy released by the chemical reaction being carried out either at constant pressure and temperature or constant volume and temperature.

Higher Heating Value (HHV) -
The water formed by combustion is assumed condensed.

Lower Heating Value (LHV) -
The water formed by combustion is assumed to remain in vapor phase.

Combustion Example Problems:

Problems in stoichiometry are divided into two main categories, depending on whether the composition of the fuel or the composition of the reaction products is known.

Example 1: Determine the stoichiometric air required for the complete combustion of 1 lbm of normal heptane C_7H_{16}. What is the percentage analysis of the products on a mass and a mole basis?

Solution: The chemical equation is

$$C_7H_{16}(\ell) + 11\ O_2(g) + 11(3.76)\ N_2(g) \rightarrow 7\ CO_2(g) + 8\ H_2O(g) + 11(3.76)\ N_2(g)$$

The air fuel ratio on a mass basis is

$$\frac{(11 \times 32) + (11 \times 3.76 \times 28)}{(7 \times 12) + (16 \times 1.008)} = 15.12 \text{ lbm air/lbm fuel}$$

Alternately, since there are (11 + 11 x 3.76) moles of air, the air fuel ratio is

$$\frac{(11 + 11 \times 3.76)(28.97)}{(100.205)} = 15.12 \text{ lbm air/lbm fuel}$$

The air fuel ratio on a mole basis is

$$\frac{11 + 11 \times 3.76}{1} = 52.3 \text{ moles air/mole fuel}$$

The analysis of the products on a mass and molal basis are as follows:

	By Mole		By Mass		
CO_2	8	12.42%	7 x 44 =	308	19.11%
H_2O	7	14.20%	8 x 18 =	144	8.94%
N_2	41.36	73.38%	41.36 x 28 =	1160	71.95%
Totals	56.36	100.00%		1612	100.00%

Example 2: Five moles of propane C_3H_8 are completely burned at atmospheric pressure in the theoretical amount of air. Determine

(a) The volume of air used in the combustion process measured at 14.7 psia and 77°F.

(b) The partial pressure of each constituent of the products.

(c) The volumetric analysis of the dry products.

Solution: (a) The chemical equation is

$$5\ C_3H_8(g) + 25\ O_2(g) + 25(3.76)\ N_2(g) \rightarrow 15\ CO_2(g) + 20\ H_2O(g) + 25(3.76\ N_2(g)$$

or

$$5\ C_3H_8(g) + 25.0_2(g) + 94.0\ N_2(g) \rightarrow 15\ CO_2(g) + 20\ H_2O(g) + 94.0\ N_2(g)$$

Thus, 25 moles of O_2 are required or 25 (1/0.2099) = 119 moles of air (25 moles of O_2 plus 94 moles of N_2).

If air is assumed a perfect gas, then

$$V = \frac{nRT}{p} = \frac{119 \times 1545(77 + 460)}{14.7 \times 144} = 46{,}700 \text{ ft}^3$$

(b) The partial pressures of the product constituents are proportional to the mole fractions. Therefore,

$$p_{CO_2} = \frac{15}{15 + 20 + 94} \times 14.7 = 1.71 \text{ psia}$$

$$p_{H_2O} = \frac{20}{129} \times 14.7 = 2.28 \text{ psia}$$

$$p_{N_2} = \frac{94}{129} \times 14.7 = \underline{10.71 \text{ psia}}$$

$$\text{Total pressure} = 14.7 \text{ psia}$$

(c) $$\frac{V_{CO_2}}{V_{CO_2} + V_{N_2}} = \frac{15}{15 + 94} = 13.75 \text{ per cent}$$

$$\frac{V_{N_2}}{V_{CO_2} + V_{N_2}} = 1 - 0.1375 = 86.25 \text{ per cent}$$

Example 3: The analysis of a sample of coal gives the following values by weight (the remainder is ash):

Carbon	80.7%
Hydrogen	4.9%
Sulfur	1.8%
Oxygen	5.3%
Nitrogen	1.1%

What is the air fuel ratio by weight if 20 per cent excess air is used in the combustion process?

Solution: For 1 lbm of fuel, the following table can be formulated:

	Number of Moles per lbm of Fuel	Moles of O_2 Required for Complete Combustion of 1bm of Fuel	Reaction Equation
C	$\frac{0.807}{12} = 0.0673$	0.0673	$0.0673\ C(s) + 0.0673\ O_2(g) \rightarrow 0.0673\ CO_2(g)$
H_2	$\frac{0.049}{2} = 0.0245$	0.01225	$0.0245\ H_2(g) + 0.01225\ O_2(g) \rightarrow 0.0245\ H_2O(g)$
S	$\frac{0.018}{32} = 0.000563$	0.000563	$0.000563\ S(s) + 0.000563\ O_2(g) \rightarrow 0.000563\ SO_2(g)$
O_2	$\frac{0.053}{32} = 0.001655$	0	
N_2	$\frac{0.011}{28} = 0.000393$	0	

O_2 required = 0.080113 mole; O_2 in fuel = 0.001655 mole;
Difference = 0.078458 mole O_2 per lbm of fuel

Stoichiometric air fuel ratio = 0.078458(100/20.99) x 28.97

= 10.83 lbm air/lbm fuel

With 20 per cent excess air, the air fuel ratio is

$$10.83 \times 1.2 = 13 \text{ lbm air/lbm fuel}$$

Example 4: The composition of a hydrocarbon fuel by mass is 85 per cent carbon, 13 per cent hydrogen, and 2 per cent oxygen. Determine

(a) The chemically correct mass of air required for the complete combustion of 1 lbm of fuel.

(b) The volumetric analysis of the products of combustion and the dew point if the total pressure is 14.7 psia.

(c) If the products of combustion are cooled to 60°F, what is the mass of the water vapor that condensed?

Solution: (a) Consider the combustion of 1 lbm of fuel composed of 0.85 lbm of carbon, 0.13 lbm of hydrogen, and 0.02 lbm of oxygen. In the complete combustion process the carbon oxidizes to CO_2 and the hydrogen to H_2O according to the equations

$$C(s) + O_2(g) \rightarrow CO_2(g)$$

$$H_2(g) + \tfrac{1}{2}\, O_2(g) \rightarrow H_2O(g)$$

The oxygen requirement as dictated by these two equations is, however, reduced by the amount of oxygen already existing in the fuel. The number of moles of

$$O_2 \text{ required} = \frac{0.85}{12.01} + \frac{0.13}{2.016 \times 2} - \frac{0.02}{32} = 0.07075 + 0.03225 - 0.000625$$

$$= 0.102375 \text{ mole of } O_2\text{/lbm of fuel}$$

$$\text{mass of air} = \frac{0.102375}{0.2099} \times 28.97 = 14.1 \text{ lbm air/lbm fuel}$$

(b) The analysis of the products of combustion by mass and volume is given in the accompanying table:

	Moles per lbm Fuel	lbm per lbm Fuel	Molal Analysis
CO_2	$\frac{0.85}{12.01} = 0.07075$	3.12	0.136
H_2O	$\frac{0.13}{2.016} = 0.0645$	1.16	0.124
N_2	$(3.76 \times 0.102375) = 0.385$	10.82	0.740
Totals	0.52025	15.1	1.000

The dew point temperature may be found from steam tables corresponding to the pressure of the water vapor in the combustion products:

Pressure of H_2O = 0.124 x 14.7 = 1.82 psia

Dew point = 122.5°F

(c) Since 60°F is below the dew point temperature, some water vapor will condense. The vapor pressure is then equal to the saturation vapor pressure corresponding to 60°F.

From steam tables, $p_{g,sat}$ at 60°F = 0.2563 psia and v_g = 1206.7 ft^3/lbm. The volume occupied by the water vapor is equal to the volume occupied by the CO_2 and the N_2 at their partial pressure of 14.7 - 0.2563 = 14.4437 psia. The volume according to perfect gas law is

$$V = \frac{nRT}{p} = \frac{(0.07075 + 0.385)\ 1545 \times 520}{14.4437 \times 144} = 176.3\ ft^3$$

$$\text{Mass of vapor in products} = \frac{V}{v_g} = \frac{176.3}{1206.7} = 0.1462\ \text{lbm}$$

Mass of vapor that condensed = 1.16 - 0.1462 = 1.0138 lbm

Example 5: A hydrocarbon fuel in the vapor state is burned with atmospheric air at 14.7 psia, 80°F, and 60 per cent relative humidity. The volumetric analysis of the dry products of combustion is

Product	Percentage
CO_2	10.00
O_2	2.37
CO	0.53
N_2	87.10 (by difference from 100)
	100.00

Calculate

(a) Ratio of hydrogen to carbon by mass for the fuel

(b) Air-fuel ratio by mass

(c) Air used as percentage of the stoichiometric value

(d) Volume of the humid air supplied per lbm of fuel

Solution: (a) The following table gives the number of atoms of carbon, oxygen, and nitrogen per mole of dry products:

	Moles (per Mole of Dry Products)	Carbon	Oxygen	Nitrogen
CO_2	0.10	0.10	0.20	--
O_2	0.0237	--	0.0474	--
CO	0.0053	0.0053	0.0053	--
N_2	0.871	--	--	1.742
Totals	1.0000	0.1053	0.2527	1.742

Assuming that H_2O is the only remaining product of combustion, the amount of hydrogen in the fuel may be determined by finding the amount of oxygen used to form the H_2O.

Total moles of O_2 supplied with 0.871 mole of N_2 $= \frac{0.871}{3.76} = 0.231.$

Moles of O_2 accounted for in the reactions considered $= 0.12635.$

Hence the difference used in formation of H_2O $= 0.10465$ moles of O_2

The formation of H_2O can thus be expressed as

$$0.2093\ H_2(g) + 0.10465\ O_2(g) \rightarrow 0.2093\ H_2O(g)$$

which indicates that 0.2093 mole of H_2O is formed per mole of dry products. The hydrogen-carbon ratio in the fuel is

$$\frac{H}{C} = \frac{2 \times 0.2093}{0.1053} = 3.97 \quad \text{(by atoms)}$$

$$= \frac{(2 \times 0.2093) \times 1.008}{(0.1053) \times 12.01} = 0.333 \quad \text{(by weight)}$$

(b) The air-fuel ratio $= \frac{[0.871 \times (1.0/0.79)]\ 28.97}{(0.1053 \times 12.01) + (0.2093) \times 2.016}$

$$= \frac{31.95}{1.687} = 18.9 \text{ lbm air/lbm fuel}$$

(c) The number of moles of oxygen required for stoichiometric combustion is

$$0.1053\ C(s) + 0.1053\ O_2(g) \rightarrow 0.1053\ CO_2(g)$$

$$0.2093\ H_2(g) + \underline{0.10465}\ O_2(g) \rightarrow 0.2093\ H_2O(g)$$

0.20995 moles of O_2 per mole of dry products

Since the oxygen-air ratio is constant in atmospheric air, the percentage of the air supplied is the same as that of the oxygen. Hence the percentage of the stoichiometric air used is

$$\frac{0.231}{0.20995} = 1.1 \text{ or } 110 \text{ per cent}$$

(d) The number of moles of dry air is

$0.87\left(\frac{1}{0.79}\right) = 1.102$ moles air/mole dry products, but the mass of fuel = 1.687 lbm fuel/mole dry products. Therefore, the number of moles of air per lbm of fuel = 1.102/1.687 = 0.653.

At 80°F

$$p_{g,sat} = 0.5069 \text{ psia}$$

and

$$p_g = 0.6 \times 0.5069 = 0.30414 \text{ psia}$$

Hence

$$V = \frac{nRT}{p} = \frac{0.653 \times 1545.33 \times 540}{(14.7 - 0.30414) \times 144} = 263 \text{ ft}^3\text{/lbm of fuel}$$

GAS DYNAMICS AND COMBUSTION 1

Air flows through a convergent-divergent nozzle with inlet area 0.8 sq. in., minimum area 0.5 sq. in., and exit area 0.60 sq. in. At the inlet the air velocity is 344 ft/sec, stream pressure 100 psia and stream temperature 160°F.

(a) Calculate the mass rate of flow through the nozzle in lbm/hr.

(b) Determine the Mach number at the minimum area.

(c) Determine the stream velocity and pressure at the exit.

Solution

*Note: It is presumed that mach function tables are available.

Assume isentropic, steady flow.

a) $\dot{m} = \rho_1 A_1 V_1$

$$\rho_1 = \frac{P_1}{RT_1} = \frac{100 \times 144}{53.34 \times 620} = 0.435 \text{ lbm/ft}^3$$

$$\dot{m} = (0.435)(344)\left(\frac{0.8}{144}\right) = \underline{\underline{0.832 \text{ lbm/sec.}}}$$

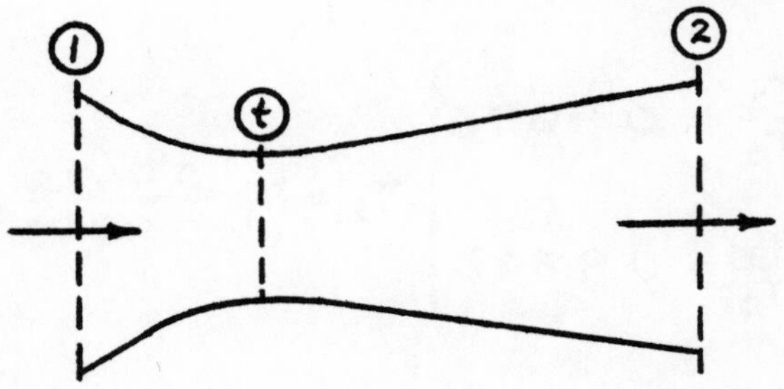

$A_1 = .8 \text{ in}^2$ $A_t = 0.5 \text{ in}^2$ $A_2 = 0.6 \text{ in}^2$

$U_1 = 344$ ft/sec.

$P_1 = 100$ psia

$T_1 = 160°F$

-continued-

b) $M_1 = \dfrac{344}{49\sqrt{620}} = 0.282$

$\dfrac{A^*}{A_1} = 0.467$ but $\dfrac{A_t}{A_1} = \dfrac{0.5}{0.8} = 0.625$

Therefore, $\dfrac{A^*}{A_t} = \dfrac{A^*/A_1}{A_t/A_1} = \dfrac{0.467}{0.625} = 0.747$

from tables, $M_t = \underline{\underline{0.5}}$

c) $\dfrac{A_2}{A_1} = \dfrac{0.6}{0.8} = 0.75$, $\dfrac{A^*}{A_2} = \dfrac{A^*/A_1}{A_2/A_1} = \dfrac{0.467}{0.75} = 0.6227$

from tables $M_2 = 0.395$ and,

$$\left.\begin{aligned}\frac{P_2}{P_0} &= 0.898\\ \frac{P_1}{P_0} &= 0.946\end{aligned}\right\} \quad P_2 = P_1\,\frac{0.898}{0.946} = 100 \times \frac{0.898}{0.946}$$

$$= \underline{\underline{94.9 \text{ psia}}}$$

$$(644\ \mathrm{KN/m^2})$$

$$\left.\begin{aligned}\frac{T_2}{T_0} &= 0.9696\\ \frac{T_1}{T_0} &= 0.9843\end{aligned}\right\} \quad T_2 = T_1\,\frac{0.9696}{0.9843} = 620 \times \frac{0.9696}{0.9843}$$

$$= 611\ ^\circ R$$

$$V_2 = C_2 M_2 = 49\sqrt{611} \times 0.395 = \underline{\underline{478 \text{ ft./sec.}}}$$

$$(145.7\ \mathrm{m/sec})$$

GAS DYNAMICS AND COMBUSTION 2

Dry air at 100 psia and 1500°R enters a converging-diverging nozzle through a line of 0.05 ft^2 area and expands to a delivery region pressure of 5 psia. Assuming isentropic expansion and a mass rate of flow of 2 lbm/sec, find:

(a) The stagnation enthalpy.

(b) The temperature and enthalpy at discharge.

(c) The Mach number and velocity of the air stream at discharge.

(d) The mass flow rate per unit area.

Solution

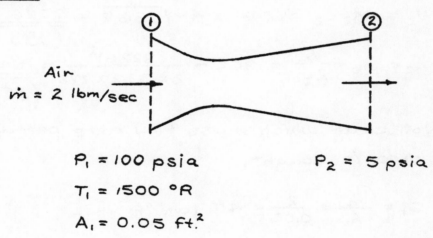

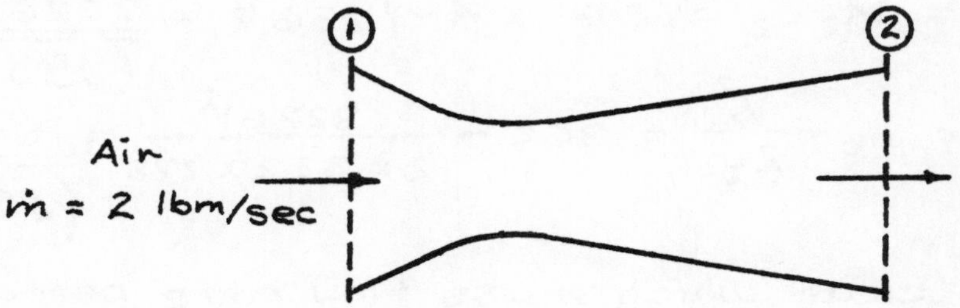

$P_1 = 100$ psia $\qquad P_2 = 5$ psia

$T_1 = 1500\ °R$

$A_1 = 0.05\ ft^2$

$$\rho_1 = \frac{P_1}{RT_1} = \frac{100 \times 144}{53.34 \times 1500} = 0.180\ lbm/ft.^3$$

$$V_1 = \frac{\dot{m}}{\rho_1 A_1} = \frac{2}{0.180 \times .05} = 222\ ft./sec.$$

$$M_1 = \frac{222}{49\sqrt{1500}} = 0.117$$

therefore $\frac{T_1}{T_o} = 0.997, \quad T_o = 1505\ °R$

$$\frac{P_1}{P_o} = 0.990, \quad P_o = 101 \text{ psia}$$

-continued-

a) $h_0 = h_1 + \frac{V_1^2}{2g_c} = c_p T_0 = 0.241 \times 1505 = 362$ Btu/lbm

b) & c)

$\frac{P_2}{P_0} = \frac{5}{101} = 0.0495$ $\quad M_2 = \underline{\underline{2.61}}$

$\frac{T_2}{T_0} = 0.423$ $\quad T_2 = \underline{\underline{636.2\ °R}}$

$V_2 = M_2 c_2 = 2.61 \times 49\sqrt{636.2} = \underline{\underline{3226 \text{ ft/sec.}}}$ (983 m/sec)

$h_2 = h_0 - \frac{V_2^2}{2g_c} = 362 - \frac{(3226)^2}{2 \times 32.2 \times 778} = \underline{\underline{154 \text{ Btu/lbm}}}$ (358 KJ/Kg)

d) Not clear which mass flow rate per unit area is sought.

$G_1 = \frac{\dot{m}}{A_1} = \frac{2}{0.05} = 40$ lbm/ft²·sec.

$G^* = \frac{\dot{m}}{A^*} = 0.532 \frac{P_0}{\sqrt{T_0}} = 0.532 \frac{101 \times 144}{38.8}$

$= 200$ lbm/ft²·sec

GAS DYNAMICS AND COMBUSTION 3

Air is flowing in a supersonic wind tunnel. At a given point, the area is 1 ft^2 and Mach number is 2.5. The reservoir conditions are 2116 lb/ft^2 and 59°F. Find the temperature pressure and density at the point where Mach number equals 2.5 and the area at the throat of the tunnel.

Solution

Assume isentropic, steady flow.

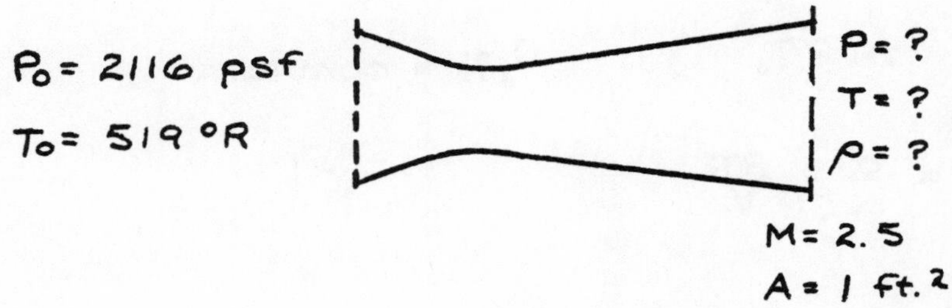

P_0 = 2116 psf

T_0 = 519 °R

P = ?
T = ?
ρ = ?

M = 2.5
A = 1 ft.^{2}

$\dfrac{T}{T_0}$ = 0.44444   T = $\underline{230.6\text{°R}}$ $\left(128.1\text{°K}\right)$

$\rho_0 = \dfrac{P_0}{R T_0}$ = 0.0764 lbm/ft^3

$\dfrac{P}{P_0}$ = 0.05853   P = $\underline{123.3 \text{ psf}}$ $\left(5.90 \text{ KH/}_{m^2}\right)$

$\dfrac{\rho}{\rho_0}$ = 0.13169 ρ = $\underline{0.01006 \text{ lbm/ft.}^3}$ $\left(1.58 \text{ N/}_{m^3}\right)$

$\dfrac{A}{A^*}$ = 2.6367 A^* = $\underline{0.3793 \text{ ft.}^2}$ $\left(0.0352 \text{ m}^2\right)$

GAS DYNAMICS AND COMBUSTION 4

A 5 ft^3 tank of compressed air discharges through a 0.500 in. diameter converging nozzle which is located in the side of the tank. If the mass flow coefficient of the nozzle, based on isentropic flow through it, is 0.95 and the gas within the tank expands isothermally, find the time for the tank to discharge from 150 psia to 50 psia if the temperature of the tank is 70°F and the surrounding pressure is 14.7 psia.

Solution

For the tank:

$$p_o V = m R T_o \qquad V, T_o = \text{constant}$$

$$\frac{dp_o}{d\tau} = \frac{dm}{d\tau}\frac{RT_o}{V}$$

For Sonic flow through nozzle, note $\frac{P_{exit}}{P_o} < 0.528$, $M_e = 1$

$$\frac{dm}{d\tau} = -\frac{p_o A}{\sqrt{T_o}} C_m \sqrt{\frac{kg_c}{R}\left(\frac{2}{k+1}\right)^{\frac{k+1}{k-1}}}$$

By elimination of $\frac{dm}{d\tau}$,

$$\frac{dp_o}{d\tau} = -\frac{p_o A}{\sqrt{T_o}} C_m \sqrt{\frac{kg_c}{R}\left(\frac{2}{k+1}\right)^{\frac{k+1}{k-1}}}\;\frac{RT_o}{V} = \text{constant}\,(C)\,p_o$$

Therefore,
$$\int_{p_i}^{p_f}\frac{dp_o}{p_o} = C\int_0^{\tau} d\tau$$

or
$$\tau = \frac{1}{C}\ln\frac{p_f}{p_i} = -\frac{V}{C_m A\sqrt{kg_c R T_o\left(\frac{2}{k+1}\right)^{\frac{k+1}{k-1}}}}\ln\frac{p_f}{p_i}$$

-continued-

for air,

$K = 1.4$

$R = 53.34$ ft-lbf/lbm°R

$$\tau = \frac{-0.0353 V \ln \frac{p_f}{p_i}}{C_m A \sqrt{T_i}} = \underline{\underline{6.49 \text{ sec.}}}$$

GAS DYNAMICS AND COMBUSTION 5

An attitude control system for a space vehicle uses nitrogen expanding through a nozzle to provide thrust vector control. At maximum thrust nitrogen is used at a rate of 360 lb/hr from an initial pressure of 500 psia and 0°F.

(a) What is the thrust of the nozzle?

(b) Calculate the throat area of the nozzle.

Solution

$\dot{m} = 360$ lbm/hr $= 0.1$ lbm/sec.

$P_o = 500$ psia $C_p = 0.25$ Btu/lbm °F

$T_o = 460$°R $K = 1.4$ $R = 55.14$ ft. lbf/lbm°R

for maximum thrust $P_{exit} = P_{atm}$

a) $$F = \frac{\dot{m}}{g_c} V_e + \underbrace{A_e (p_e - p_a)}_{0 \text{ for max. thrust}}$$

but, $$V_e = \sqrt{2 g_c C_p (T_o - T_e)} = \sqrt{2 g_c C_p T_o \left(1 - \left(\frac{P_e}{P_o}\right)^{\frac{K-1}{K}}\right)}, \quad C_p = \frac{KR}{K-1}$$

-continued-

$$\therefore F_{max} = \frac{0.1}{32.2}\sqrt{\frac{2 \times 1.4 \times 1545.44 \times 460 \times 32.2}{28 \times 0.4}\left(1-\left(\frac{14.7}{500}\right)^{0.285}\right)}$$

$$= \underline{\underline{5.91 \text{ lbf}}}$$

b) $$\frac{\dot{m}}{A_t} = \sqrt{\frac{K g_c}{R/M_{(\text{molecular wt.})}}\left(\frac{2}{K+1}\right)^{\frac{K+1}{K-1}}}\ \frac{P_o}{\sqrt{T_o}} \approx 0.523\ \frac{P_o}{\sqrt{T_o}}$$

$$= 0.523\ \frac{500 \times 144}{\sqrt{460}} = 1757 \text{ lbm/ft}^2\text{sec}$$

$$\therefore A_t = \frac{0.1 \times 144}{1757} = \underline{\underline{0.00820 \text{ in.}^2}} \quad (5.29 \text{ mm}^2)$$

GAS DYNAMICS AND COMBUSTION 6

A turbine test area hopes to obtain 1/4 lb/sec of cooling air through a 29 ft 1/2 in. I.D. pipe.

Assuming no heat transfer, what is the available maximum flow rate from a 100°F 200 psia source if the outlet is maintained at 50 psia?

Friction factor in pipe is f = 0.01 ($\Delta P = \frac{4fl}{D}\frac{\rho V^2}{2g}$)

List all assumptions and references.

Solution

Assume:

a) fanno line solution for a perfect gas

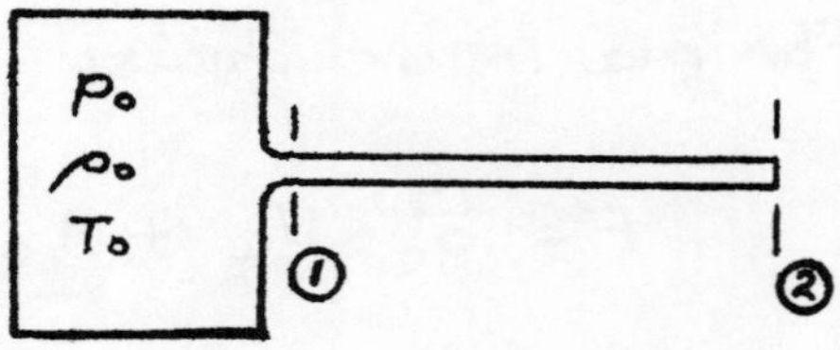

b) pipe entrance velocity low, such that $T_o \simeq T_1$, $\rho_o \simeq \rho_1$ (check subsequently)

-continued-

For desired flow rate,

$$V_1 = \frac{\dot{m}}{\rho_0 A_1} = \frac{0.25 \times 53.34 \times 560 \times 4}{200 \times 144 \times \pi \times 0.25} = 190 \text{ ft/sec.}$$

For this velocity, assumption b) is not seriously violated.

$$\therefore \quad M_1 = \frac{190}{49\sqrt{560}} = 0.16$$

$$\frac{4fL_{1-2}}{D} = \frac{0.04 \times 29}{1/24} = 27.84$$

For $M_1 = 0.16$ from tables $\frac{4fL_{max}}{D} = 22.96$

Then conclude M_1 is too large.

Further $\frac{p_1}{p^*} = 6.6$ from tables, but $\frac{p_1}{p_2} = 4$ from problem gives additional support $M_1 < 0.164$

<u>This means $\dot{m}$ will be below desired flow rate</u>

To estimate $\dot{m}$, assume $M_1 = 0.15$ Then $\left.\frac{4fL_{max}}{D}\right)_1 = 27.93$

$$\left.\frac{4fL_{max}}{D}\right)_2 = \left.\frac{4fL_{max}}{D}\right)_1 - 27.84 = 0.09$$

$\therefore M_2 = 0.78$ from tables, $\frac{p_1/p^*}{p_2/p^*} = \frac{p_1}{p_2} = \frac{7.29}{1.326} = 5.5$

$\therefore M_1$ is slightly lower than 0.15 to give $p_1/p_2 = 4$

$$V_1 \simeq 190 \times \frac{0.15}{0.16} = 174 \text{ ft/sec.}$$

$$\dot{m} = 0.25 \times \frac{0.15}{0.164} = \underline{0.229 \text{ lbm/sec.}}\ (0.104 \text{ Kg/sec})$$

GAS DYNAMICS AND COMBUSTION 7

Gas flows through a duct that extends from station "a" through station "d" and it has a cross-sectional area of 100 in^2. At station "a" the gas has a velocity of 880 ft/sec, a static pressure of 20.0 psia and a total temperature of 670°R. Between stations "b" and "c" heat is transferred to/from the flow. If the gas has a specific heat at constant pressure of 0.275 Btu/lb - °R and a gas constant of 70 ft-lbf/lbm°R, what heat flow rate is necessary for a Mach number of 0.300 at station "d"? Is the heat added or removed? Neglect velocity distribution and all real gas effects.

<u>Solution</u>

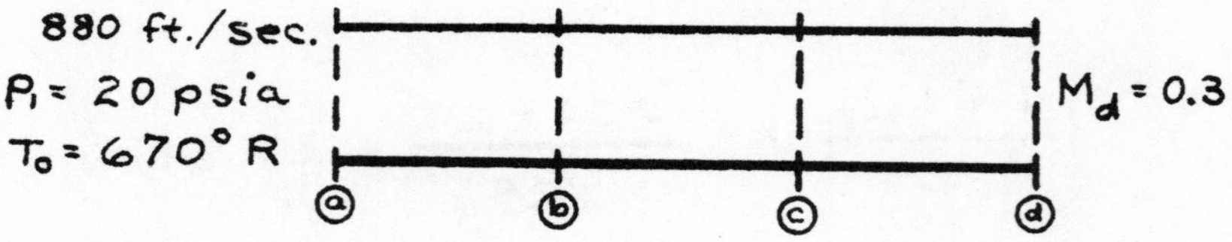

$A = 100$ in^2

$C_p = 0.275$ Btu/lbm °F

$R = 70$ ft·lbf/lbm°R

$$M_a = \frac{880}{\sqrt{K g_c R T_a}}$$

$$C_p - C_v = R$$

$$C_v = 0.275 - \frac{70}{778} = .185$$

$$K = \frac{C_p}{C_v} = \frac{.275}{.185} = 1.49$$

$$T_a = T_o - \frac{V^2}{2 g_c C_p} = 670 - \frac{(880)^2}{2 \times .275 \times 778 g_c} = 670 - 56.2 = 613.8°R$$

Therefore,

$$M_a = \frac{880}{\sqrt{1.49 \times 32.2 \times 70 \times 613.8}} = 0.613$$

Since $M_a > M_d$ and both are subsonic

Therefore, <u>heat is removed.</u>

$$\frac{T_{o2}}{T_{o1}} = \frac{M_2^2(1+kM_1^2)^2(1+\frac{K-1}{2}M_2^2)}{M_1^2(1+KM_2^2)^2(1+\frac{K-1}{2}M_1^2)}$$

$$= \frac{.09(1+1.49 \times .376)^2(1+.245 \times .09)}{.376(1+1.49 \times .09)^2(1+.245 \times .376)} = 0.447$$

-continued-

Therefore, $T_{o2} = 0.447 \times 670 = 300°R$

Heat removed $= c_p(T_{o1} - T_{o2}) = .275(670-300) = 101.9$ Btu/lbm

Total $Q = \left(\frac{20 \times 144}{70 \times 613.8} \times \frac{100}{144} \times 880\right)(101.9) = 4167$ Btu/sec

GAS DYNAMICS AND COMBUSTION 8

A 16-inch pipe line is transporting natural gas across the desert. The pressure and temperature as it leaves pumping station A is 400 psia and 97°F. One hundred miles further along the pipe line at B, measurements reveal a pressure of 175 psia and a temperature of 110°F.

How much could the temperature be reduced at B by insulating the 100 miles of pipe?

Solution

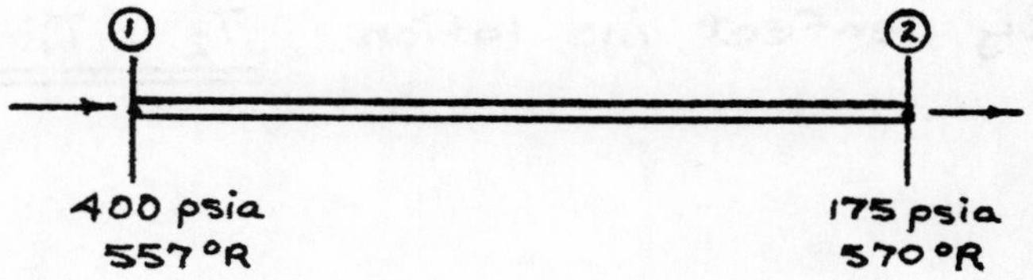

If one assumes no heat transfer, the fanno line would represent the solution.

For $M < 1$, $T_2 > T_1$

$\therefore$ There is heat supplied to pipe.

From 1st Law,

$$h_1 + \frac{V_1^2}{2g_c} + q = h_2 + \frac{V_2^2}{2g_c}$$

$$c_p(T_2 - T_1) = q + \frac{V_1^2 - V_2^2}{2g_c}$$

-continued-

Check to see if $\Delta\left(\frac{V^2}{2g_c}\right)$ is negligible

$$\Delta p \simeq f \frac{\ell}{D} \frac{\bar{V}^2}{2g_c} \bar{\rho} \simeq .01 \times \frac{5.28 \times 10^5}{1.25} \frac{\bar{V}^2}{64.4} \frac{\bar{p} M}{R_o \bar{T}}$$

* for check, use arithmetic average pressure and temp.

Then,

$$\bar{V}^2 = \frac{225 \times 144 \times 1.25 \times 1544 \times 563 \times 64.4}{.01 \times 5.25 \times 10^5 \times 288 \times 144 \times 16}$$

$$\bar{V} = 25.5 \text{ ft./sec.}$$

∴ neglect vel. effect.

By perfect insulation $T_2 \simeq T_1$ (97°F) (36.1°C)

GAS DYNAMICS AND COMBUSTION 9

A residual fuel oil used in central station steam power plant has the following ultimate analysis:

Constituent	Per Cent by Weight
Carbon	84.00
Hydrogen	13.10
Oxygen	.75
Nitrogen	1.00
Sulfur	1.10
Ash	.05

Assume that this fuel is burned completely with nine per cent excess air by volume.

The regulations of the air pollution authorities require that sulfur products calculated as sulfur dioxide not exceed 0.2 per cent by volume of the exhaust products, and that ash particles discharged not exceed 0.2 grain per cubic foot of exhaust gas at 60°F and atmospheric pressure. Assume that only the original fuel ash remains in the solid state after combustion.

Will the regulations of the authorities be complied with?

Show data and calculations necessary to support your answer.

Solution

Moles/lbm of fuel	Moles of O_2 required for complete combustion
C $\frac{0.84}{12} = 0.07$	0.07
H_2 $\frac{0.13}{2} = 0.065$	0.0325
O_2 $\frac{.0075}{32} = 0.000234$	—
N_2 $\frac{0.01}{28} = 0.000357$	—
S $\frac{0.011}{32} = 0.000344$	0.000344
Ash —	—
	0.102844 O_2 required
	0.000234 O_2 in fuel
	0.102610 (moles of O_2/lbm fuel)

$$\text{air-fuel ratio} = 0.102610 \times \frac{100\%}{20.99\%}(1.09)$$

= 0.53 moles of air/lbm fuel

= 15.35 lbm air/lbm fuel

-continued-

Product Analysis

		Moles/lbm fuel
CO_2	0.07	= 0.07
H_2O	0.065	= 0.065 Assume all H_2O in vapor state
O_2	0.102610 x 0.09	= 0.00922
N_2	3.76 x .102610 x 1.09	= 0.420
SO_2	0.000344	= 0.000344
		0.565

$$SO_2 = \frac{.000344}{.565} = \underline{\underline{0.06\% < 0.2\%}} \quad (OK)$$

mass of ash = 0.0005 x 7000 = 3.5 grains/lbm fuel

$$V_{products} = \frac{nRT}{p} = 214 \ ft^3/lbm \ Fuel,$$

$$ash = \frac{3.5}{214} = \underline{\underline{0.0163}} \ grains/ft.^3 \ product \quad (OK)$$

$$(37.3 \times 10^{-6} Kg/m^3)$$

GAS DYNAMICS AND COMBUSTION 10

An air pollution regulation restricts the discharge to atmosphere of sulfur compounds calculated as sulfur dioxide (SO_2) to 0.2 per cent by volume.

Fuel having the following analysis on a weight basis is available as boiler fuel:

> Ash - 0.10, Carbon - 85.5, Hydrogen - 9.50
> Nitrogen - 1.10, Oxygen - 0.80, Sulfur - 3.0
> Dry bulb temperature - 80°F
> Web bulb temperature - 70°F
> Exit flue gas temperature - 300°F
> Atmospheric pressure - 14.7 psi

(a) What would the sulfur dioxide (SO_2) content be in the flue gas when burning this fuel with 20 per cent excess air?

(b) Would this fuel meet requirements of sulfur discharge?

<u>Solution</u>

	moles / pound fuel	moles O_2 required
C	$0.855/12 = 0.0713$	0.0713
H_2	$0.095/2 = 0.0475$	0.0238
N_2	$0.0110/28 = 0.0039$	—
O_2	$0.008/32 = 0.0003$	—
S	$0.030/32 = 0.0009$	0.0009
	(less O_2 in fuel)	-0.0003
		$\Sigma = 0.0957$

$$\text{moles air} = 4.76 \times 0.0957 \times 1.2 = 0.5466$$

$$\text{moles } N_2 = \frac{3.76}{4.76} \times 0.5466 = 0.4318$$

-continued-

moles exhaust:

$$CO_2 = 0.0713$$

$$H_2O = 0.0475$$

$$N_2 = 0.0039 + 0.4318 = 0.4357$$

$$O_2 = .0957 \times 0.20 = 0.0191$$

$$SO_2 = \underline{0.0009}$$

$$\Sigma = 0.5745$$

since water will be in vapor form.

$$\frac{n_{SO_2}}{n_{total}} = \frac{\nu_{SO_2}}{\nu_{total}} = \frac{0.0009}{0.5745} = \underline{\underline{0.16\%}} \quad \text{level is OK}$$

GAS DYNAMICS AND COMBUSTION 11

Determine the horsepower required to drive the forced draft fans on a pressurized furnace type boiler unit with output at 500,000 pounds of steam per hour at 1250 psig and 1000°F at superheater outlet. Efficiency of the boiler unit is 86.0 per cent when burning fuel oil of the following ultimate analysis:

Carbon	- 85.5 per cent by weight	
Hydrogen	- 9.5 per cent by weight	
Oxygen	- 0.80 per cent by weight	
Nitrogen	- 1.10 " " " "	
Sulfur	- 3.00 " " " "	
Ash	- 0.10 " " " "	Btu/pound 18,500

Firing condition is 20% excess air and 21" water total pressure difference.

Solution

Assume excess air is on a mass base

At 1250 psig & 1000°F, $h = 1497$ Btu/lbm

At 410 °F & 1250 psig, $h_f = 374.97 + 1.13 = 376.1$ Btu/lbm

-continued-

$$Q = 500{,}000\,(1497 - 376.1) = 560.45 \times 10^6 \text{ Btu/hr.}$$

$$Q_{fuel} = \frac{560.45 \times 10^6}{0.86} = 652 \times 10^6 \text{ Btu/hr.}$$

$$\text{Mass of fuel} = \frac{652 \times 10^6}{18500} = 35200 \text{ lbm/hr.}$$

To determine air required/lbm of fuel :

	Moles/lbm of fuel		Moles of O_2 required
C	$\frac{.855}{12}$ =	.0713	.0713
H_2	$\frac{.095}{2}$ =	.0475	.0238
O_2	$\frac{.008}{32}$ =	.0003	—
N_2	$\frac{.011}{28}$ =	.0004	—
S	$\frac{.03}{32}$ =	.0009	.0009
			.0960
			−.0003 O_2 in fuel
			.0957

$$A/F \text{ ratio} = .0957 \times \frac{1.20}{0.21} \times 28.97 = 15.84 \text{ lbm air/lbm fuel}$$

$$\text{Total mass of air} = 35200 \times 15.84 = 5.58 \times 10^5 \text{ lbm/hr}$$

$$H_a\text{, head in ft. of same fluid} = H_w \frac{\rho_w}{\rho_a}$$

$$\rho_a = \frac{14.7 \times 144}{53.34 \times 540} = .0735 \text{ lbm/ft}^3$$

Therefore,

$$HP = \frac{\dot{m}_a H_a}{33000\,\eta}$$

$$= \frac{5.58 \times 10^5}{60}\,\frac{21}{12} \times \frac{62.4}{.0735} \times \frac{1}{33000 \times 0.7}$$

$$= \underline{\underline{598 \text{ HP}}}$$

(445.9 KW)

GAS DYNAMICS AND COMBUSTION 12

A tube solid propellant grain is bonded to its case and has inhibitor on the grain ends. The case I.D. is 40.0 inches in diameter. The grain I.D. is 26.0 inches in diameter and has a 70.0 inch length. The propellant density is 0.062 lbs/cu.in. and its burning rate is characterized by the following equation:

$$b = ap^n \text{, in/sec, where } a = 0.001 \text{ and } n = 0.75$$

p - chamber pressure, psia

The products of combustion have a constant specific heat ratio of 1.25, a gas constant of 70.0 ft-lbf/lbm°R, and a flame temperature of 3950°F. At zero time the chamber pressure is 190 psia. What is the chamber pressure at the end of 40.0 seconds? Neglect grain erosion effects. State all assumptions.

Solution

* Assume: Stagnation Conditions in combustion zone
radial burning only.

Initial burning rate

$$b_i = 0.001 \times 190^{0.75} = 0.0512 \text{ in/sec}$$

$$\therefore \dot{m}_i = \pi D \ell b_i \rho = \pi \times 26 \times 70 \times 0.0512 \times 0.062$$

$$= 18.15 \text{ lbm/sec.}$$

* Assume: Sonic Flow of perfect gas at nozzle throat.

$$\dot{m}/A^* = \left[\frac{K g_c}{R}\left(\frac{2}{K+1}\right)^{\frac{K+1}{K-1}}\right]^{1/2} \frac{p_{o,i}}{T_{o,i}^{1/2}}$$

$$A^* = \frac{18.15}{190}\left[\frac{4410}{\frac{1.25 \times 32.2}{70}\left(\frac{2}{2.25}\right)^{\frac{2.25}{0.25}}}\right]^{1/2}$$

$$= 14.2 \text{ in}^2$$

—continued—

*Assume: burning rate remains constant for 40 sec.

$$\therefore \dot{m}_{40\,sec.} = \pi(D_i + 2 b_i t)\, \ell\, b_i \rho$$

$$= \pi(26 + 2 \times 0.0512 \times 40)\, 70 \times .0512 \times .062$$

$$= 21.0 \text{ lbm/sec}$$

*Assume: $T_{o,i} = T_{o,40\,sec.}$

Then, $$P_{o,40\,sec.} = \frac{\dot{m}}{A} \frac{T_o^{1/2}}{\left[\frac{K g_c}{R}\left(\frac{2}{K+1}\right)^{\frac{K+1}{K-1}}\right]^{1/2}}$$

$$= \underline{\underline{220 \text{ psia}}}$$

$$(1.52 \times 10^3 \text{ KN/m}^2)$$

For more accurate value, use $\bar{P}_o$ to get new $\bar{b}$ & hence a better value of $P_{o,40\,sec.}$

9

Hydraulic Machines

R. IAN MURRAY

In this chapter the performance characteristics of pumps, fans and turbines of the hydrodynamic type are reviewed. Emphasis is placed on the determination of the operating point for specific conditions and on the selection of appropriate machines.

Dimensional analysis leads to the definition of the following dimensionless variables:

Capacity Coefficient	$C_Q = \dfrac{Q}{ND^3}$
Head Coefficient	$C_H = \dfrac{Hg}{N^2D^2}$
Power Coefficient	$C_P = \dfrac{P}{\rho N^3 D^5}$
Efficiency	$\dfrac{Q\gamma H}{P}$
Machine Reynolds Number	$\dfrac{ND^2}{\nu}$

N is the machine speed in revolutions per unit of time; D is a characteristic dimension of the machine, such as the impeller or runner diameter. H is the change of total head across the machine. For fans, H is generally replaced by $\frac{\Delta p}{\gamma}$ where Δp is the increase in total pressure produced by the fan. Other symbols are as defined in Chapter 6.

When dealing with low viscosity fluids such as water and air, the effects of changes in the machine Reynolds number are negligible for the machine sizes and speeds generally encountered. For this reason the machine Reynolds number is omitted from the following relationships.

PUMPS

For geometrically similar pumps operating under conditions which ensure that the pressure is everywhere greater than the vapor pressure of the fluid, the head coefficient is a function of the capacity coefficient only. The significance of this statement is that the relationship among H, Q, N, and D for an entire family of geometrically similar pumps operating at different speeds can be represented by a single curve of C_H vs. C_Q. This knowledge makes it possible, for example, to predict the H vs. Q characteristic at speed N_2 from test data at speed N_1.

Corresponding points are points that plot as a single point on the C_H vs. C_Q curve. Thus for corresponding points

$$C_Q = \frac{Q_1}{N_1 D_1^3} = \frac{Q_2}{N_2 D_2^3}$$

$$C_H = \frac{H_1 g}{N_1^2 D_2^2} = \frac{H_2 g}{N_2^2 D_2^2}$$

Furthermore,

$$\frac{C_Q^2}{C_H} = \frac{Q^2}{H\,gD^4} = \text{constant}$$

is the locus of corresponding points on an H vs. Q plot. The constant, of course, is different for each point on the C_H vs. C_Q curve.

From a hydraulic point of view alone, and under the same assumptions as above, the power coefficient and the efficiency are each functions of the capacity coefficient only. However, since the hydraulic analysis does not take mechanical friction into account, the above statement for power and efficiency is only a first approximation.

The above relationships are used to predict one set of performance characteristics from another. They should not be used directly to predict one operating point from another, for in general two operating points are not corresponding points. An operating point is determined by the intersection of a "head rise required" vs. flow rate curve with the H vs. Q characteristic curve of the pump. The "head rise required" (or system) curve is determined by the principles reviewed in Chapter 6. It depends on the system the pump must supply, not on the pump.

All of the above is predicated on the assumption that the pressure is everywhere greater than the vapor pressure of the liquid being pumped, i.e., that the pump does not cavitate. To ensure this, the "net positive suction head" (NPSH) must be greater than a critical value that is a function of the design, the size, the flow rate and the speed of the pump. The NPSH is defined as the total head at inlet in feet of liquid absolute, less the vapor pressure of the liquid in feet of liquid. The critical value of NPSH scales in accordance with the head coefficient.

To select the proper type of pump for a given application, the notion of "specific speed" (N_S) is introduced. If C_Q and C_H are combined in such a way as to eliminate the size factor D, the following dimensionless variable is obtained:

$$\frac{NQ^{1/2}}{(Hg)^{3/4}}$$

When this is evaluated at the point of maximum efficiency, it becomes a useful parameter to identify the type or design of pump. In the United States it is common practice to omit g and to specify the units of H in feet, the units of Q in gpm and the units of N in rpm. The parameter is then known as specific speed. Its value ranges from 500 for large centrifugal pumps to 15,000 for propeller pumps.

TURBINES

Speed, head and flow rate are independent variables in the performance of turbines. Thus, under the same assumptions as for pumps, the power coefficient and the efficiency are each functions of the capacity coefficient and the head coefficient. In the operation of a turbine, however, the speed is generally fixed, and the head is approximately constant or a function of the flow rate. Under these conditions the power is a function of the flow rate only (density also being fixed) and the efficiency can be expressed as a function of the power only.

Specific speed is again a useful parameter for selecting the proper type of turbine for a given application. The definition, however, is different than that for a pump:

$$N_S = \frac{NP^{1/2}}{H^{5/4}} \quad \text{(P is in hp)}$$

Small values of N_S are associated with impulse turbines; large values, with axial flow.

Energy is extracted from a fluid as it passes through the rotor of a turbine. Corresponding to this energy extraction either the velocity or the pressure of the fluid decreases, or both may decrease. The ratio of the energy extraction associated with a pressure drop to the total energy extraction is called "reaction". An "impulse" turbine is one in which the "reaction" is zero.

HYDRAULIC MACHINES 1

You are to select a boiler-feed pump for condensate at a maximum temperature of 130°F and maximum vacuum of 25 inches of mercury. Proposed are a horizontal centrifugal pump and a vertical turbine pump. Each requires 8 feet of net positive suction head at its suction inlet for water at 72°F. Your design has the suction inlet 4 feet below the condensate surface. It is suggested the suction condition is improved by lowering the suction inlets 10 feet. This would be done by lengthening the horizontal pump's suction pipe (friction = 10 feet per 100 feet) and lengthening the vertical pump's column pipe (friction = 15 feet per 100 feet).

REQUIRED: (a) Are the inlets all right as is?

(b) What do you think of the suggestion?

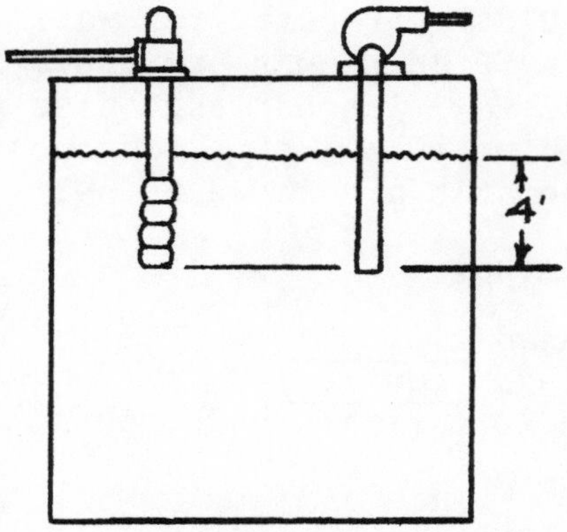

Your Design

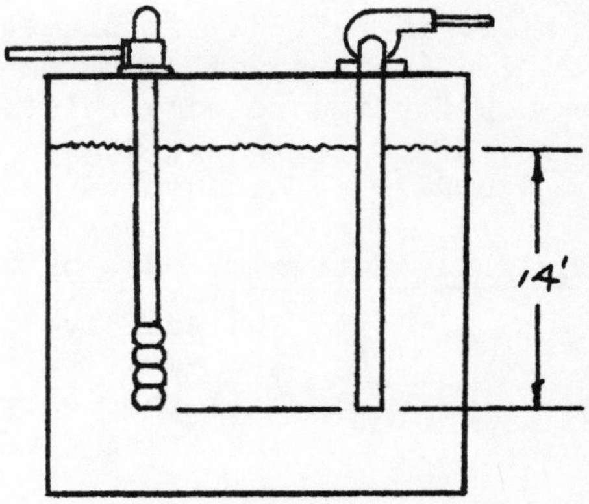

Suggestion

Solution

a) The design is not satisfactory. The pressure at the surface of the condensate is equal to the vapor pressure. Thus the net positive suction head (NPSH) on the verticle turbine pump is approximately 4 ft.; on the horizontal centrifugal pump it is zero.

b) The suggestion for the horizontal centrifugal pump is useless, for the NPSH would remain zero.

—Continued—

The suggestion for the verticle turbine pump is excessive, for the NPSH would then be approximately 14 feet.

HYDRAULIC MACHINES 2

Water at the boiling point of 228°F is pumped at a rate of 120 gallons per minute from a deaerator into a boiler. Pressure in the deaerator is 5.3 psig. The water level in the tank is controlled by a high-low device. The water level varies in the tank from a maximum height of 17.0 feet to a minimum height of 15.0 feet above the centrifugal pump center line. The piping is 2-1/2" steel pipe, Sched. 40, 18 feet long. There is one 2-1/2" gate valve, and one 2-1/2" long radius elbow. The pump operates at sea level. Pump impeller is 8-31/32" in diameter. At 228°F, viscosity for water is 0.26 centipoises and specific weight is 59.43 pounds per cubic foot. K = 0.5 for square edge inlet; K = 0.22 for flanged wedge disc gate valve; K = 0.27 for 90 degree large radius elbow; pipe roughness $\frac{\varepsilon}{D}$ = 0.000729. The pump curves and schematic are shown below.

REQUIRED: Determine each of the following:

(a) Net positive suction head available.
(b) Net positive suction head required.
(c) Will the centrifugal pump cavitate under these operating conditions?

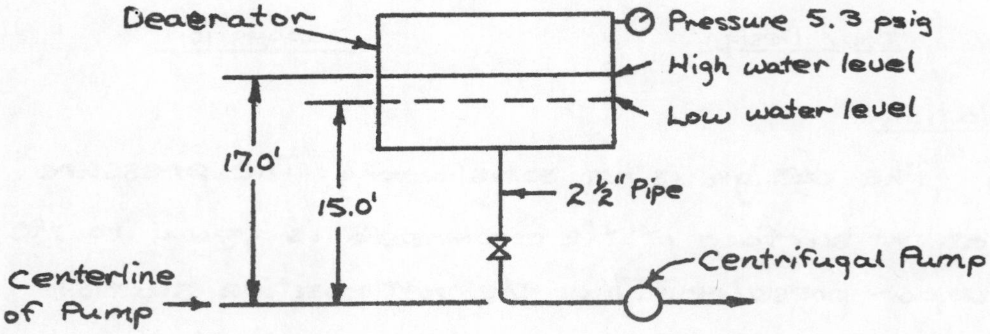

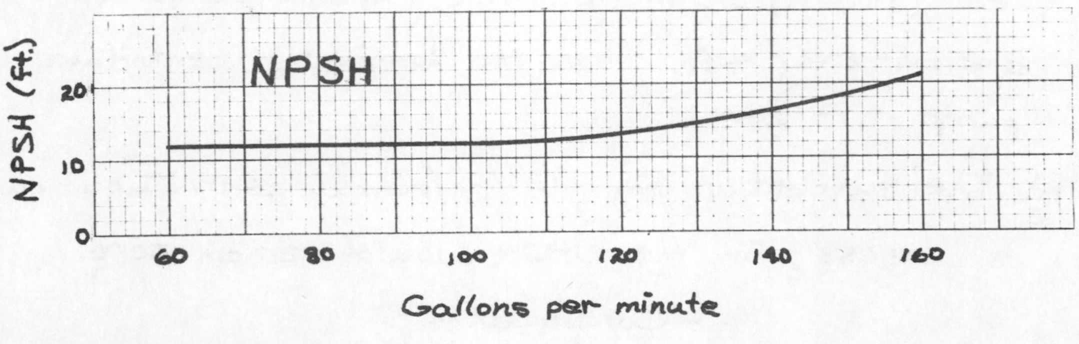

Solution

a) $NPSH = \begin{Bmatrix} 17' \\ 15' \end{Bmatrix} - H_L$

$$V = \frac{Q}{A} = \frac{120 \text{ gal./min.} \times \frac{ft^3}{7.48 \text{ gal.}} \times \frac{min.}{60 \text{ sec.}}}{\frac{\pi}{4}\left(\frac{2.47}{12}\right)^2 \ ft^2} = 8.04 \text{ ft/sec.}$$

$$\frac{V^2}{2g} = 1 \text{ ft.}$$

$$N_R = \frac{vd\rho}{\mu} = \frac{(8.04 \text{ ft./sec.})\left(\frac{2.47}{12} \text{ ft.}\right)\left(\frac{59.4}{32.2} \ \frac{lb.sec^2}{ft.^4}\right)}{(0.26 \text{ centipoise})\left(\frac{lb. \ sec.}{4.79 \times 10^4 \ ft.^2 \text{ centipoise}}\right)}$$

$$N_R = 5.6 \times 10^5$$

$$f = 0.0185 \text{ (from Moody Diagram)}$$

$$H_L = \left(0.0185 \times \frac{18 \times 12}{2.47} + 0.50 + 0.22 + 0.27\right)\frac{V^2}{2g} = 2.6'$$

NPSH ranges from 12.4' to 14.4' ←

b) NPSH required = 13' (from graph)

c) Pump will cavitate

HYDRAULIC MACHINES 3

You need to increase the pressure of a hydrocarbon fuel 125 psi at 450 gallons per minute. The hydrocarbon has a specific gravity of 0.72 and a vapor pressure of 35 psia, and a temperature of 90°F. You are considering a centrifugal pump whose characteristics are shown by a 1770 rpm test with water at 72°F to be:

G.P.M.	Head (feet)	N.P.S.H. Required (feet)
0	436	
100	400	3.9
200	377	5.0
300	346	6.5
400	287	8.3
500	200	11.0

REQUIRED: (a) What speed must the pump be operated for your needs?

(b) What suction pressure must be supplied the pump?

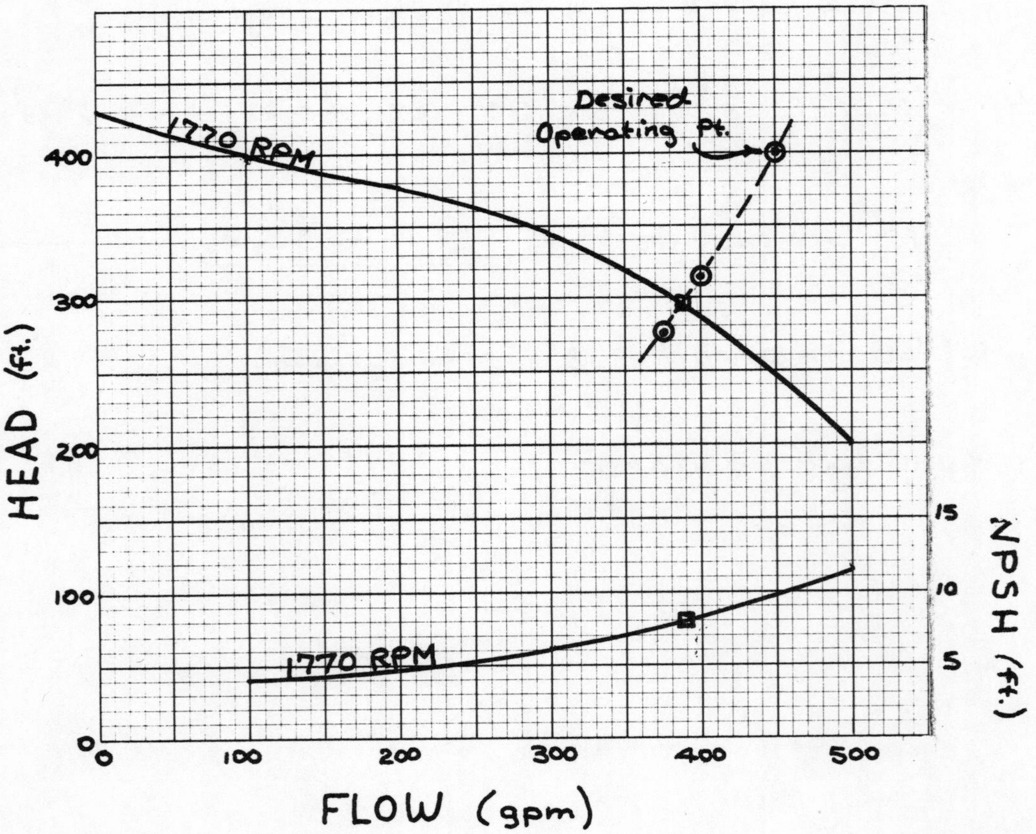

Solution

$$H = \frac{\Delta P}{\gamma} = \frac{125 \times 144}{0.72 \times 62.4} = 400 \text{ ft. @ } 450 \text{ gpm}$$

* For corresponding points to desired operating points,

$$\frac{Q^2}{H} = \text{constant}$$

$$= \frac{(450)^2}{400} = \frac{(400)^2}{316} = \frac{(375)^2}{277}$$

* Corresponding point at 1770 RPM,

$$H_1 = 295 \text{ ft.}$$

$$Q_1 = 385 \text{ gpm}$$

$$NPSH_1 = 8 \text{ ft.}$$

$$N_2 = N_1 \frac{Q_2}{Q_1}$$

$$N_2 = 1770 \times {}^{450}/_{385} = \underline{\underline{2070 \text{ rpm}}} \leftarrow$$

* Since the NPSH scale's in accordance with the head coeficient,

$$NPSH_2 = 8 \text{ ft.} \times \left({}^{450}/_{385}\right)^2 = 10.9 \text{ Ft} = \frac{P_s + P_a - P_v}{\gamma}$$

$$P_s = \frac{0.72 \times 62.4 \times 10.9}{144} + 35 - 14.7 = 3.4 + 35 - 14.7$$

$$= \underline{\underline{23.7 \text{ psig}}} \leftarrow$$

P_s is the total suction pressure.

HYDRAULIC MACHINES 4

An electric-motor-driven (1770 rpm) pump is used to raise water from one open tank to another, as shown. The pump data presented to you is:

Flow (gpm)	Head (feet)	Overall Efficiency
0	635	0
500	570	41.5
1000	474	66.2
1250	415	71.0
1500	334	69.5
1750	235	65.0
2000	120	49.5

The system curve, before installing the automatic flow control valve, is:

Flow (gpm)	Head (feet)
0	127
400	133
800	151
1200	180
1600	222
2000	275

The valve allows a flow of 1400 gpm. What is the pressure differential across the valve?

If another pump is available with the same head-capacity curve and 10% less power consumption, how much could economically be paid for it, assuming pumping to be 1000 acre-feet per year, for 5 years, power cost 1.8¢ per kwh, and money worth 7%?

Calculate the pump cost with and without the flow control valve.

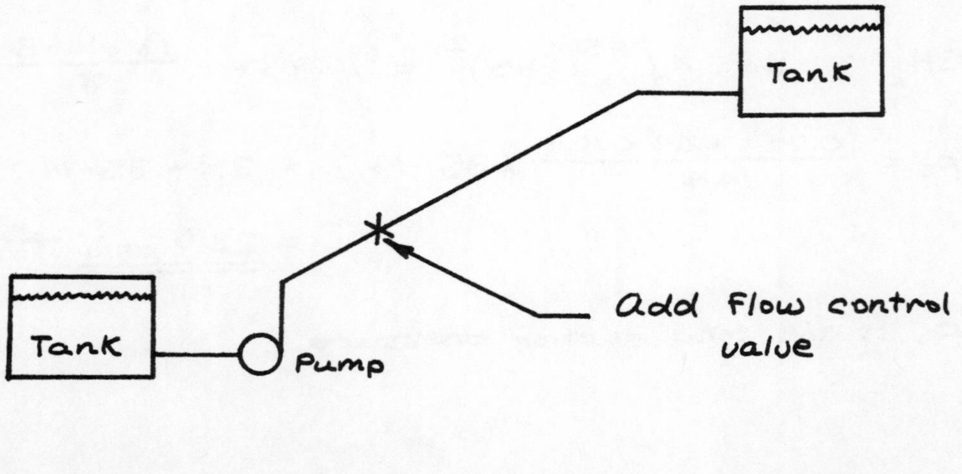

Solution

Plot the given characteristics for both pump and system (see next page for plot). At 1400 gpm, observe the pump head to be 370 ft. and the system head to be 200 ft.. Thus, the head loss at the control valve is 170 ft.

$$\Delta p = \gamma \Delta H = \left(62.4 \frac{lb}{ft^3}\right)(170 \text{ ft})\left(\frac{ft^2}{144 \text{ in}^2}\right)$$

$$= \underline{\underline{73.7 \text{ psi}}}$$

Operation without control valve:

$$\left.\begin{array}{l} Q = 1730 \text{ gpm} \\ H = 240 \text{ ft.} \\ \eta = 66\% \end{array}\right\} P = \frac{Q\gamma H}{\eta} = 118.5 \text{ kw}$$

$$P = \left(1730 \frac{gal}{min.}\right)\left(\frac{ft^3}{7.48 \text{ gal}}\right)\left(62.4 \frac{lb}{ft^3}\right)(240 \text{ ft})\left(\frac{1}{.66}\right)\left(\frac{kw \cdot min.}{44{,}254 \text{ ft} \cdot lb}\right)$$

$$\underset{(\text{per year})}{\text{Pumping time}} = \frac{1000 \text{ acre ft.}}{1730 \text{ gpm}} \times \frac{326 \times 10^3 \text{ gal.}}{\text{acre} \cdot \text{ft}} \times \frac{hr}{60 \text{ min}} = 3140 \text{ hr.}$$

$$\text{Energy Cost per year} = \frac{\$0.018}{kwh} \times 118.5 \text{ kw} \times 3140 \text{ hr}$$

$$= \$6700$$

Annual Savings with other pump $= \$670$

Present value @ 7% for 5 yrs. $= 4.1 \times \$670 = \underline{\underline{\$2740}}$

Operation with control valve:

$$\left.\begin{array}{l} Q = 1400 \text{ gpm} \\ H = 370 \text{ ft} \\ \eta = 71\% \end{array}\right\} P = 1.16 \times 118.5 \text{ kw}$$

$$\underset{(\text{per year})}{\text{Pumping time}} = \frac{1730}{1400} \times 3140 \text{ hr} = 1.235 \times 3140 \text{ hr.}$$

Energy cost per year = 1.16 x 1.235 x \$6700 = \$9600

Annual savings with other pump = \$960

Present value* @ 7% for 5 yrs. = 4.1 x \$960 = **\$3940**

*See Chapter 12.

However, control valve should not be used unless it is needed for reasons not indicated in the problem statement.

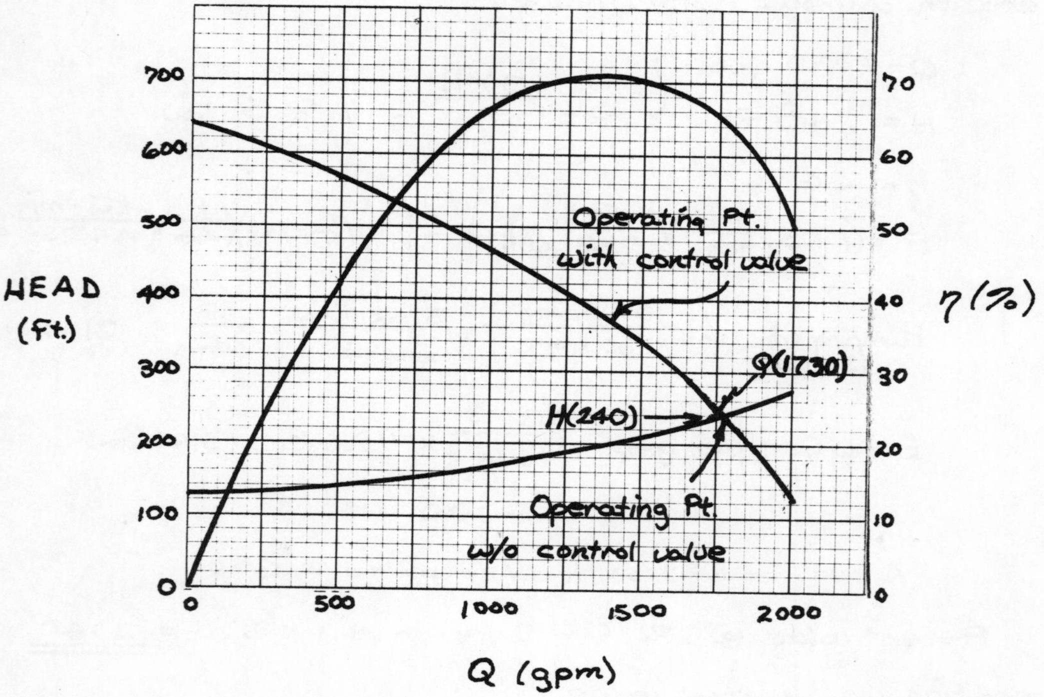

HYDRAULIC MACHINES 5

A variable-speed centrifugal water pump has the following H - Q curve at its maximum speed of 1680 rpm:

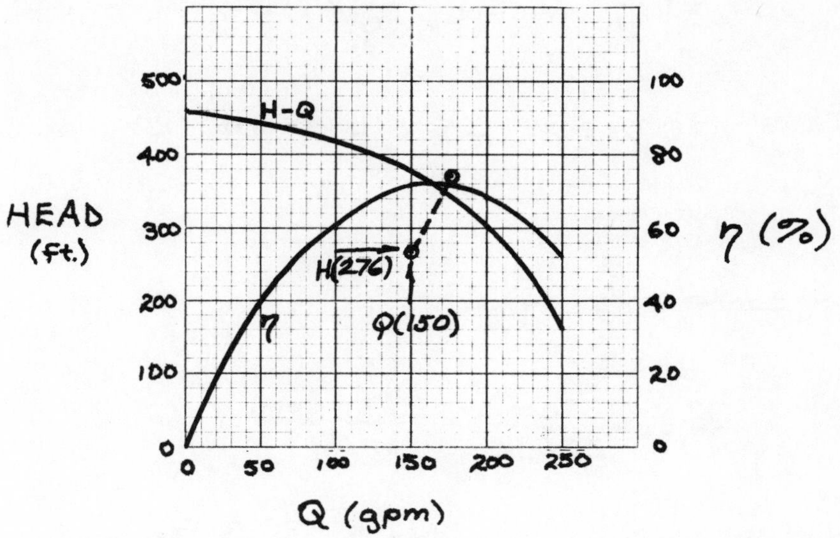

Suction pressure varies from 23 to 40 psig. The system demand varies from 0 to 200 gpm.

(a) What speed range must the pump have to maintain a constant discharge pressure of 150 psig?

(b) What speed would the pump be running when pumping 150 gpm with a discharge pressure of 150 psig and suction pressure of 30 psig?

Solution

Suction pressure is minimum at maximum flow.

$$H = \frac{(150-40)\times 144}{62.4} = 253 \text{ ft. } @ \; Q=0 \qquad N = ?$$

$$H = \frac{(150-23)\times 144}{62.4} = 293 \text{ ft } @ \; Q = 205 \text{ gpm} \quad N = 1680 \text{ rpm}$$

$$H = \frac{(150-30)\times 144}{62.4} = 276 \text{ ft } @ \; Q = 150 \text{ gpm} \quad N = ?$$

- continued -

a) At shut off :

$$N_2 = \sqrt{\frac{H_2}{H_1}}\, N_1 = \sqrt{\frac{253}{460}} \times 1680 \text{ rpm} = 1240 \text{ rpm}$$

Thus, the speed range is from 1240 rpm to 1680 rpm.

b) For corresponding points :

$$\frac{Q^2}{H} = \frac{(150 \text{ gpm})^2}{276 \text{ ft.}} = \frac{(175 \text{ gpm})^2}{H_3}$$

$$H_3 = 375 \text{ ft.}$$

Sketch curve of corresponding points on graph and estimate that @ 1680 rpm, $Q_5 = 168$ gpm.

Then,

$$N_4 = \frac{Q_4}{Q_5} N_5 = \frac{150}{168} \times 1680$$

$$= \underline{\underline{1500 \text{ rpm}}}$$

HYDRAULIC MACHINES 6

A pump takes water from the Mad River and delivers it through 4000 feet of pipe line to a paper plant water tank. The static lift is 200 feet. The pump when new had the following performance:

Flow in gpm	Head in feet	Efficiency of Pump and Electric Motor Combined
0	785	0
1000	685	43.5
1500	620	57.8
2000	579	67.1
2500	539	73.0
3000	476	75.0
3500	389	71.3
4000	260	61.9

The pump has worn so that it presently pumps to the tank 2000 gpm at a head of 323 feet, while using power at the rate of 328 kilowatts. You are to decide whether to leave the pump as is for awhile, repair it to original condition, or replace it. Power costs 1.5¢ per kwh. Repair of the present pump to new condition would cost $3000. Replacement would cost $40.00 per nominal motor h.p. The plant uses 1.1 billion gallons annually, operating 363 days. The pump should deliver 20% more than the annual average, to allow for future wear.

What is the economical solution?

Solution

$$\text{Use rate} = \frac{1.1\times10^9 \text{ gal.}}{363 \text{ days}} \times \frac{\text{day}}{24 \text{ hrs.}} \times \frac{\text{hr}}{60 \text{ min}}$$

$$= 2100 \text{ gpm}$$

Thus, one alternative is eliminated; the present pump will not do the required job.

The pump should deliver $1.2 \times 2100 \text{ gpm} = 2530 \text{ gpm}$. Assume that the head loss in the system is proportional to the square of the flow rate.

$$H = 200 \text{ ft} + KQ^2$$

$$H = 200 \text{ ft} + KQ^2$$

$$323 = 200 + K(2000)^2 \longrightarrow K = \frac{123}{4 \times 10^6}$$

$$H = 200 + \frac{123}{4 \times 10^6}(2530)^2 = 396 \text{ ft.}$$

$$P_{\text{new pump}} = \frac{Q\gamma H}{\eta} = \frac{2530 \text{ gpm}}{7.48 \text{ gal/ft}^3} \cdot \frac{62.4 \text{ lb/ft}^3 \; 396 \text{ ft}}{33{,}000 \frac{\text{ft}\cdot\text{lb}}{\text{hp}\cdot\text{min.}}} \cdot \frac{1}{\eta}$$

$$= \frac{253.5}{\eta} \text{ hp}$$

Assume $\eta = 75\%$, then $P_{\text{new pump}} = 338 \text{ hp} = 252 \text{ Kw}$

∴ minimum replacement cost $= 40 \times 338 = \$13{,}500$

$$\underset{\text{(new pump)}}{\text{Power cost}} = 252 \text{ Kw}\left(363 \times 24 \times \frac{2100}{2530} \text{ hr}\right) \$0.015/\text{kwh}$$

$$= \$27{,}300/\text{yr.}$$

Operating point if present pump were repaired would be 3000 gpm @ 476 ft. head and 75% efficiency. This is obtained by inspection of the pump characteristic data and trying $Q = 3000$ gpm in the head required equation developed above. It could also be obtained by plotting the pump characteristic data and the head required equation and observing their intersection.

$$\$27{,}300 \times \frac{476 \text{ ft.}}{396 \text{ ft.}} = \$32{,}800$$

Additional annual power cost for repaired pump is $5,500. Thus, the incremental cost of a new pump would be recoved in two years. Buy a new pump!

HYDRAULIC MACHINES 7

A water purveyor pumps from the Sacramento River, as shown, schematically.

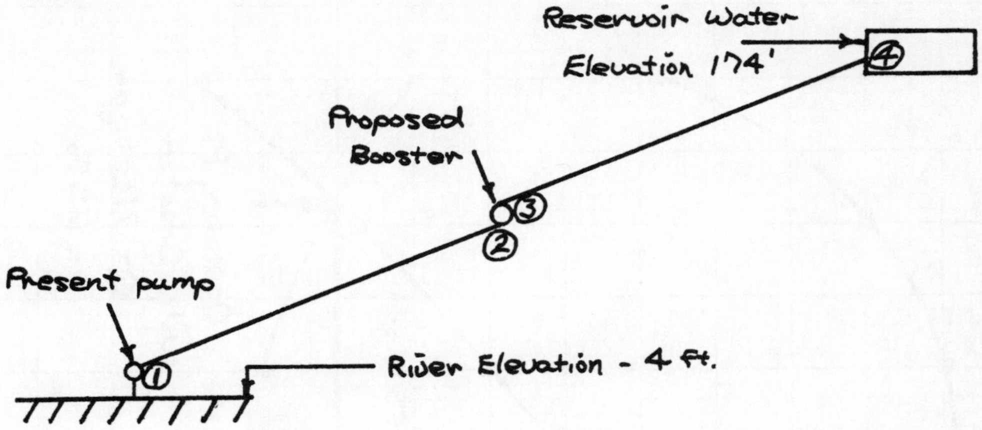

The existing booster is a pump that has the following performance:

Flow - gpm	Head in feet	Efficiency of Pump and Electric Motor Combined
0	615	0
1000	590	27.3
2000	555	49.8
3000	500	66.5
4000	465	75.8
5000	415	81.7
6000	345	81.4
7000	265	75.3
8000	155	56.1

This pump presently pumps 5,500 gpm through 16,500 feet of 16" I.D. concrete-lined steel pipe from elevation 4' to elevation 174'. To increase the system flow, the addition of a booster is proposed at elevation 94', 8,000 feet from the existing booster. The proposed booster is to take suction directly from the pipe line.

REQUIRED: (a) What is the maximum flow for which the new booster can be designed?

(b) A second proposal is to construct a parallel pipe line and booster to give the same increase in flow as the first proposal. Assume friction in the new pipe line to be 1.5 feet per 100 feet, and velocity to be 6.0 feet per second. What would be the annual savings in power cost for 2,000 hours operation at 1.4¢ per kilowatt hour?

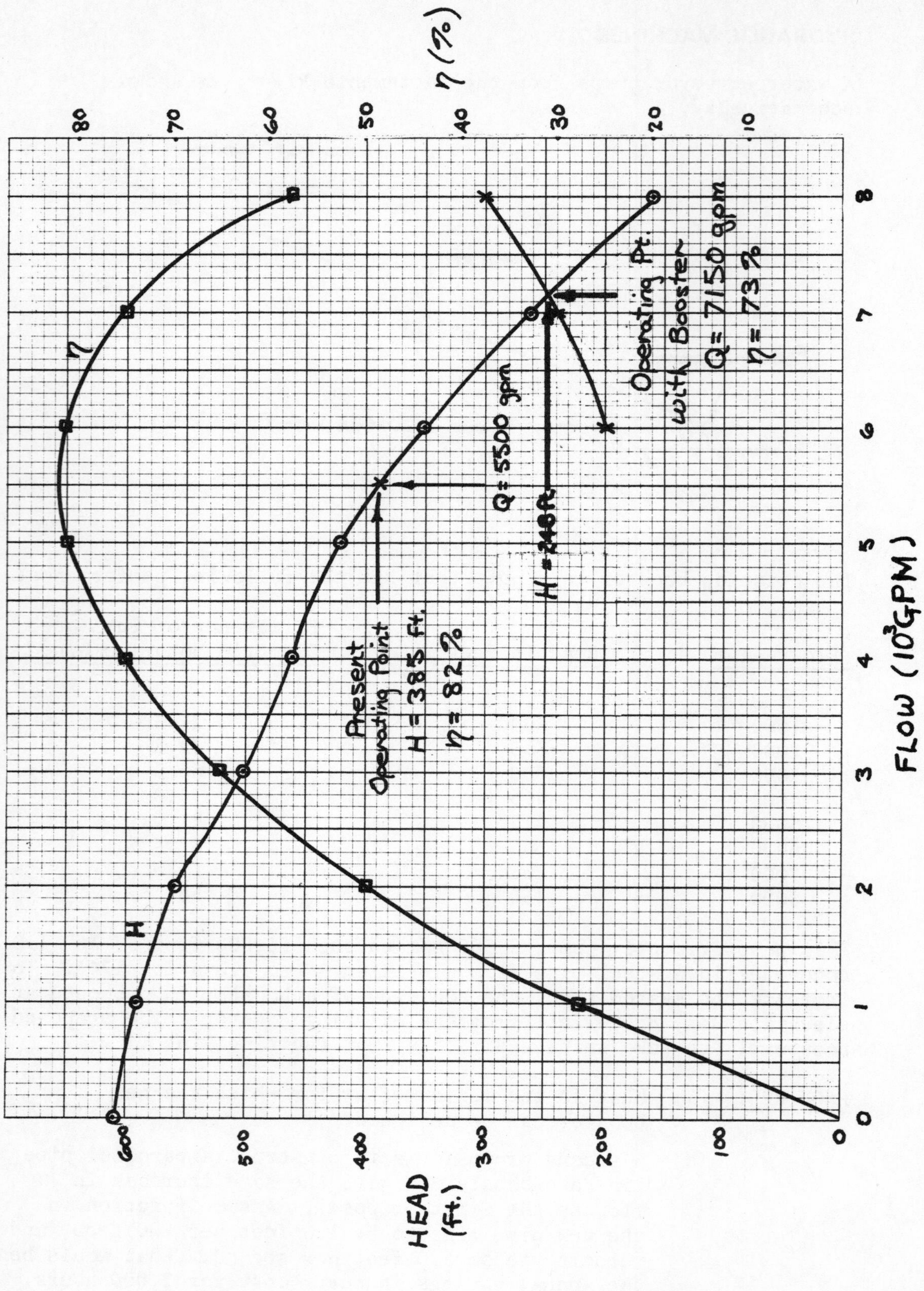
η (%)
80
70
60
50
40
30
20
10
η
H
Present
Operating Point
H = 385 ft.
η = 82 %
Q = 5500 gpm
Operating Pt.
with Booster
Q = 7150 gpm
η = 73 %
0
1
2
3
4
5
6
7
8
FLOW (10³GPM)
600
500
400
300
200
100
0
HEAD
(Ft.)

Solution

Present Operation:

$$H_1 + H_P = H_4 + H_{L_{1-4}}$$

$$4' + 385' = 174' + H_{L_{1-4}} \longrightarrow H_{L_{1-4}} = 215 \text{ ft.}$$

assume $H_L = K\ell Q^2 \longrightarrow K = 0.43 \times 10^{-9} (\text{gpm})^{-2}$

Operation with Booster:

$$\text{For } \ell = 8000', \; K\ell = 3.44 \times 10^{-6} \frac{\text{ft}}{(\text{gpm})^2}$$

$$\text{Thus, } H_{L_{1-2}} = 3.44 \times 10^{-6} \frac{\text{ft}}{(\text{gpm})^2} Q^2$$

Since no information is given about NPSH, assume a suction head of −15 ft. at the inlet to the booster.

*Note: a zero suction head would be a better assumption; but the problem asks for the maximum flow, and −15 ft. is a common rule of thumb.

Then, $H_2 = 94' - 15' = 79 \text{ ft.}$

$$H_1 + H_P = H_2 + H_{L_{1-2}}$$

$$4' + H_P = 79' + 3.44 \times 10^{-6} Q^2$$

$$\longrightarrow H_P = 75 + 3.44 \times 10^{-6} Q^2$$

Plot the curve of H vs. Q for the pump and the curve of H_P vs. Q. for the system. From the intersection of these curves obtain:

$Q = 7150$ gpm ← max. flow

$H_P = 248$ ft.

$\eta = 73\%$

Booster requirements:

For $\ell = 8{,}500'$, $K\ell = 3.65 \times 10^{-6} \dfrac{ft}{(gpm)^2}$

$$H_{L_{3-4}} = (3.65 \times 10^{-6})(7150)^2 = 186.5 \text{ ft.}$$

$$H_2 + H_B = H_4 + H_{L_{3-4}}$$

$$79' + H_B = 174' + 186.5' \longrightarrow H_B = 282 \text{ ft.}$$

Power Requirements

$$P = \frac{Q\gamma H}{\eta} = \frac{QH}{\eta}\left(62.4 \frac{lb}{ft^3} \times \frac{ft^3}{7.48 \, gal.} \times \frac{2.26 \times 10^{-5} \, kw\text{-}min}{ft \cdot lb}\right)$$

$$= \frac{QH}{\eta}\left(1.88 \times 10^{-4} \frac{kw \cdot min}{ft \cdot gal}\right)$$

Present Operation:

$$P = \frac{5500 \times 385}{0.82}(1.88 \times 10^{-4}) = 485 \text{ kw}$$

Pump (Original) when Booster is used:

$$P = \frac{7150 \times 248}{0.73}(1.88 \times 10^{-4}) = 457 \text{ kw}$$

Booster (assuming $\eta = 80\%$):

$$P = \frac{7150 \times 282}{0.80}(1.88 \times 10^{-4}) = 474 \text{ kw}$$

Total Power 931 kw

Parallel installation (assuming $\eta = 80\%$):

$$Q = 7150 - 5500 = 1650 \text{ gpm}$$

$$H = 170 + 1.5 \times 165 = 418 \text{ ft}$$

$$P = \frac{1650 \times 418}{0.80}(1.88 \times 10^{-4}) = 162 \text{ kw}$$

Original Inst. → 485

647 kw Total

$$\Delta P = 931 - 647 = 284 \text{ kw}$$

—continued—

Annual power cost savings for parrallel installation:

$$284 \text{ Kw} \times 2000 \text{ hr.} \times \frac{\$0.014}{\text{Kw-hr.}} = \underline{\underline{\$7950}}$$

HYDRAULIC MACHINES 8

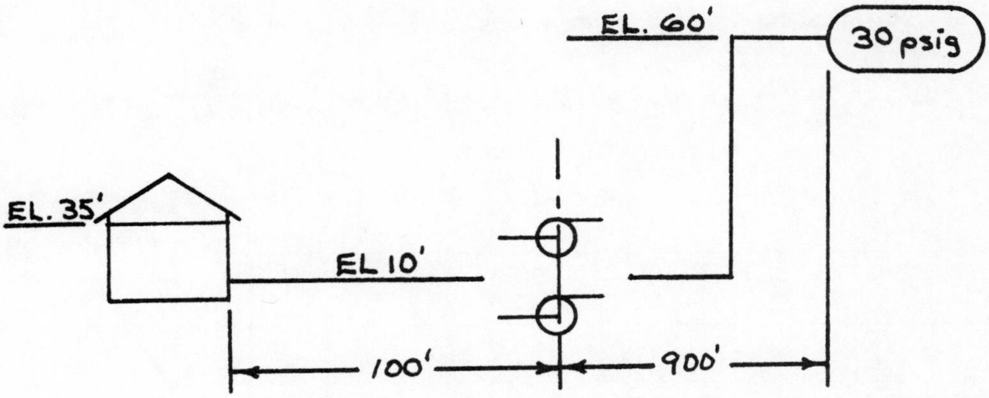

Gasoline is to be pumped from a feed tank whose liquid surface is at elevation 35' to a closed vessel operating at 30 psig pressure and located at elevation 60'. Two identical centrifugal pumps are available, located at elevation 10'. They can be operated either singly, in series, or in parallel to pump over a pipe line from the feed tank to the vessel. The pipe line consists of 100 equivalent feet of 4 inch pipe from the feed tank to the pumps, and 900 equivalent feet of 4 inch pipe from the pumps to the vessel. At the pumping temperature, the gasoline has a specific gravity of 0.72 and a vapor pressure of 6 psia.

Data for constructing the pump head capacity curves and the pipe line friction curve are given in the following tabulation:

G.P.M.	Pump Head Feet of Liquid	Friction Drop in Pipe Feet of Liquid per 100 Feet of Pipe
0	252	0
200	252	4.29
400	235	15.5
600	187	32.8
800		55.8
1000		84.3

(a) Find the maximum pumping rate.

(b) Find the available NPSH at this pumping rate.

Solution

To obtain the H vs. Q characteristic curve for pumps in series, add the heads corresponding to common flow rates; for pumps in parallel, add the flow rates corresponding to common heads.

To obtain the piping system curve:

$$H_P = H_2 - H_1 + H_L = (60-35)\text{ ft.} + \frac{30 \times 144}{0.72 \times 62.4}\text{ ft.} + H_L$$

$$= 121 \text{ ft} + H_L \quad \left(\begin{array}{c}\text{neglecting velocity}\\ \text{head at discharge}\end{array}\right)$$

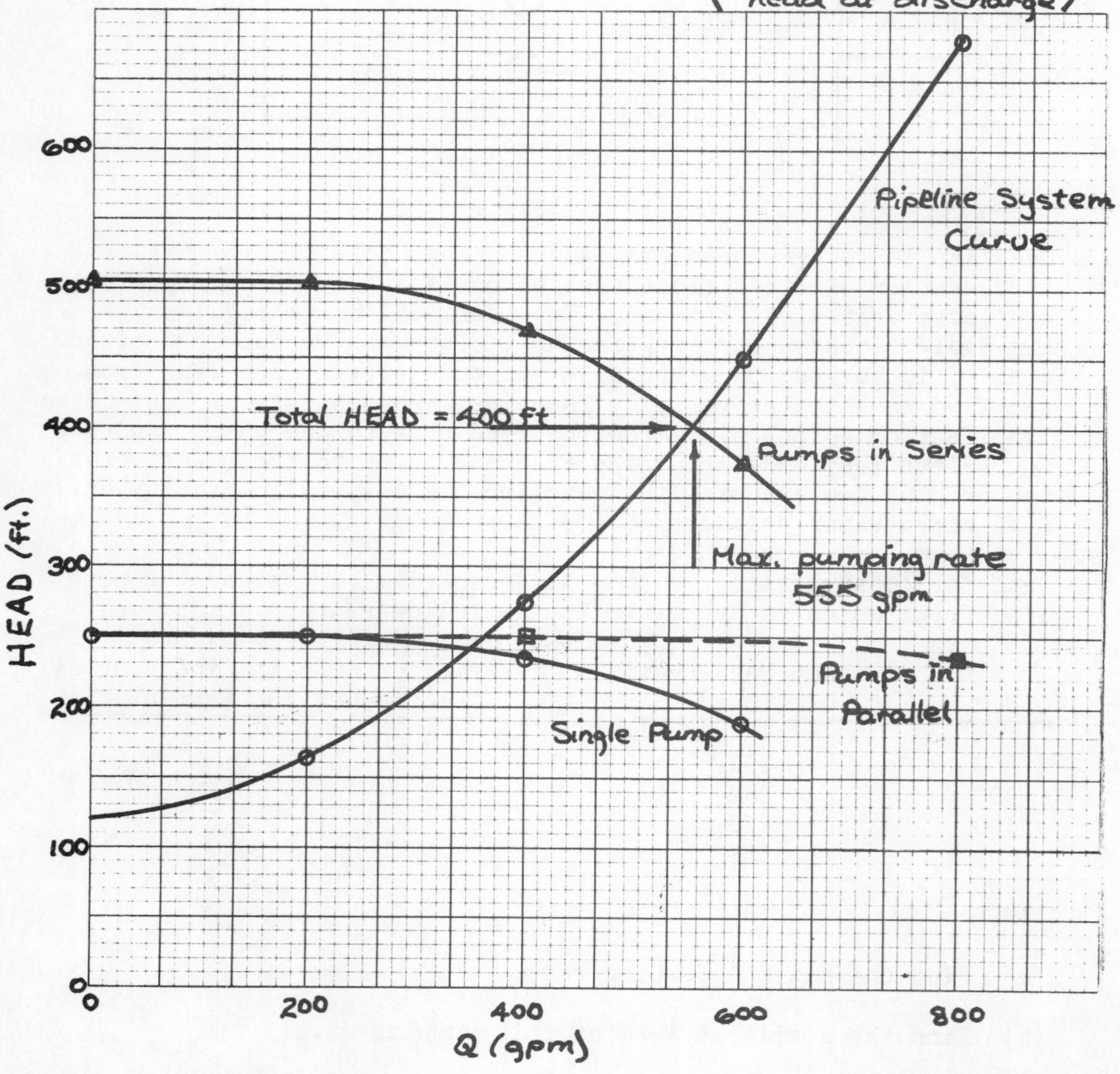

From the intersection of the pumps in series curve and the piping system curve the maximum pumping rate is 555 gpm at a total pumping head of 400 ft.

Thus, the head loss is 400' − 121' = 279'

→ 27.9 ft/100 ft.

$$NPSH = (35' - 10') - 27.9' + \frac{(14.7 - 6.0) \times 144}{0.72 \times 62.4} \text{ ft}$$

$$= \underline{25 \text{ ft.}}$$

HYDRAULIC MACHINES 9

An engine-driven deep well pump takes water from a well, and puts it through 1300 feet of pipe of unknown size to irrigate a field of sugar beets. The level beet field is 75 feet higher than ground level at the well. The water is dispersed through a number of sprinkler heads. Flow is 700 gallons per minute and pressure at the pump is 115 psig.

The water level in the well is 175 feet below ground level when pumping 700 gpm and 140 feet when not pumping. Assume the well water level to be a linear relationship with the flow up to 2000 gpm.

The pump and engine are running at 1500 rpm, which is the limit of the governor. The engine at 1500 rpm is rated at 150 continuous bhp and 190 bhp intermittent. At 1800 rpm, it is rated at 175 bhp continuous and 225 bhp intermittent.

The farmer needs a greater flow. You are an engineer for a pump company with pumps for all conditions. Assume any pump you have has an efficiency of 75%.

(a) What is the efficiency of the existing pump? Should it be pumping more?

(b) What would be the flow with your best pump for the existing engine, pipe, and sprinkler system? Run the engine at 1800 rpm.

(c) What would you recommend be done?

Solution

a) Assume the velocity head to be negligible.

Pump head $H = 175 \text{ ft.} + \dfrac{115 \times 144}{62.4} = 175 + 265 = 440 \text{ ft.}$

Power supplied to Flow $= Q\gamma H$

$= 700 \text{ gpm} \times \dfrac{\text{ft}^3}{7.48 \text{ gal}} \times 62.4 \dfrac{\text{lb}}{\text{ft}^3} \times 440 \text{ ft.} \times \dfrac{\text{hp}\cdot\text{min.}}{33{,}000 \text{ ft}\cdot\text{lb}}$

$= 77.9 \text{ hp}$

Efficiency cannot be determined since the actual Engine power is not known. Correspondingly, pumping performance cannot be evaluated.

b) Assume the head loss in the pipe and sprinkler heads is proportional to Q^2.

* System head curve equation : H is in ft.
Q is in GPM

$$H = 140' + \frac{35'}{700\,gpm} Q + 75' + \frac{265' - 75'}{(700\,gpm)^2} Q^2$$

$$H = 215 + 5Q + 3.88Q^2 \quad (Q \text{ in } 100 \text{ GPM units})$$

* System power curve equation : P in hp
H in ft.
Q in 100 GPM

$$P = \frac{Q\gamma H}{\eta} = \frac{62.4\,QH}{7.48 \times 330 \times 0.75} = 0.03375\,QH$$

Q (100 gpm)	H (ft.)	P (Hp)
7	440	104
8	503	136
9	574	174

← New Pump @1800 rpm, 900 gpm (continuous)

Computed Values from equations above

c) Present pump @ 1800 rpm.

Corresponding point with present operating point :

$$Q_2 = \frac{N_2}{N_1} Q_1 = \frac{1800}{1500} \times 700 = 840\,gpm$$

$$H_2 = \left(\frac{N_2}{N_1}\right)^2 H_1 = \left(\frac{1800}{1500}\right)^2 \times 440 = 634 \text{ ft.}$$

The new operating point can not be determined without knowing the pump characteristic curve, but it can be approximated by sketching typical curve segments through known points.

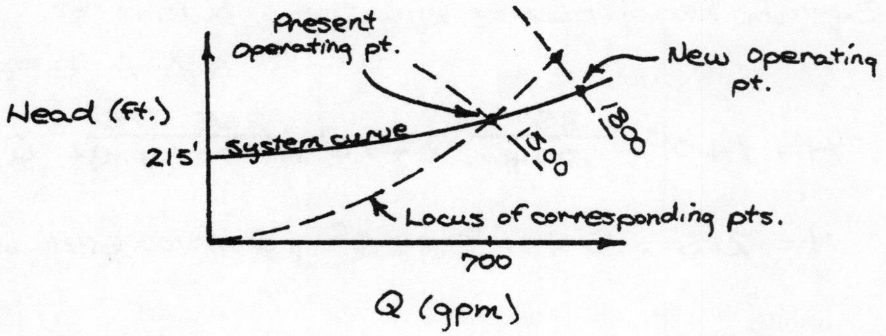

By comparing approximate new operating point with what a new pump would do, it appears that the present pump is still in good condition.

Recommendation :

1) Keep Present Pump

2) Reset governor for 1800 rpm.

HYDRAULIC MACHINES 10

A Francis-type hydroelectric turbine and generator unit is to be installed at a plant where the operating head is 200 feet. The unit is to have a rating of 10,000 hp and the power system operates at 60 cps. Assume the plant elevation is 500 feet above sea level and ambient temperature is 70°F.

REQUIRED: Answer the following questions, giving full reference for all formulas, charts, grpahs, etc., you use:

(a) What specific speed would you recommend for the runner?

(b) What is the maximum recommended runner speed in rpm?

(c) What is the closest synchronous speed to the maximum recommended runner speed that is still less than the maximum recommended runner speed?

(d) How many pole pieces (not pairs) will the generator have at this synchronous speed?

(e) What runner discharge diameter would be recommended for the given head and power rating?

(f) What setting (vertical distance between the bottom of the runner and tail water elevation) would you recommend?

(g) What maximum efficiency would you expect the turbine (exclusive of the generator) to have?

(h) Across what range of percentages of full load will this peak efficiency be expected to occur?

Solution

Ref: Baumeister + Marks, "Std. Handbook for M.E.", seventh ed., pp 9-(183-200)

a) $n_s = 675/\sqrt{H} = 675/\sqrt{200} = \underline{\underline{47.7}}$

n_s - specific speed

b) $n_s = n\sqrt{P}/H^{5/4} \quad \therefore n = n_s H^{5/4}/\sqrt{P}$

$n = \dfrac{47.7(200)^{5/4}}{\sqrt{10^4}} = \underline{\underline{359 \text{ rpm}}}$

H in ft.
P in hp.
n in rpm

c) 360 and $\underline{\underline{327 \text{ rpm}}}$ are synchronus speeds.

d) $n = 120 \times f/p$ $\qquad p = \frac{120 \times 60}{327} = \underline{\underline{22}}$

* Note: the above answers the question as worded, but does not represent good design. The 360 rpm synchronus speed should be chosen which would mean 20 poles and a specific speed of 47.9. These values will be used.

e) $\varphi = \frac{\pi D n}{720\sqrt{2gH}} = 0.73$ $\quad \left\{ \begin{array}{l} D \text{ in inches} \\ n \text{ in rpm} \\ H \text{ in ft.} \end{array} \right.$

$$D = \frac{0.73 \times 720 \times 8.03\sqrt{200}}{\pi \times 360} = \underline{\underline{52.8 \text{ in.}}}$$

f) $\sigma = \frac{H_b - H_v - H_s}{H} \geq \frac{n_s^{3/2}}{2000} = 0.1658$ (if model test data not available)

$H_s \leq H_b - H_v - 33.16$ ft.

Assume $p_{atm} = 14.7 - 0.3 = 14.4$ psia

$p_v = 0.4$ psia

$$H_b - H_v = \frac{14.0 \times 144}{62.4} = \underline{32.31 \text{ ft}}$$

$H_s \leq \underline{\underline{-0.85 \text{ ft.}}}$ (Negative value means submergence)

g) + h) $\underline{\underline{90\text{-}92\%}}$ $\qquad$ $\underline{\underline{\text{From } 60\text{-}100\% \text{ of rated load}}}$

HYDRAULIC MACHINES 11

A steam turbine stage has a blade root velocity of 800 ft/sec and a root radius of 2.0 feet. Swirl velocity is to be zero at exit to blading along its total length (6") while axial velocity remains constant along length.

Assume free vortex swirl velocity distribution, "0" reaction at root, and blade entrance angle = blade exit angle = 45° at root.

REQUIRED: Find:

(a) The blade entrance and exit angles at tip.

(b) The amount of reaction at tip.

Solution

For the conditions of this problem the energy transfer associated with a pressure difference is equal to the change of kinetic energy associated with the relative velocity of the fluid with respect to the blade. Thus,

$$\text{Reaction} = \frac{V_{R_2}^2 - V_{R_1}^2}{V_1^2 - V_2^2 + V_{R_2}^2 - V_{R_1}^2}$$

where: V_R – relative velocity

V – absolute velocity

V_B – blade velocity

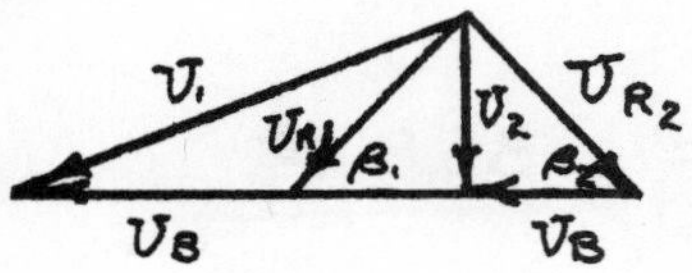

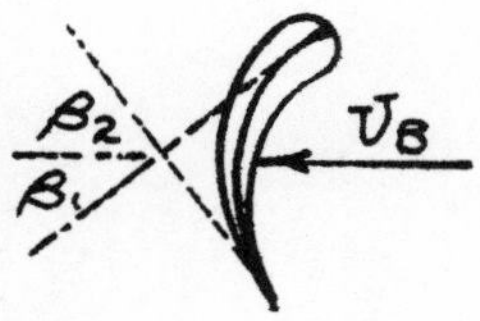

Condition at root:

$V_{R_1} = V_{R_2}$ for zero reaction

$\beta_1 = \beta_2 = 45°$

$V_2 = V_B = 800$ ft/sec

V_S – tangential component of $V_1 = 2V_B = 1600$ ft/sec

$V_S\, r = \text{const.} = 3200$ ft²/sec

$\omega = \frac{V_B}{r} = 400\ \text{sec}^{-1}$

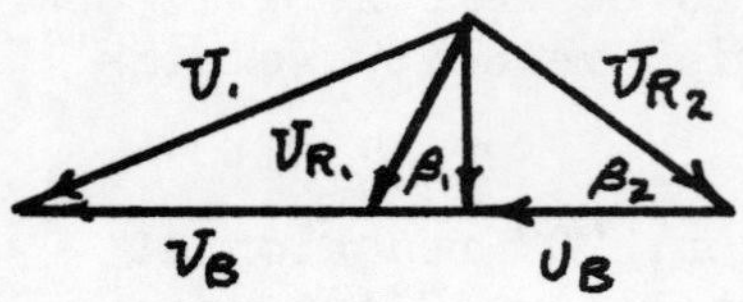

Condition at tip:

$$V_B = r\omega = 2.5 \times 400 = 1000 \text{ ft/sec}$$

$$V_S = \frac{3200}{2.5} = 1280 \text{ ft/sec.}$$

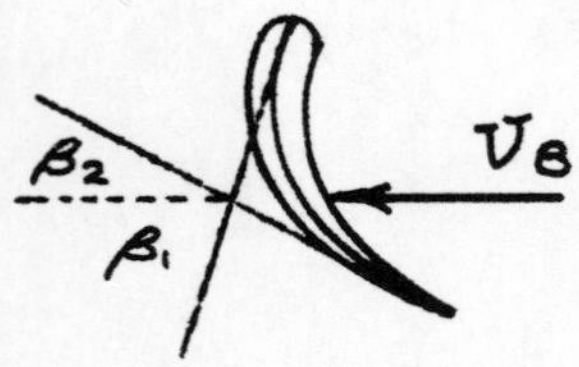

$$\beta_1 = \tan^{-1}\frac{V_2}{V_S - V_B}$$

$$= \tan^{-1}\frac{800}{1280-1000}$$

$$\beta_1 = \underline{\underline{70.7^\circ}} \longleftarrow$$

$$\beta_2 = \tan^{-1}\frac{V_2}{V_B} = \tan^{-1}\frac{800}{1000} = \underline{\underline{38.7^\circ}}$$

$$V_{R_2} = \frac{V_2}{\sin\beta_2} = 1280 \text{ ft/sec}$$

$$V_{R_1} = \frac{V_2}{\sin\beta_1} = 848 \text{ ft/sec}$$

$$V_1 = \sqrt{V_2^2 + V_S^2} = 1510 \text{ ft/sec}$$

$$V_1^2 - V_2^2 = V_S^2 = 1.64 \times 10^6 \text{ ft}^2/\text{sec}^2$$

$$V_{R_2}^2 - V_{R_1}^2 = 0.919 \times 10^6 \text{ ft}^2/\text{sec}^2$$

$$\text{Reaction} = \frac{0.919}{1.64 + 0.919} = \underline{\underline{35.9\ \%}}$$

10

Power Plants

RICHARD K. PEFLEY

Reciprocating Engine Closed Cycle Models:

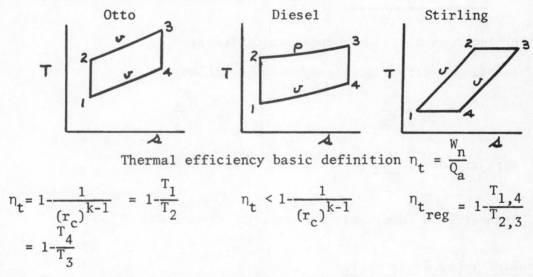

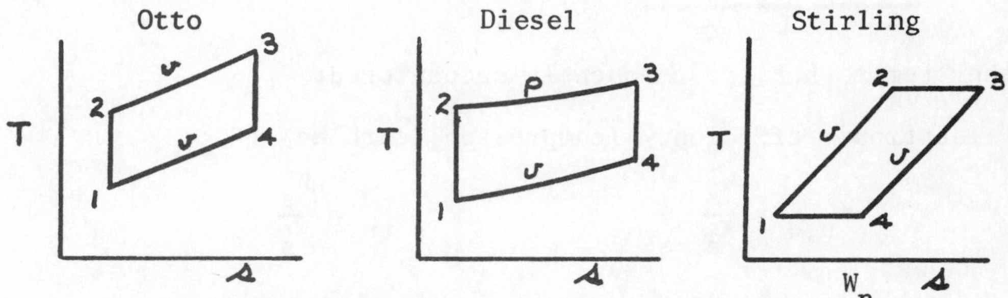

Thermal efficiency basic definition $\eta_t = \frac{W_n}{Q_a}$

$\eta_t = 1 - \frac{1}{(r_c)^{k-1}} = 1 - \frac{T_1}{T_2} = 1 - \frac{T_4}{T_3}$ $\qquad \eta_t < 1 - \frac{1}{(r_c)^{k-1}}$ $\qquad \eta_{t_{reg}} = 1 - \frac{T_{1,4}}{T_{2,3}}$

Other terms that are frequently encountered:

Volumetric efficiency: $\eta_v = \left.\frac{\nu_a}{\nu_d}\right]_{p,T}$

Compression ratio: $r_c = \frac{\nu_d + \nu_c}{\nu_c}$

Mean effective pressure (brake, indicated)

$$\text{bmep} = \frac{w_{b,cc}}{A\ell} \qquad \text{imep} = \frac{w_{i,cc}}{A\ell}$$

Specific fuel consumption (brake, indicated)

$$\text{bsfc} = \frac{\dot{m}_f}{\text{bhp}} \qquad \text{isfc} = \frac{\dot{m}_f}{\text{ihp}}$$

Horsepower (brake, indicated)

$$\text{bhp} = \frac{2\pi NT}{33{,}000} = \frac{(\text{bmep})\ell AnN}{33{,}000}$$

$$\text{ihp} = \text{bhp} + \text{fhp} = \frac{(\text{imep})\ell AnN}{33{,}000}$$

Gas Turbines - Brayton Cycle:

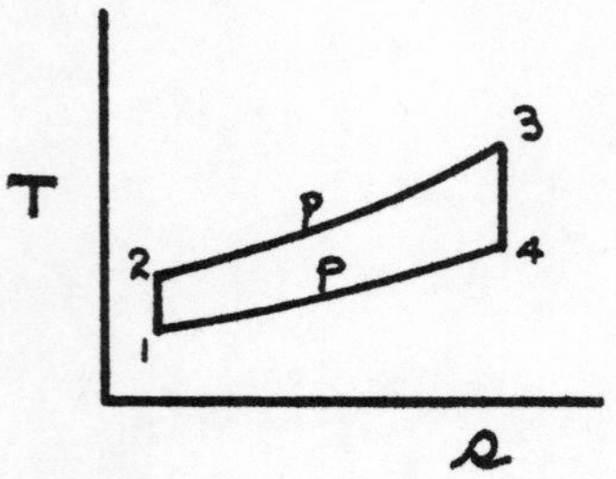

$$\eta_t = 1 - \frac{1}{(r_p)^{\frac{k-1}{k}}} = 1 - \frac{T_1}{T_2} = 1 - \frac{T_4}{T_3}$$

Other terms that are frequently encountered:

Isentropic efficiency (compressor, turbine)

$$\eta_{s,c} = \frac{W_s}{W_a} \qquad \eta_{s,tu} = \frac{W_a}{W_s}$$

Work ratio: $\dfrac{W_{tu}}{W_c}$

Regenerative ideal heat exchange: $\eta_t = 1 - \dfrac{T_2}{T_3}$

Rankine Vapor Power Cycle:

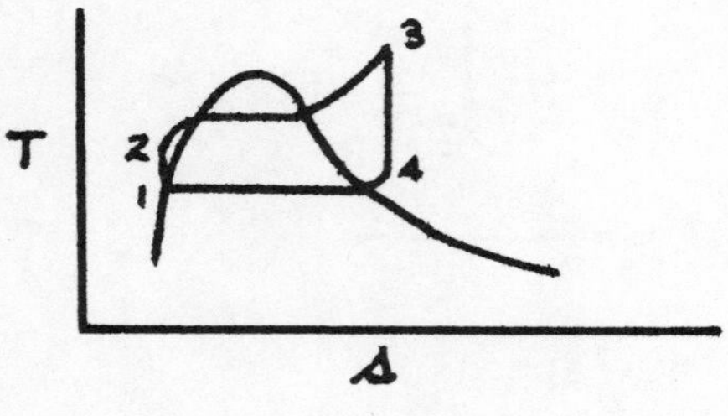

$$\eta_t = \frac{(h_3\ h_4) - (h_2 - h_1)}{(h_3 - h_2}$$

Other terms that are frequently encountered:

Heat rate: $\text{HR} = \dfrac{Q_a\,(\text{Btu})}{W_n\,(\text{kw-hr})}$

Boiler hp = h_{fg} x 34.5 lbm/hr at 212°F = 33,475 Btu/hr

Terminology for Power Plants Review

Symbols:

A	piston area
bhp	brake horsepower
bmep	brake mean effective pressure
bsfc	brake specific fuel consumption
fhp	friction horsepower
HR	heat rate
ihp	indicated horsepower
imep	indicated mean effective pressure
isfc	indicated specific fuel consumption
k	specific heat ratio
ℓ	stroke length
$\dot{m}$	mass flow rate
N	revolutions per minute
n	number of cylinders
p	pressure
r	ratio
T,t	temperature
ν	specific volume
W	work
η	efficiency

Subscripts:

a	actual, added
b	brake
c	clearance, compressor, compression
cc	cylinder-cycle
d	displacement
i	indicated
n	net
t	thermal
tu	turbine
reg	regeneration
v	volume
f	fuel

NEW VEHICLE EMISSIONS STANDARDS SUMMARY
(Taken from California Air Resources Board Fact Sheet 72-6 5000 6-72)
LIGHT-DUTY VEHICLES UNDER 6,000 lbs

Year	Standard	Cold Start Test	Hydrocarbons	Carbon Monoxide	Oxides of Nitrogen
Prior to Controls			850 ppm (11 gm/mi)*	3.4% (80 gm/mi)	1000 ppm (4 gm/mi)
1966-67	State	7-mode	275 ppm	1.5%	no std.
1968-69	State & Federal	7-mode			
		50-100 CID	410 ppm	2.3%	no std.
		101-140 CID	350 ppm	2.0%	no std.
		over 140 CID	275 ppm	1.5%	no std.
1970	State & Federal	7-mode	2.2 gm/mi	23 gm/mi	no std.
1971	State	7-mode	2.2 gm/mi	23 gm/mi	4 gm/mi
	Federal	7-mode	2.2 gm/mi	23 gm/mi	-
1972	State	7-mode	1.5 gm/mi	23 gm/mi	3 gm/mi
		or CVS-1	3.2 gm/mi	39 gm/mi	3.2 gm/mi**
	Federal	CVS-1	3.4 gm/mi	39 gm/mi	-
1973	State	CVS-1	3.2 gm/mi	39 gm/mi	3 gm/mi
	Federal	CVS-1	3.4 gm/mi	39 gm/mi	3 gm/mi
1974	State	CVS-1	3.2 gm/mi	39 gm/mi	2 gm/mi
	Federal	CVS-1	3.4 gm/mi	39 gm/mi	3 gm/mi
1975	State	CVS-1	1 gm/mi	24 gm/mi	1.5 gm/mi
	Federal	CVS-2	0.41 gm/mi	3.4 gm/mi	3 gm/mi
1976	State	CVS-1	1 gm/mi	24 gm/mi	1.5 gm/mi
	Federal	CVS-2	0.41 gm/mi	3.4 gm/mi	0.4 gm/mi

ppm — parts per million concentration

gm/mi — grams per mile

7-mode — is a 137 second driving cycle test

CVS-1 — is a constant volume sample cold start test

CVS-2 — is a constant volume sample cold start test average with a constant volume sample hot start test, both with the Federal 22-minute driving cycle

* — The values in parentheses are approximately equivalent values

NEW VEHICLE EMISSIONS STANDARDS SUMMARY
(Taken from California Air Resources Board Fact Sheet 72-6 5000 6-72)

HEAVY-DUTY VEHICLES OVER 6,000 lbs

Year	Standard	Hydrocarbons	Carbon Monoxide	Oxides of Nitrogen
1969-71	State-gasoline	275 ppm	1.5%	no std.
1972	State-gasoline	180 ppm	1.0%	no std.
1973-74	State-gasoline & diesel	$HC + NO_X = 16$ gm/BHP hr. $CO = 40$ gm/BHP hr.		
1975 & later	State-gasoline & diesel	$HC + NO_X = 5$ gm/BHP hr. $CO = 25$ gm/BHP hr.		

gm/BHP hr. grams per brake horsepower-hour

POWER PLANTS 1

A space vehicle in an earth distance orbit about the sun has a solar panel array which is its electrical power generator. The solar cells are of the P-N type with a "red-blue" filter and have 50 mil. thick protective glass covers. The solar panels are sun oriented. If the radiant solar energy in this orbit is 1400 watts/m^2, and the array is 10% efficient and develops 460 watts, what power will it develop in a Mars distance orbit about the sun? Assume that the earth and Mars orbits are 1.495×10^8 km and 2.276×10^8 km, respectively, from the sun.

Solution

* Assume temperature of panels and hence efficiency are constant.

* Note that radiation intensity varies inversely as the square of the distance from the sun.

Then, $P \propto \frac{1}{r^2}$ $\quad \therefore \frac{P_e}{P_m} = \frac{r_m^2}{r_e^2}$

$$P_m = 460\left(\frac{1.495}{2.278}\right)^2 = \underline{\underline{200 \text{ watts}}}$$

POWER PLANTS 2

From a test of a 7 x 10 inch, single-acting, four-stroke-cycle gas engine, the following data was obtained: speed = 340 rpm, average area of indicator cards = 1.13 sq. in.; length of cards = 2.90 in.; scale of indicator spring = 200 lb/sq.in./in.; net brake load = 37.1 lb.; length of brake arm = 4.0 ft.

Determine:
- (a) The indicated mean effective pressure (mep).
- (b) The indicated horsepower (ihp).
- (c) The brake horsepower (bhp).
- (d) The mechanical efficiency.
- (e) The brake torque.

Solution

* Note : no. of cylinders not specified

a) $\bar{h} = \left.\frac{A}{l}\right)_{card} = \frac{1.13}{2.90} = 0.39 \text{ in}$

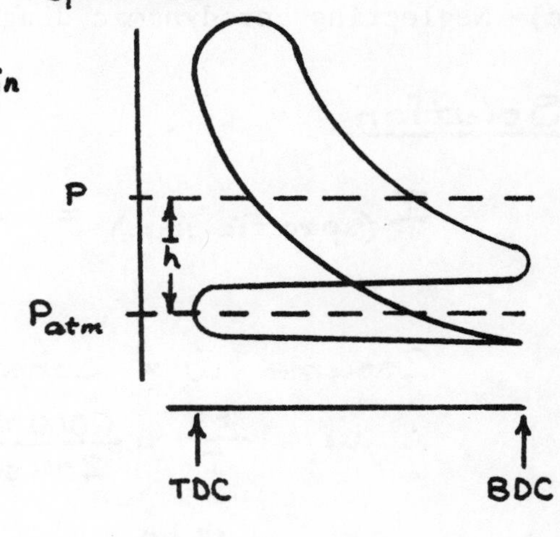

$imep = \bar{h} k_{spring}$

$= 0.39 \times 200$

$= \underline{\underline{78 \text{ psi}}}$

b) $ihp/cyl = imep \times A_p \times stroke \times \frac{rpm}{2}$

$$= \frac{78 \times \pi \times 49 \times 10 \times 170}{4 \times 33{,}000 \times 12} = \underline{\underline{12.9 \text{ hp}}}$$

c) $bhp = \frac{2\pi \times rpm \times T}{33{,}000} = \frac{2\pi \times 340 \times 4 \times 37.1}{33{,}000} = \underline{\underline{9.61 \text{ hp}}}$ (7.17 kw)

∴ single cyl. engine since bhp is a reasonable % of ihp/cyl.

d) $\eta_{mech.} = \frac{9.61}{12.9} = \underline{\underline{74.5\%}}$

e) $T = F \times L = 37.1 \times 4 = \underline{\underline{148.4 \text{ ft}\cdot\text{lbf}}}$

(2.012 Nm)

POWER PLANTS 3

A vertically-launched rocket vehicle has an initial weight of 3,000 lbm. The rocket has 1,500 lbm of first-stage propellant and a variable expansion ratio nozzle with a constant system delivered specific impulse of 240 sec.

If the liftoff thrust-to-weight ratio is 2.00 lbf/lbm, then at first-state burnout:

(a) What is the time elapsed from launch?

(b) What is the thrust-to-weight ratio?

(c) Neglecting aerodynamic drag, what is the velocity?

Solution

$$I_{\text{(specific imp.)}} = \frac{F_t \text{ (thrust)}}{\dot{W} \text{ (prop. flow rate)}}$$

Assume $\dot{w}$ = constant

$$\therefore \dot{w} = \frac{F_t}{I} = \frac{6000 \text{ lbf}}{240 \text{ sec.}} = 25 \text{ lbf/sec.}$$

a) $t = \frac{W}{\dot{w}} = \frac{1500}{25} = \underline{\underline{60 \text{ sec}}}$

b) $\frac{F_t}{W_T \text{ (total)}} = \frac{6000}{W_{T,o} - \dot{W}t} = \frac{6000}{1500} = \underline{\underline{4:1}}$

c) at any instant,

$$F_t = \frac{1}{g_c} \frac{d\, m_T V_r \text{ (rocket)}}{dt} + W_T$$

-continued-

$$m_T = m_{T,0} - \dot{m}t$$

$$F_t = \frac{1}{g_c}\frac{d[(m_{T,0} - \dot{m}t)V_r]}{dt} + (W_{T,0} - \dot{\omega}t)$$

$$F_t = \frac{1}{g_c}(m_{T,0} - \dot{m}t)\left(\frac{dU_r}{dt} + g\right)$$

$$\int_0^V dU_r = \int_0^t F_t\, g_c \frac{dt}{(m_{T,0} - \dot{m}t)} - \int_0^t g\, dt$$

$$U_r = \frac{F_t g_c}{\dot{m}} \ln \frac{m_{T,0}}{(m_{T,0} - \dot{m}t)} - gt$$

$$= \frac{6000 \times 32.2}{25} \ln 2 - 32.2 \times 60 = 3428 \frac{ft}{sec}$$

(1.045 KM/SEC)

POWER PLANTS 4

Consider the ideal isotope powered space power system shown below.

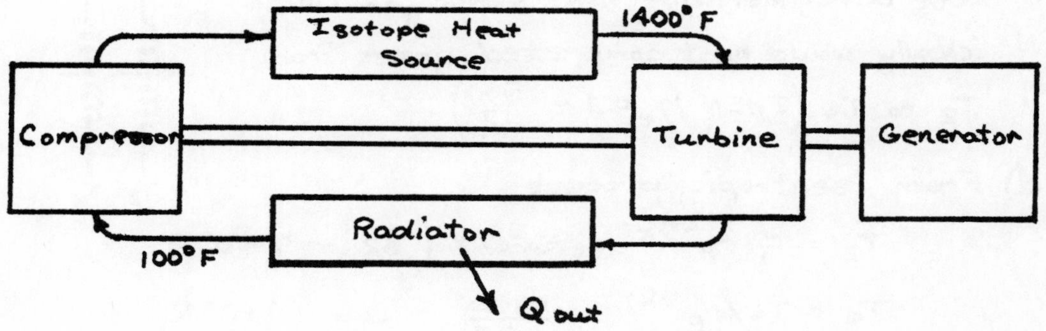

Assume: (1) A simple Brayton Cycle.

(2) There are no losses in the system.

(3) The working fluid is a perfect monatomic gas.

(4) The compressor pressure ratio is two to one.

-continued-

Then:

(a) What is the cycle efficiency?

(b) How does the cycle efficiency compare with that of a Carnot engine operating between the same temperature limits?

(c) What would you do to greatly improve the cycle efficiency of the Brayton Cycle while still operating between the same temperature limits?

(d) What is the cycle efficiency of this improved cycle assuming ideal conditions?

Solution

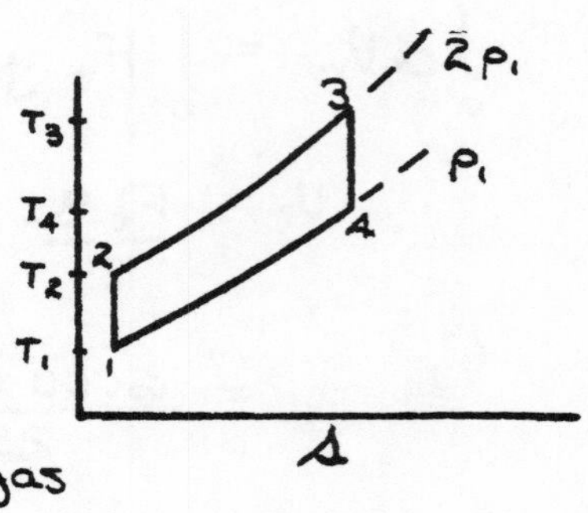

a) For the Brayton Cycle

$$\eta_t = 1 - \frac{1}{r_p^{k-1/k}}$$

From kinetic theory for monatomic gas

$$C_v = \frac{3}{2}\frac{R_0}{M} \quad \text{and} \quad C_p = C_v + \frac{R_0}{M} \qquad \therefore k = 5/3 = 1.67$$

$$\eta_t = 1 - \frac{1}{2^{0.67/1.67}} = \underline{\underline{24.2\ \%}}$$

b) $$\eta_{t\ carnot} = 1 - \frac{T_{min}}{T_{max}} = 1 - \frac{T_1}{T_3} = 1 - \frac{560}{1860} = \underline{\underline{69.9\ \%}}$$

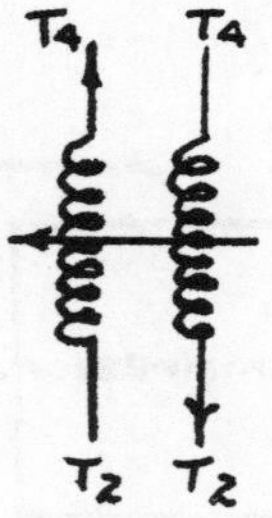

c) Use a regenerative heat exchanger which ideally would heat compressed gases from T_2 to T_4. (THEN $\eta_t = 1 - T_2/T_3$)

d) From isentropic process

$$T_2 = T_1\, r_p^{k-1/k} = 560 \times 1.32 = 740°R$$

$$T_4 = T_3/r_p^{k-1/k} = \frac{1860}{1.32} = 1410°R$$

For original cycle $\eta_{to} = \dfrac{W_{net}}{Q_{add_o}} = \dfrac{W_{net}}{C_p(T_3 - T_2)}$

For regenerator $\eta_{tr} = \dfrac{W_{net}}{Q_{add_r}} = \dfrac{W_{net}}{C_p(T_3 - T_4)}$

Since W_{net} same for both cases,

$$\eta_{tr} = \eta_{to}\frac{T_3 - T_2}{T_3 - T_4} = 24.2\,\frac{1860 - 740}{1860 - 1410} = \underline{\underline{60.3\ \%}}$$

POWER PLANTS 5

A 55,000 lb truck (loaded weight) has a power train which delivers 200 horsepower at the wheels. When the loaded truck is traveling at 60 mph on a level road, 100 horsepower is required to overcome wind drag and the remaining 100 horsepower is required to overcome rolling resistance.

Can the same truck and load maintain a speed of 20 mph on a 6 per cent uphill grade? Support your answer with calculations. Show all assumptions.

Solution

P_z (Climbing power required) = W(weight) V_z (vert. vel.)

$$V_z = 20 \text{ mph} \times \frac{5280}{3600} \times .06 = 1.76 \frac{\text{ft.}}{\text{sec.}}$$

$$P_z = \frac{55{,}000 \times 1.76}{550} = 176 \text{ hp}$$

P_a (air drag power) $\propto V^3$ (veh. vel.) $\quad P_{a_{20mph}} = 100\left(\frac{20}{60}\right)^3 = 3.7$ hp

P_r (rolling drag power) $\propto V^*$ $\quad P_{r_{20mph}} = 100\left(\frac{20}{60}\right) = 33.3$ hp

$P_z + P_a + P_r = 213$ hp (159 kW) higher than 200 hp available

* actually rolling drag increases slowly with speed so that $P_r \propto V^{1+}$. This depends on tire pressure, type of road, etc.. Thus, it is possible that the truck could maintain 20 mph on grade.

POWER PLANTS 6

The compressed air motor shown consists of two identical chambers (A) to which nozzles are attached. Air is admitted through expansion valves (B). The chambers are mounted on a hub which rotates freely about the supply line (C). The air in the chambers (A) has negligible velocity.

Pressure in supply line	500 psig
Pressure in chambers (A)	100 psig
Exhaust pressure	14.7 psia

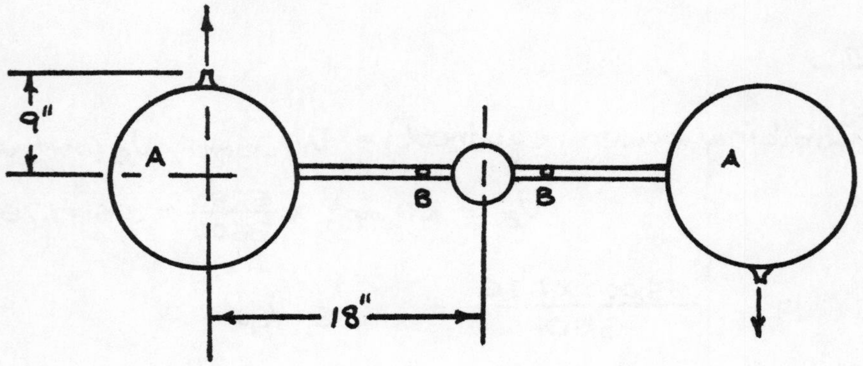

Temperature of supply air	100°F
Temperature of exhaust	20°F
Discharge rate of air	2,000 lb/hr
Distance from center of bearing to center lines of nozzles	18 in
Radius of gyration of motor	15 in
Weight of motor (all rotating parts)	200 lb
Distance from center of chambers (A) to ends of nozzles	9 in

Find: (a) If the motor is originally held by a brake which is suddenly released, what speed will the motor attain in 6 seconds?

(b) What will be the maximum speed of the motor, assuming no frictional or air resistance?

Give both answers in revolutions per minute.

Solution

Assume adiabatic nozzle flow & stagnation in spheres.

From 1st law of thermo,

$$V_r \text{ (rel. exh.)} = \sqrt{2 g_c C_p (T_0 - T_{exh})}$$

$$= (64.4 \times 0.241 \times 778 \times 80)^{1/2}$$

$$= 980 \text{ ft/sec.}$$

From impulse-momentum principle for angular momentum,

$$\Sigma T + \frac{1}{g_c} \dot{m} (\vec{r} \times \vec{V}_{abs})_{in\ flow} = \frac{1}{g_c} m r_g^2 \frac{d\omega}{dt}$$

for free rotation $\Sigma T = 0$. Let cc. moments be (+).

$$\frac{1}{g_c} \dot{m} (\vec{r} \times \vec{V}_{abs}) = -\frac{1}{32.2}\left(\frac{2000}{3600}\right)(1.5)(-V_r + r\omega)$$

$$= -\frac{1}{32.2}\left(\frac{2000}{3600}\right)(1.5)(-980 + 1.5\omega)$$

$$= 25.9 - .039\,\omega$$

$$\frac{1}{g_c} m r_g^2 \frac{d\omega}{dt} = \frac{200}{32.2}(1.25)^2 \frac{d\omega}{dt} = 9.68 \frac{d\omega}{dt}$$

Assume ω small for first 6 seconds.

$$\therefore 9.68\, d\omega/dt = 25.9 \qquad \frac{d\omega}{dt} = 2.68 \text{ rad/sec}^2$$

a) in six seconds, $\omega = 6 \times \frac{d\omega}{dt} = 16 \frac{\text{rad}}{\text{sec}} = 2.5 \text{ rps}$

b) At runaway $d\omega/dt = 0$

$$.039\,\omega = 25.9 \qquad \omega = 666 \frac{\text{rad}}{\text{sec}} = 106 \text{ rps}$$

POWER PLANTS 7

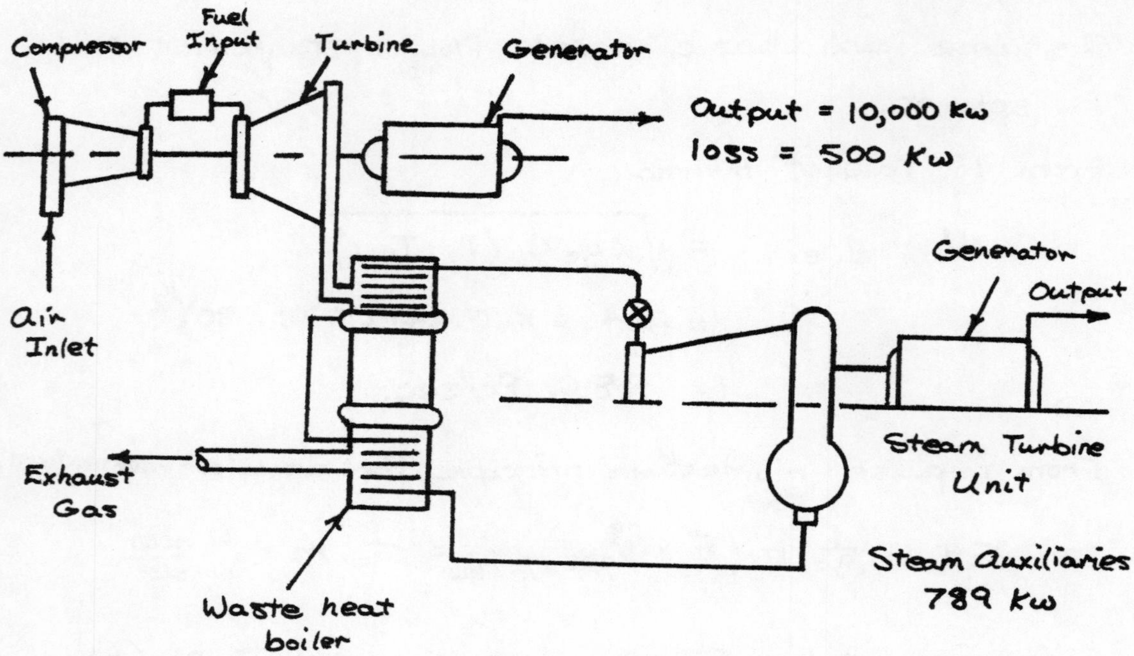

The above diagram shows a combined cycle gas and steam turbine electric generating unit. At 10,000 kw output on the gas turbine unit the electrical and mechanical loss is 500 kw. The overall thermal efficiency of the gas turbine generator unit is 20 per cent. The gas turbine exhausts to a waste heat steam generator which is 75 per cent efficient. The steam from the steam generator drives a steam turbine electric generator which has an overall heat rate of 11,000 Btu per kwh. Steam turbine unit auxiliaries require 789 kw.

Determine:
(a) The output of the steam turbine generating unit in kw.
(b) The net heat rate in Btu per kwh of the combined cycle.
(c) The thermal efficiency.

Solution

Assume 10,000 kw output is net output

$$\dot{E}_{\text{supplied to gas turbine}} = \frac{P_{net}}{\eta_t} = \frac{10,000}{0.20} = 50,000 \text{ kw}$$

$$\dot{E}_{boiler} = \dot{E}_{turbine} - (P_{net} + loss) = 39,500 \text{ kw}$$

$$\dot{E}_{steam} = \dot{E}_{boiler} \times \eta_{boiler} = 29,600 \text{ kw}$$

$$\text{boiler heat rate} = \frac{11,000 \text{ Btu/kwh}}{3413 \text{ Btu/kwh}} = 3.22 \frac{\text{kw input}}{\text{kw output}}$$

a) Assume boiler heat rate based on $\dot{E}_{boiler}$.

$$\therefore P_{\text{steam turb.}} = \frac{39,500}{3.22} = \underline{\underline{12,200 \text{ kw}}}$$

b)
$$\text{Heat rate} = \frac{50,000 \times 3413}{10,000 + 12,200} = \underline{\underline{7700 \frac{\text{Btu}}{\text{kwh}}}}$$

c)
$$\eta_{t,overall} = \frac{22,200}{50,000} = \underline{\underline{44.5\ \%}}$$

POWER PLANTS 8

A turbo-generator produces 8500 kw in a refinery power plant. Generator efficiency is 95%. Turbine inlet steam is 815 psia, 750°F; exhaust steam is 145 psia, 440°F at A. If the turbine trips out, steam would be depressed and desuperheated to the 145 psia header. Steam flow rate, pressure and temperature are to be the same at A and B.

(a) What is the turbine engine efficiency?

(b) What is the flow rate of A?

(c) What is the flow rate at C?

(d) What is the steam temperature at C?

Assume line losses and heat losses are negligible.

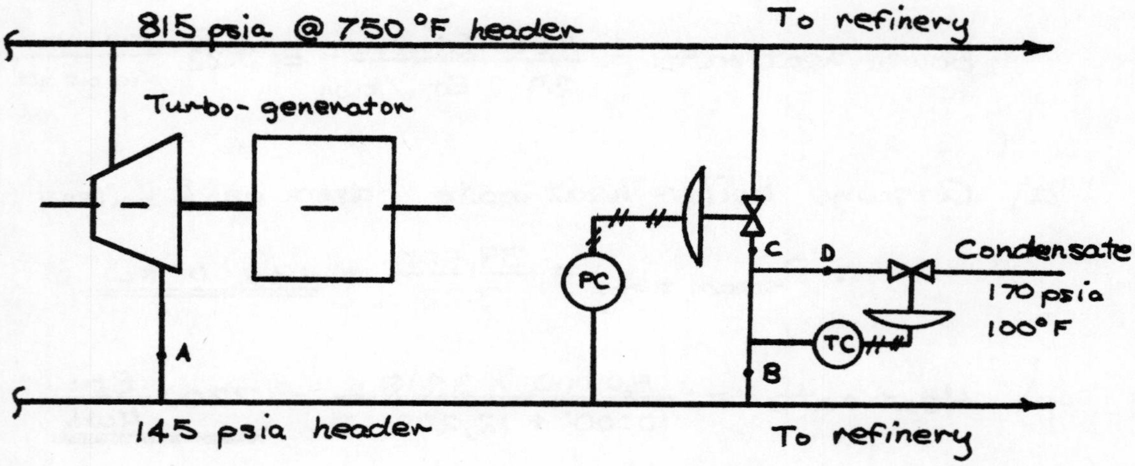

Solution

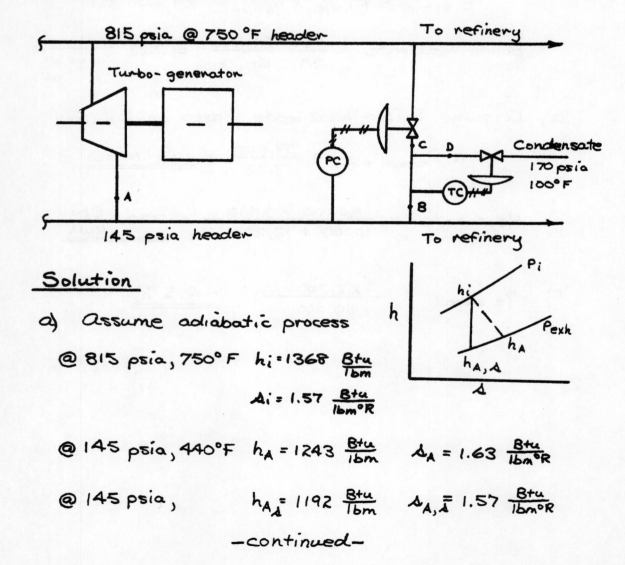

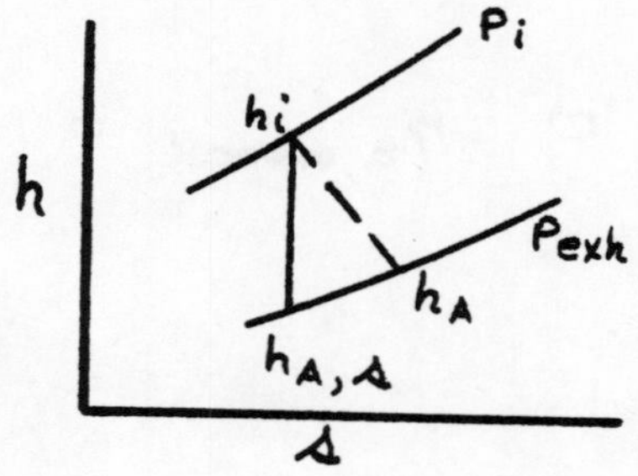

a) Assume adiabatic process

@ 815 psia, 750°F $h_i = 1368 \frac{Btu}{lbm}$

$s_i = 1.57 \frac{Btu}{lbm°R}$

@ 145 psia, 440°F $h_A = 1243 \frac{Btu}{lbm}$ $s_A = 1.63 \frac{Btu}{lbm°R}$

@ 145 psia, $h_{A,s} = 1192 \frac{Btu}{lbm}$ $s_{A,s} = 1.57 \frac{Btu}{lbm°R}$

—continued—

$$\eta_{engine} = \frac{\Delta h_{actual}}{\Delta h_s} = \frac{1368-1243}{1368-1192} = \underline{\underline{71\%}}$$

b) From 1st Law of Thermo.

$$\dot{m}(h_i - h_A) = P_{turbine} = \frac{P_{gen.}}{\eta_{gen.}} = \frac{8500\,\text{KW} \times 3413\,\frac{\text{Btu}}{\text{KWh}}}{0.95}$$

$$\dot{m}_A = \frac{8500 \times 3413}{0.95 \times 125} = \underline{\underline{2.44 \times 10^5 \ \frac{\text{lbm}}{\text{hr}}}}$$

(111 Kg/HR)

c) @ 170 psia, 100°F $h_D = 68 \ \frac{\text{Btu}}{\text{lbm}}$

assume adiabatic mixing and negligible K.E. change of streams.

$$\dot{m}_C(h_i - h_{B=A}) = \dot{m}_D(h_B - h_D) \quad \& \quad \dot{m}_D = \dot{m}_A - \dot{m}_C$$

$$\dot{m}_C(125) = (\dot{m}_A - \dot{m}_C)(1243 - 68)$$

$$\dot{m}_C = \frac{1175}{1300}\dot{m}_A = \underline{\underline{2.21 \times 10^5 \ \frac{\text{lbm}}{\text{hr}}}}$$

(100 Kg/HR)

d) Assume steam is throttled

$h_C = h_i \qquad P_C = 145$ psia

$$\underline{\underline{T = 680\,°\text{F}}}$$

(360°C)

POWER PLANTS 9

A stationary diesel engine generates 2,000 brake horsepower with a specific fuel rate of 0.42 lb/bhp-hr. The heating value of the fuel is 19,000 Btu/lb, and the air-fuel ratio is 16/1. Assume surroundings of 90°F, 14.7 psia and the exhaust temperature 800°F.

Estimate the cooling water flow rate (gpm) necessary if the water cooling apparatus can produce a 30°F temperature drop in the water.

Solution

$$\dot{E}_{shaft} = 2000 \times 2544 = 5.09 \times 10^6 \ \frac{\text{Btu}}{\text{lbm}}$$

$$\dot{m}_f = \text{bsfc} \times \text{bhp} = 0.42 \times 2000 = 840 \ \text{lbm/hr.}$$

$$\dot{E}_{supplied} = LHV \dot{m}_f = 18{,}000 \ \frac{\text{Btu}}{\text{lbm}} \times 840 \ \frac{\text{lbm}}{\text{hr}}$$

$$= 15.2 \times 10^6 \ \frac{\text{Btu}}{\text{hr}}$$

* Note: Assume HV given was HHV & LHV ≈ HHV − 1000

$$\dot{E}_{exh} = \dot{m} c_p (T_{exh} - T_{atm}) = (\dot{m}_{air} + \dot{m}_{fuel}) c_p (T_e - T_a)$$

$$= \dot{m}_f (16 + 1)(0.26)(800 - 90)$$

$$= 2.63 \times 10^6 \ \frac{\text{Btu}}{\text{hr}}$$

allow 5% $\dot{E}_{supplied}$ for incomplete comb. & heat loss to surroundings

$$\dot{E}_{coolant} = (\dot{E}_{su} - \dot{E}_{sh} - \dot{E}_{ex} - .05 \dot{E}_{su})$$

$$= 6.73 \times 10^6 \ \text{Btu/hr}$$

$$\text{gpm} = \frac{6.73 \times 10^6 \ \text{Btu/hr}}{60 \frac{\text{min}}{\text{hr}} \times 30°\text{F} \times 1 \frac{\text{Btu}}{\text{lbm °F}} \times 8.35 \frac{\text{lbm}}{\text{gal}}} = \underline{\underline{445}}$$

$(0.028 \ \text{m}^3/\text{SEC})$

POWER PLANTS 10

A boat, powered by a normally aspirated diesel engine, has been given a pier (stationary) test in San Francisco Bay. The resulting required engine horsepower curve and the maximum engine horsepower curve versus speed are shown on the graph below. If this boat is moved to the Great Salt Lake in Utah, its performance will not be satisfactory. Explain why (credit will be based on the quality of your analysis using the data supplied) and recommend changes that will improve the performance.

	San Francisco	Great Salt Lake
Elevation - ft	0	4,000
Baro pressure - inches of Hg	29.9	26.6
Air Temp. - °F	65.0	65.0
Water Temp. - °F	55.0	55.0
Density of water - lb/ft^3	65.0	78.0
Viscosity of water - $lb\ sec/ft^2$	2.65×10^{-5}	4.15×10^{-5}

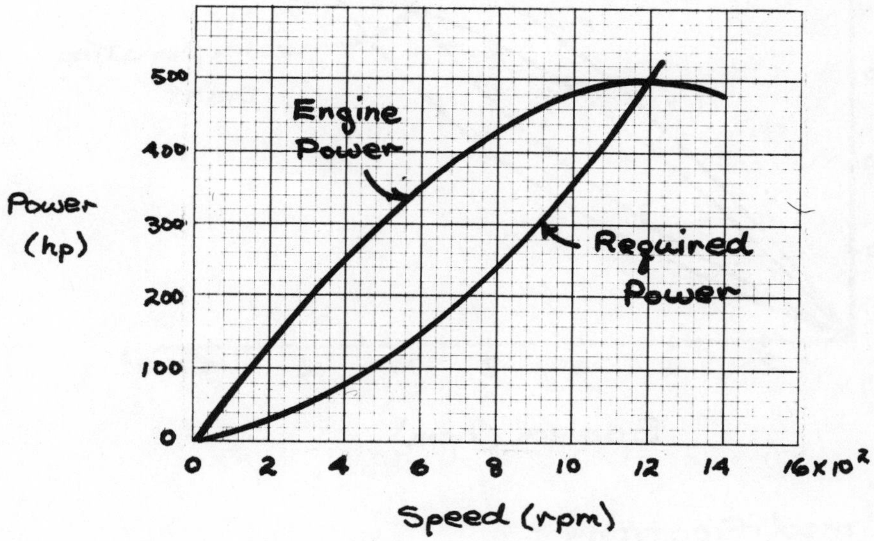

Solution

Max. Engine Power $\propto \rho_{air\ intake} = \frac{p}{RT}\Big)_{intake}$

Required Power $\propto \rho_{water}$

-continued-

$$\therefore \quad \frac{P_{eng.,\,S.L.}}{P_{eng.,\,Sea\,L.}} = \frac{\rho_{atm,\,S.L.}}{\rho_{atm,\,sea\,L.}} \quad \text{plot new engine curve}$$

$$\frac{P_{prop.,\,S.L.}}{P_{prop.,\,sea\,L.}} = \frac{\rho_{water,\,S.L.}}{\rho_{water,\,Sea\,L.}} \quad \text{plot new required curve}$$

new intersection shows engine cannot come up to rated speed.

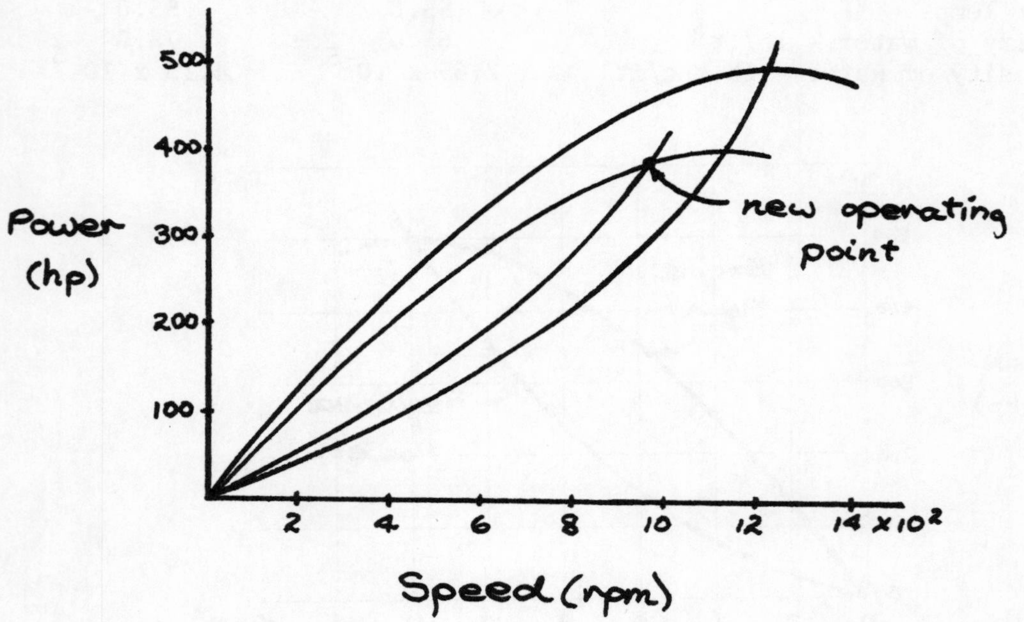

Possible modifications :

1) Reduce prop. dia or change prop.

2) Supercharge engine or change engine

POWER PLANTS 11

From the flow diagram below of a steam turbine generator unit, calculate the output in kw, the throttle steam flow and the unit heat rate in Btu per kwh with steam flow from the turbine to the condenser of 50,000 pounds per hour of 2.0 inches Hg absolute.

Mechanical efficiency of the unit 98.0% and generator efficiency 97.0%. Throttle conditions 750 psia and 760 FTT.

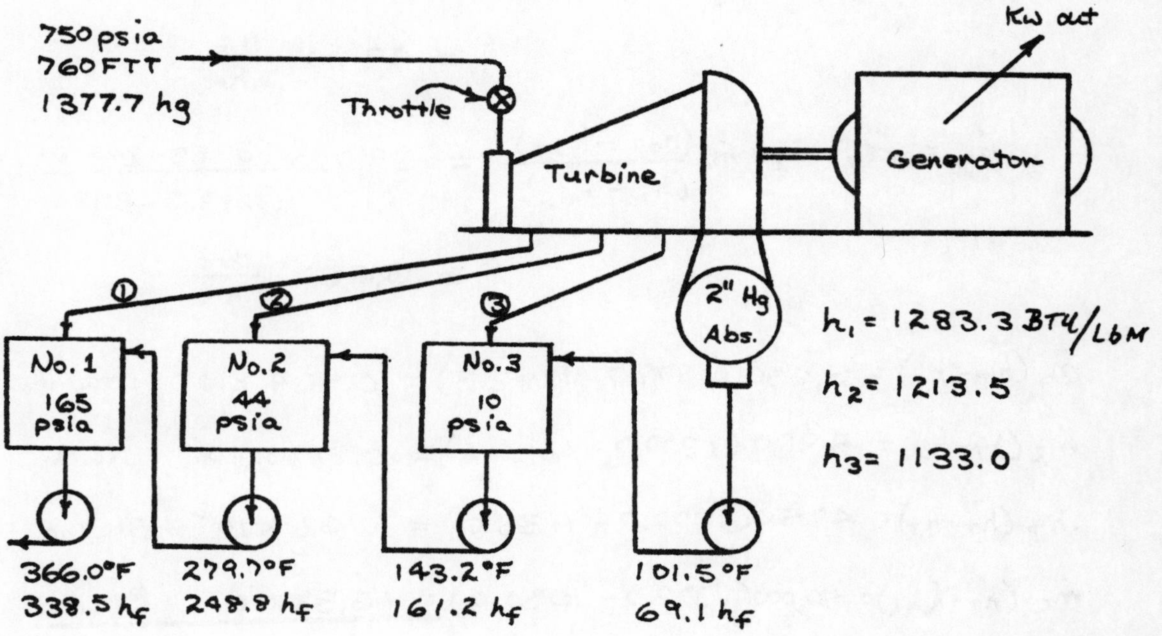

Solution

$$\text{Power output} = \eta_{mech}\,\eta_{gen.}\,\frac{\dot m_1(h_T-h_1)+\dot m_2(h_T-h_2)+\dot m_3(h_T-h_3)+\dot m_c(h_T-h_{exh})}{3413\ \text{Btu/kwh}}$$

* Note: Data for all enthalpies given except h_{exh}. Based on plot of intermediate turbine states on Mollier diag. suggests $h_{exh} \simeq 1050\ \frac{\text{Btu}}{\text{lbm}}$.

-continued-

To find $\dot{m}_{1,2,3}$ perform energy balences on feed water heaters.

$$\dot{m}_3 (h_3 - h_{out,3}) = \dot{m}_c (h_{o,3} - h_c)$$

$$\dot{m}_3 = 50{,}000 \frac{\text{lbm}}{\text{hr.}} \frac{(161.2 - 69.1)}{(1133.0 - 161.2)} = 4{,}750 \text{ lbm/hr.}$$

$$\dot{m}_2 = \dot{m}_3 + \dot{m}_c \frac{(h_{o,2} - h_{o,3})}{(h_2 - h_{o,2})} = 54{,}750 \frac{(248.8 - 161.2)}{(1213.5 - 248.8)}$$

$$= 4{,}970 \frac{\text{lbm}}{\text{hr.}}$$

$$\dot{m}_1 = \dot{m}_3 + \dot{m}_2 + \dot{m}_c \frac{(h_{o,1} - h_{o,2})}{(h_1 - h_{o,1})} = 59{,}720 \frac{(338.5 - 248.8)}{(1283.3 - 338.5)}$$

$$= 5{,}650 \frac{\text{lbm}}{\text{hr.}}$$

$$\dot{m}_1 (h_T - h_1) = 5{,}650 (1377.7 - 1283.3) = 0.534 \times 10^6 \text{ Btu/hr.}$$

$$\dot{m}_2 (h_T - h_2) = 4{,}970 (1377.7 - 1213.5) = 0.817 \times 10^6 \text{ Btu/hr.}$$

$$\dot{m}_3 (h_T - h_3) = 4{,}750 (1377.7 - 1133.0) = 1.162 \times 10^6 \text{ Btu/hr.}$$

$$\dot{m}_c (h_T - h_{exh}) = 50{,}000 (1377.7 - 1050.0) = \underline{16.38 \times 10^6 \text{ Btu/hr.}}$$

$$18.89 \times 10^6 \text{ Btu/hr.}$$

$$P_{out} = 0.98 \times 0.97 \frac{18.89 \times 10^6}{3.413 \times 10^3} = \underline{\underline{5261 \text{ Kw}}}$$

$$\dot{m}_t = \dot{m}_1 + \dot{m}_2 + \dot{m}_3 + \dot{m}_c = \underline{\underline{65{,}370 \frac{\text{lbm}}{\text{hr.}}}}$$

$$\text{heat rate} = \frac{\dot{m}_t (h_T - h_{o,1})}{P_{out}} = \frac{65{,}370}{5261} (1377.7 - 338.5)$$

$$= \underline{\underline{12{,}900 \frac{\text{Btu}}{\text{Kwh}}}}$$

$$(3.79 \text{ Kw/Kw})$$

POWER PLANTS 12

Many boiler explosions occur during the "light-off" period due to unorthodox operational procedures.

Program a safe procedure for igniting a single burner using natural gas as the fuel. Consider the burner equipped with adjustable register, pilot and igniter. Account for all permissive requirements prior to initial light-off.

(a) Program "light-off" sequence.

(b) List all factors which may cause fuel valve closure.

(c) Do you recommend "post purge"? Discuss this for normal shut-downs and after loss of flame.

Solution

a)
1. Check for presence of gas
2. Arm pilot light gas circuit
3. Operate igniter control
4. Check for automatic arming of main gas supply to burner.
5. Observe main burner ignition
6. Adjust register for correct flame quality

b)
1. Loss of pilot flame
2. Loss of fuel pressure
3. Flame out due to abnormal draft
4. Excessive flue or combustion zone temperature
5. Normal Temperature controller shutdown

c) Post purge is helpful in a safety and re-ignition sense because it eliminates combustion products, unburned fuel, and hot gases prior to restart thus assuring smooth ignition.

POWER PLANTS 13

A power plant supplies an auxiliary header with steam at 600 psig and 700°F. The total steam requirement for the auxiliaries is 25,000 lb/hr at 300 psig and 450°F.

REQUIRED: (a) Design a reducing and desuperheating station to achieve the above results. Water is available at 350 psig and 105°F.

(b) Compute the amount of spray water required and state your reasons for selecting the number and type of spray nozzles.

(c) Sketch the control and piping complex supporting your design.

Solution

a), c)

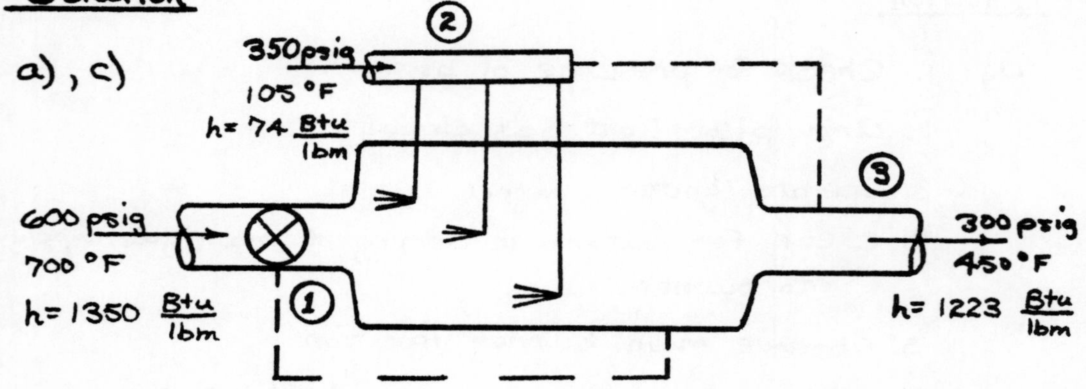

① main steam line throttle value, holds down stream pressure @ 300 psig

② Temperature controlled nozzle actuating value holds downstream temperature by supplying tempering water to appropriate number of spray nozzles.

-continued-

b) From 1st Law (assume ΔKE. negligible, adiabatic)

$$\dot{m}_w (h_3 - h_2) = \dot{m}_s (h_1 - h_3)$$

$$\dot{m}_s = 25{,}000 \text{ lb/hr} - \dot{m}_w$$

$$\therefore \dot{m}_w (h_3 - h_2) + \dot{m}_w (h_1 - h_3) = 25{,}000 (h_1 - h_3)$$

$$\dot{m}_w = 25{,}000 \frac{(h_1 - h_3)}{(h_1 - h_2)} = \underline{\underline{2490 \text{ lbm/hr}}}$$

(1,129 Kg/hr)

Since flow rate of water is only about 5 gpm, four to six orifice type nozzles should be able to handle the requirement.

VEHICLE EMISSIONS

SPECIAL PROBLEM 1

The NO_X is found to be 550 ppm equivalent of NO_2 in a vehicle exhaust sample using California Standard Smog Test. Determine the gr/mi if the exhaust is 700 ft^3/mi at standard pressure and temperature.

$$gr/mi = \frac{ppm}{10^6} \times \left(\rho_{NO_X}\right)_{P_{std},\ T_{std}} \nu$$

Assume $NO_X = NO_2$

$$\rho_{NO_2} = \frac{pM}{R_oT} = 54.16\ gr/ft^3$$

$$gr/mi = \frac{500}{10^6} \times 54.16 \times 700 = \underline{\underline{19.8}}$$

Note: CO(gr/mi) is determined in the same way.

SPECIAL PROBLEM 2

The hydrocarbons are found to be 1000 ppm equivalent of hexane in a vehicle exhaust sample using California Standard Smog Test. Determine the gr/mi if the exhaust volume is 600 ft^3/mi at standard pressure and temperature.

$$\rho_{HC} = 16.33\ gr/ft^3 \text{ assumes C:H = 1:1.85}$$

$$gr/mi_{\text{carbon equivalent}} = 1.8^* \times \frac{ppm}{10^6} \times \rho_{HC} \times c_{eq} \times \nu$$

$$= 1.8 \times \frac{10^3}{10^6} \times 16.33 \times 6 \times 600 = \underline{\underline{106}}$$

---

* 1.8 is correction factor for NDIR instruments - not necessary for FID instruments.

11

Heating, Ventilating and Air Conditioning

RICHARD K. PEFLEY

Ventilation of an enclosed space is usually for the purposes of controlling the air temperature, humidity, and/or chemical composition. When humans occupy the space (room), the ventilation is necessary to provide adequate breathing oxygen and to transport the waste heat and moisture produced by the occupants. Normally, the waste heat and moisture far exceed the breathing oxygen ventilation requirement.

The heat and moisture release rates by the occupants are functions of their activity level, age, size, and sex. The fraction of energy associated with heat transferred to the air by convection and surrounding surfaces by radiation is called sensible heat while that associated with moisture evaporation is called latent heat. These fractions vary, depending on activity level and environmental state. In addition to the occupant heat and moisture load there may be an equipment and lighting load ventilation requirement.

It is normally assumed that the room air is homogeneous and is the same as the room air exhaust state.

<u>Basic Refrigeration Cycles</u>:

Carnot

$$C.O.P._{hp} = \frac{Q_r}{W_{net}} = \frac{1}{1 - \frac{T_c}{T_h}}$$

$$C.O.P._{ref} = \frac{Q_{abs}}{W_{net}} = C.O.P._{hp} - 1$$

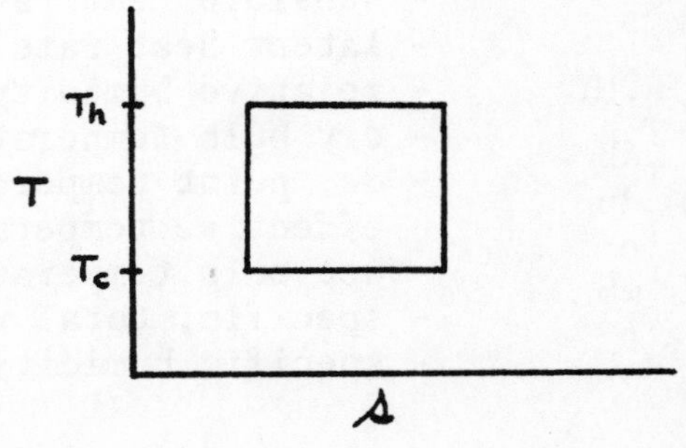

Reference - ASHRAE Handbook of Fundamentals.

Rankine-Vapor

$$C.O.P._{hp} = \frac{Q_r}{W_{net}} = \frac{h_2 - h_3}{h_2 - h_1}$$

$$C.O.P._{ref} = C.O.P._{hp} - 1$$

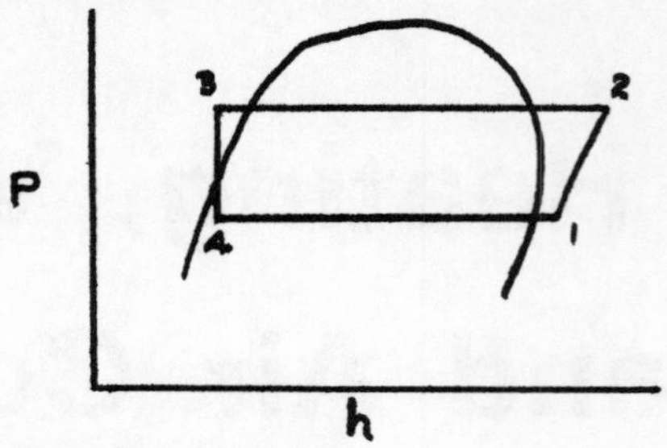

Gas Compression-Expansion Refrigeration Cycle:

$$C.O.P._{hp} = \frac{Q_r}{W_{net}} = \frac{1}{1 - \frac{1}{r_p^{\frac{k-1}{k}}}}$$

$$C.O.P._{ref} = C.O.P._{hp} - 1$$

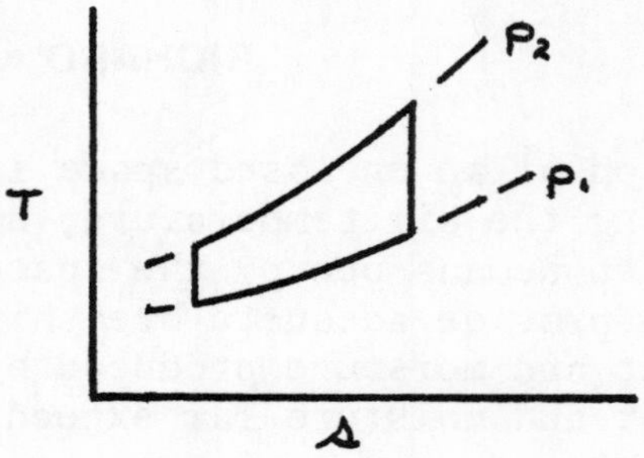

Human Factors and Constants:

Sedentary heat output rate	- 400 Btu/hr-person
Normal sedentary fresh air requirement	- 10 cfm/person
Minimum fresh air breathing requirement	- 3 cfm/person
Nominal comfort state ($q_s/q_\ell \simeq 3$)	- $75°F_{db}$ $60°F_{wb}$
1 ton refrigeration = 12000 Btu/hr	

Terminology:

c, c_p, c_v - specific heats
$C.O.P._{hp}$ - coefficient of performance of a heat pump
$C.O.P._{ref}$ - coefficient of performance of a refrigerator
h_a, H - enthalpy of air water mixture per lbm dry air, enthalpy or head
$m, \dot{m}$ - mass, mass flow rate
P - pressure
q_s - sensible heat rate
q_ℓ - latent heat rate
R.H. - relative humidity
T_{db} - dry bulb temperature
T_{dp} - dew point temperature
T_e - effective temperature
T_{wb} - wet bulb temperature
v, V - specific, total volume
w - specific humidity

Psychrometric Chart

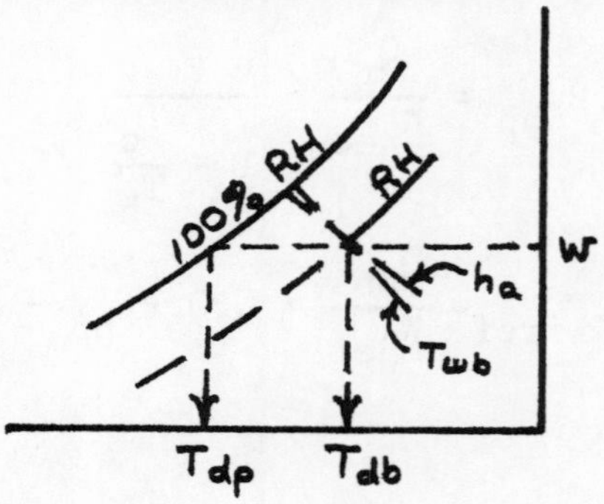

Subscripts:

wv - water vapor
O.A. - outside air
R.A. - room, exhaust air
M.A. - adiabatically mixed air
S.A. - supply air
C.A. - conditioned air
e - exit, exhaust
i - initial, in
s - steam
a,da - air, dry air

HEATING, VENTILATING AND AIR CONDITIONING 1

A steam line at 100 psia carrying dry saturated steam develops a leak as it passes through a sealed room, 20 ft by 20 ft by 12 ft. The air in the room is initially at 70°F, 50% relative humidity, 14.7 psia. The steam leaks into the room at the rate of 10 lbs per hour, diffuses thoroughly.

Assuming no heat exchange between the room and its surroundings, what is the relative humidity in the room at the end of one hour?

Solution

Treating room as a control volume and applying 1st Law:

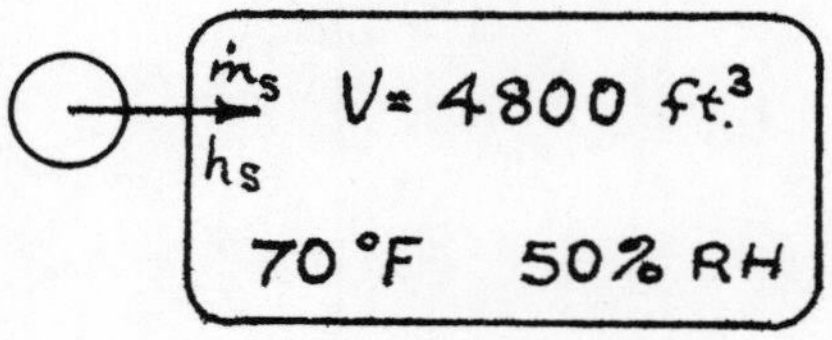

$$m_s h_s = m_s u_f + m_a u_f - m_a u_i$$

By rearrangement:

$$m_s(h_s - u_f) = m_a(u_f - u_i) \quad \text{or} \quad m_s[h_s - h_f + (pv)_f] = m_a(u_f - u_i)$$

Assume steam and air behave as perfect gases

$$m_s[c_p(T_s - T_f) + RT_f] = m_a c_v(T_f - T_i)$$

Let: $c_{p_s} = 0.45 \frac{\text{Btu}}{\text{lbm °F}}$, $R_s = 0.11 \frac{\text{Btu}}{\text{lbm °F}}$, $c_{v_a} = 0.17 \frac{\text{Btu}}{\text{lbm °F}}$

$$m_a = \frac{V}{v} = \frac{4800 \text{ ft}^3}{13.5 \text{ ft}^3/\text{lbm}_{da}} = 356 \text{ lbm}_{da}$$

Substituting values:

$$10[0.45(787.8 - T_f) + 0.11\, T_f] = 356 \times 0.17(T_f - 530)$$

$$T_f = 556 \text{ R or } 96 \text{ F}$$

Initially, $m_{wv} = \omega \frac{V}{v} = 0.0078 \times 356 = 2.78$ lbm

-continued-

Finally $m_{wv} = 10 + 2.78 = 12.78$ lbm

$$\text{Then,}\quad p_{wv} = \frac{mRT}{V} = \frac{12.78 \times 86 \times 556}{4800 \times 144} = 0.885 \text{ psia}$$

$$RH = \left.\frac{p_{wv}}{p_{wv_{sat}}}\right]_T = \frac{0.885}{0.840} \quad \text{(cannot be)}$$

∴ air will be saturated and some water will be in liquid form. T_{room} will be somewhat higher, because vapor will have condensed.

HEATING, VENTILATING AND AIR CONDITIONING 2

A cooling tower is to supply 3500 kw of cooling under the following conditions. Air enters at 100°F DB and 70°F WB and leaves at 83°F DB 95% RH. Recirculating water enters at 105°F and leaves at 80°F. Make up water temperature is 65°F. Assume a reasonable figure for blowdown and windage losses.

Determine:

(a) The total air horsepower if the expected draft loss through the described tower is 0.40" H_2O excluding the fan, and the net fan disc area is 140 ft^2.

(b) The rate of gallons of make up water required under the above conditions.

Solution

Assume steady state cooling tower.

$\Sigma \dot{m}e)_{in} = 0$

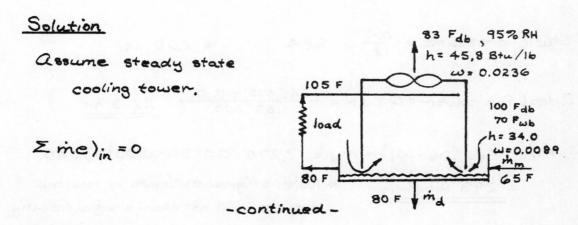

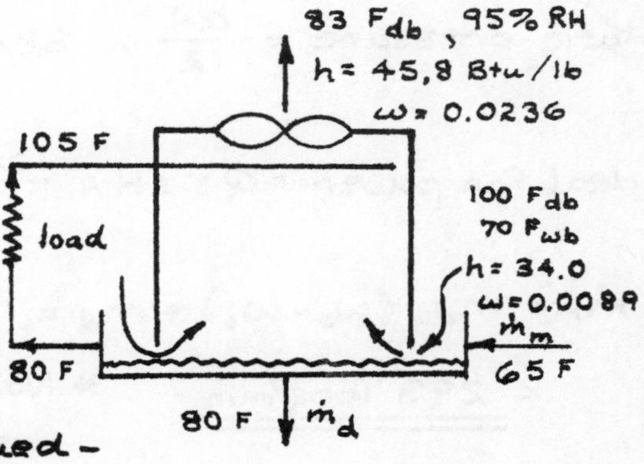

-continued-

$$\dot{m}_{da}(h_i - h_e) + \dot{m}_c(h_i - h_e) + \dot{m}_m h_i - \dot{m}_d h_e = 0$$

From Load,

$$\dot{m}_c = \frac{3500 \times 3413}{60 \times 1 \times 25} = 7960 \text{ lbm/min}$$

Since cooling comes essentially from evaporation, the make-up water can be estimated:

$$h_{fg}\,\dot{m}_m = \text{load} \quad \dot{m}_m = \frac{3500 \times 3413}{60 \times 1000} = 200 \text{ lbm/min}$$

Allowing 1/2% of circulation rate for blow down,

$$\dot{m}_d = 40 \text{ lbm/min}$$

Then,

$$\dot{m}_{da}(h_e - h_i) = \dot{m}_c C(T_i - T_e) + \dot{m}_m C(T_m - 32F) - \dot{m}_d C(T_d - 32F)$$

$$\dot{m}_{da}(45.8 - 34.0) = \text{load} + 200 \times 1(65 - 32) - 40 \times 1(80 - 32)$$

$$\dot{m}_{da} = 17{,}270 \text{ lbm/min} \quad \text{cfm} = 17{,}270 \times 14.2 = 2.45 \times 10^5 \text{ cfm}$$

$$V_e)_{\text{through fan}} = \frac{\text{cfm}}{60\,A} = 29.2 \text{ ft/sec}$$

$$\text{Dynamic pressure} = \frac{V^2}{2 g_c v} = \frac{851}{64.4 \times 14.2} = 0.93 \text{ psf}$$

$$\text{Static pressure} = \frac{0.4}{12} \times 62.4 = 2.08 \text{ psf}$$

$$\text{Ideal fan power} = Q\gamma\Delta H = \frac{245{,}000 \times 3.01}{33{,}000} = \underline{\underline{22.3 \text{ hp}}}$$

$$\dot{m}_m = \dot{m}_{da}(\omega_o - \omega_i) + \dot{m}_d = 17{,}270\,(0.0236 - 0.0089) + 40$$

$$= \underline{\underline{293 \text{ lbm/min}}}$$

* Note: original estimate is low but results will not change significantly

HEATING, VENTILATING AND AIR CONDITIONING 3

A room which is to be maintained at 76°F DB and 40% RH has an exterior wall consisting of exterior stucco on 1/2" gypsum sheathing with metal lath and plaster on the interior, all supported by two by four studding 16 inches on center. There are two 3'-0" by 4'-6" single glazed windows. Determine the external temperature at which condensation will begin to appear on the interior.

Solution

Since glass offers least resistance to heat transfer, its surface will be first to condense moisture.

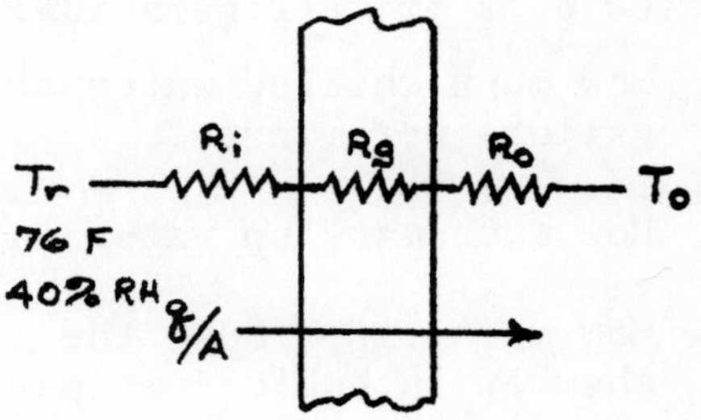

Room $T_{dp} = 49°F$

When inner surface of glass reaches 49°F, condensing will occur.

Assume: $h_i = 1.0 \frac{Btu}{hr\ ft^2\ °F}$ $\qquad R_i = \frac{1}{h_i} = 1.0 \frac{hr\ ft^2\ °F}{Btu}$

$h_o = 2.5 \frac{Btu}{hr\ ft^2\ °F}$ $\qquad R_o = \frac{1}{h_o} = 0.4 \frac{hr\ ft^2\ °F}{Btu}$

$k_{glass} = 0.40 \frac{Btu}{hr\ ft^2\ °F}$ $\qquad R_g = \frac{L}{k} = \frac{0.25}{12 \times .40} = 0.052 \frac{hr\ ft^2\ F}{Btu}$

glass thickness = 1/4"

no radiation, steady state

$$q/A = \frac{T_r - T_o}{R_i + R_g + R_o} = \frac{T_r - T_{dp}}{R_i}$$

$$\frac{T_r - T_o}{1.45} = \frac{T_r - T_{dp}}{1.0}$$

$$T_o = 1.45\ T_{dp} - 0.45\ T_r = 71 - 34.2 = \underline{\underline{36.8\ F}}$$

HEATING, VENTILATING AND AIR CONDITIONING 4

A vacuum refrigeration system consists of a large insulated flash chamber kept at low pressure by a steam ejector which pumps vapor to a condenser. Condensate is removed by a condensate pump, and an air ejector discharges air from the condenser to an air vent. Warm return water enters the flash chamber at 55°F; chilled water comes out of the flash chamber at 40°F. Vapor leaving the flash chamber has a quality of 0.97, and the temperature in the condenser is 90°F.

For 100 tons of refrigeration:

(a) How much chilled water at 40°F does this system provide in gallons per minute?

(b) How much make up water is needed in pounds per minute?

(c) How much vapor must the steam ejector remove from the flash chamber in cubic feet per minute?

<u>Solution</u>

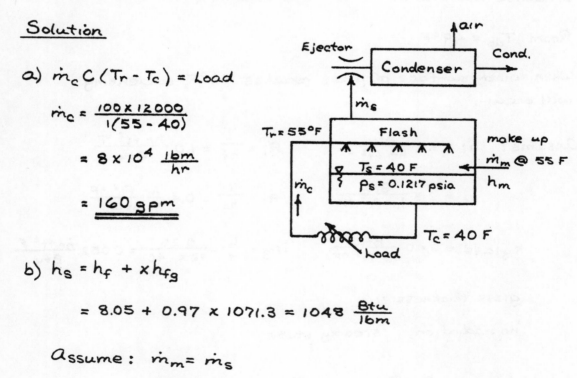

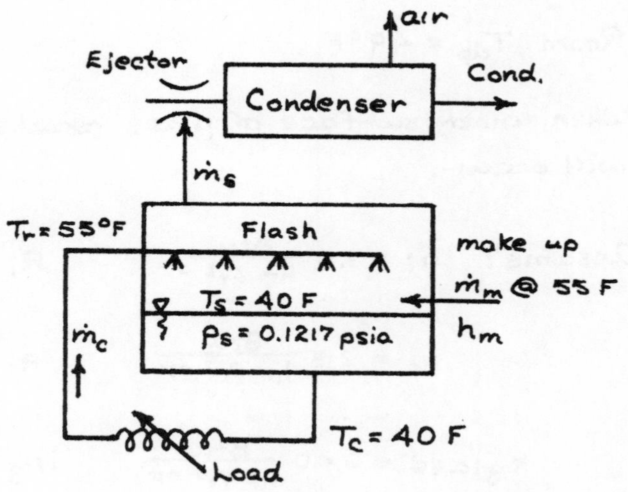

a) $\dot{m}_c C (T_r - T_c) = \text{Load}$

$$\dot{m}_c = \frac{100 \times 12000}{1(55-40)}$$

$$= 8 \times 10^4 \ \frac{\text{lbm}}{\text{hr}}$$

$$= 160 \text{ gpm}$$

b) $h_s = h_f + x h_{fg}$

$$= 8.05 + 0.97 \times 1071.3 = 1048 \ \frac{\text{Btu}}{\text{lbm}}$$

Assume: $\dot{m}_m = \dot{m}_s$

$T_m = 55$ F

From 1st Law applied to flash chamber,

$$\dot{m}_s (h_s - h_m) = \dot{m}_c (T_r - T_c) = \text{Load}$$

$$\dot{m}_s = \frac{1.2 \times 10^6}{(1048 - 23)} = 1.16 \times 10^3 \text{ lbm/hr}$$

c) $v_s + v_f + x v_{fg} \simeq x v_{fg} = 0.97 \times 2444 = 2360\ ft^3/lbm$

$$Q_s = \frac{\dot{m}_s v_s}{60} = \frac{1.16 \times 10^3 \times 2.36 \times 10^3}{60} = \underline{\underline{4.55 \times 10^4\ cfm}}$$

HEATING, VENTILATING AND AIR CONDITIONING 5

An air conditioning system serving a hospital operating room requires 100% outside air as a safety precaution, in view of the hazard relating to the use of anesthetics. Outside design is 100°F DB and 79°F WB, while inside design is 80°F DB and 65°F WB. The air is supplied to the room at 65°F DB to minimize chilling drafts on the patient. The dehumidifying coil is supplied with the chilled water and controlled so as to provide a surface temperature of 50°F.

The operating room is a separate interior zone on the second floor of a three-story completely air conditioned building. Ten persons are present when the room is in use. There are ten kilowatts of lighting, and surgical and medical equipment add 12,600 Btuh latent, and 23,900 Btuh sensible load.

(a) What CFM of air should be supplied the room?

(b) What wet bulb temperature of the supply air as it leaves the dehumidifier would maintain the desired room condition?

(c) What is the coil by-pass factor?

(d) What should be the wet bulb temperature of the supply air to the room?

(e) What is the cooling load on the coil, in tons of refrigeration?

(f) Under design conditions, how many gallons of water must be supplied if the temperature difference between water entering and leaving is 10°F for chilled water and 20°F for hot?

Note: Base all calculations on standard air. An accurate graphic solution will be acceptable.

Solution

assume:

1) human load @ 400 Btuh/person, $q_l = q_s$ due to extra clothing.

– continued –

2) Air leaves dehumidifier saturated at apparatus temp.

*Sensible load

34,130
23,900
2,000
60,030 Btu/hr

*Latent load

2,000
12,600
14,600 Btu/hr

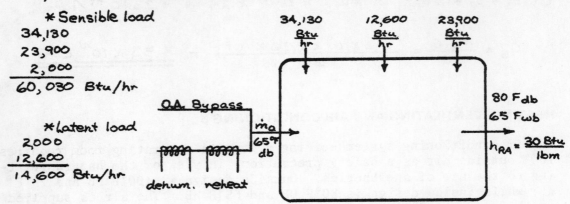

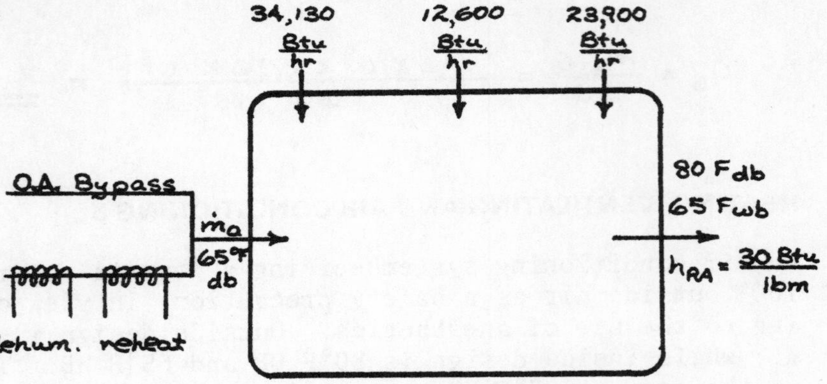

$$\frac{Q_{sensible}}{Q_{total}} = \frac{60{,}030}{74{,}630} = 0.805$$

Using psychrometric protractor, exit state and inlet dry bulb temperature locates inlet state as 65°F$_{db}$ and 69% RH, $H_i = 25.5 \frac{Btu}{lbm}$.

a) $Q/v\,(h_{RA} - h_{SA})$ = total load $\quad Q = \dfrac{74{,}630 \times 13.4}{60\,(30 - 25.5)} = \underline{\underline{3710\ cfm}}$

b) Assuming air is saturated $T_{wb} = \underline{\underline{50\,°F}}$

c) $(\dot{m}\Delta h)_{OA} = (\dot{m}\Delta h)_{coil} \quad \dfrac{\dot{m}_{coil}}{\dot{m}_{OA}} = \dfrac{42.6 - 25.5}{25.5 - 22.6} = \underline{\underline{5.90}}$

∴ 14.5% of supply air bypasses humidifier cell.

d) $\underline{\underline{58.5\,°F}}$

e) Ref. load $= \dot{m}_{coil}\,(h_{OA} - h_{SA}) = \dfrac{3710 \times 0.86}{13.4}(42.6 - 20.4)\left(\dfrac{60}{12000}\right)$

$= \underline{\underline{26.5\ tons}}$

f) $\dot{m}_{water\ hum.}\, c\,(\Delta T)$ = Ref. load $\quad \dot{m}_{water\ hum.} = \dfrac{26.5 \times 12000}{1 \times 10 \times 60} = 530 \dfrac{lbm}{min}$

$= \underline{\underline{63.0\,gpm}}$

$\dot{m}_{water\ reheat}\, c\,(\Delta T) = \dot{m}_{coil}\, C_p\, \Delta T \quad \dot{m}_{water\ reheat} = \dfrac{3710 \times 0.86 \times 0.24\,(59 - 50)}{13.4 \times 20 \times 8.4}$

$= \underline{\underline{3.1\ gpm}}$

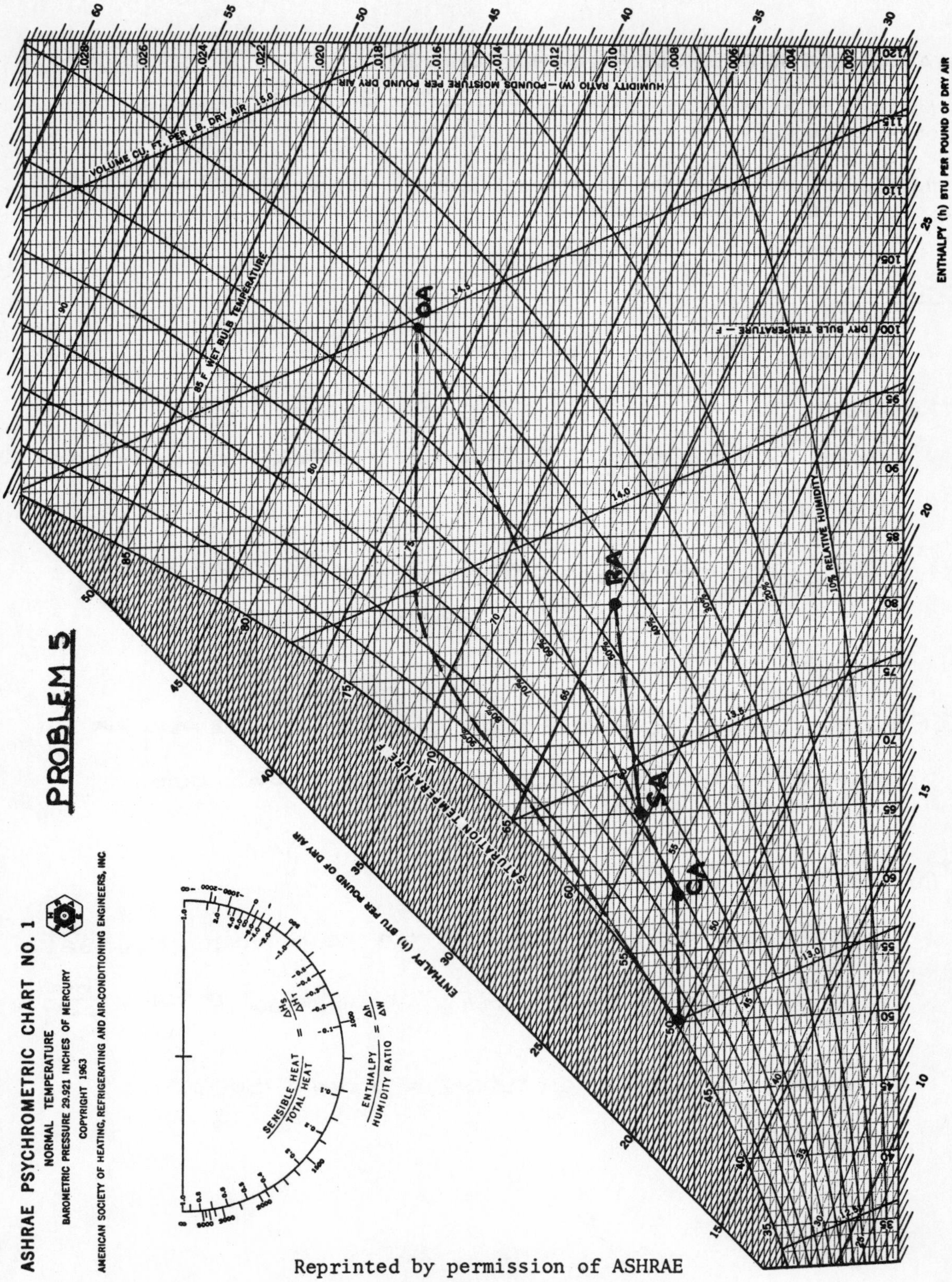

Reprinted by permission of ASHRAE

HEATING, VENTILATING AND AIR CONDITIONING 6

A cubic yard of standard air with dew point of 50°F is compressed isothermally until its volume is one cubic foot.

Find the dew point of the compressed air and the weight of water vapor, if any, that is condensed.

Solution

Assume T_{db} std. air is 70 F

Therefore,

have 27 ft^3 air @ 70 F_{db} 50 F_{dp} (50% RH) compressed to 1 ft^3 @ T=C.

Now, $p_{wv} = RH\, p_{sat} = 0.50 \times 0.363 \text{ psia} = 0.182 \text{ psia}$

$$m_{wv} = \frac{V}{v}\omega = \frac{27}{13.5} \times 0.0078 = 0.0156 \text{ lbm wv}$$

If there were no condensation $p_{wv} = 27 \times 0.182$ psia on compression, but p_{wv} cannot exceed 0.363 psia.

Therefore,

condensation will occur. From steam tables $m_{wv} = \frac{1}{868}$

$$m_{wv} = 0.00115 \frac{\text{lbm}}{\text{ft}^3}$$

Condensate = 0.0156 − 0.0012 = 0.0144 lbm

HEATING, VENTILATING AND AIR CONDITIONING 7

A 10,000 square foot church auditorium seating 1000 people is to be air conditioned. Outside design is 95°F DB and 76°F WB, inside design is 76°F DB and 64°F WB. Transmission load is 400,000 Btu/hr; lighting averages 3 watts per square foot; supply air at 55°F is to be 25% outside air.

Calculate:

(a) Sensible, latent and total heat loads.

(b) Tons of refrigeration and supply cfm.

(c) Coil entering and leaving conditions.

(d) Plot appropriate work on psychrometric chart. (Ask the proctor for a copy if you wish to work this problem.)

<u>Solution</u>

Assume for humans, $q_s/q_e = 3$

a)

Sensible Load	Transmission load		400,000 Btuh
	Lighting	$3 \times 3.413 \times 10^4$	102,000 Btuh
	Human	1000×300	300,000 Btuh
			802,000 Btuh
Latent Load	Human	1000×100	100,000 Btuh
		Total	902,000 Btuh

$$q_s/q_t = 0.89$$

b) Assume outside air & return air mixed ahead of coil. Then air to coil $T_{db} = 81$ F $T_{wb} = 67^{+}$ F (see psyc. chart) Room process line known from q_s/q_t.

– continued –

∴ Since T_{db} supply air = 55 °F, it appears that supply air is approximately saturated.

$$\dot{m}_{SA}(h_{RA} - h_{SA}) = \text{Load}$$

$$\dot{m}_{SA} = \frac{902,000}{29.2 - 23.4} = 156,000 \text{ lbda/hr}$$

$$\therefore \dot{m}_{OA} = 39,000 \text{ lbda/hr or } \underline{\underline{9300 \text{ cfm}}}$$

$$\dot{m}_{RA} = 117,000 \text{ lbda/hr or } \underline{\underline{26,700 \text{ cfm}}}$$

$$\text{Ref. Load} = \dot{m}_{SA}(h_{Ret.A} - h_{SA}) = \frac{1.56 \times 10^5 (31.8 - 23.4)}{12000}$$

$$= \underline{\underline{109 \text{ tons}}}$$

c) Ret. A T_{db} = 81 F T_{wb} = 67 F

SA. approximately sat. @ 55 F

d) see chart

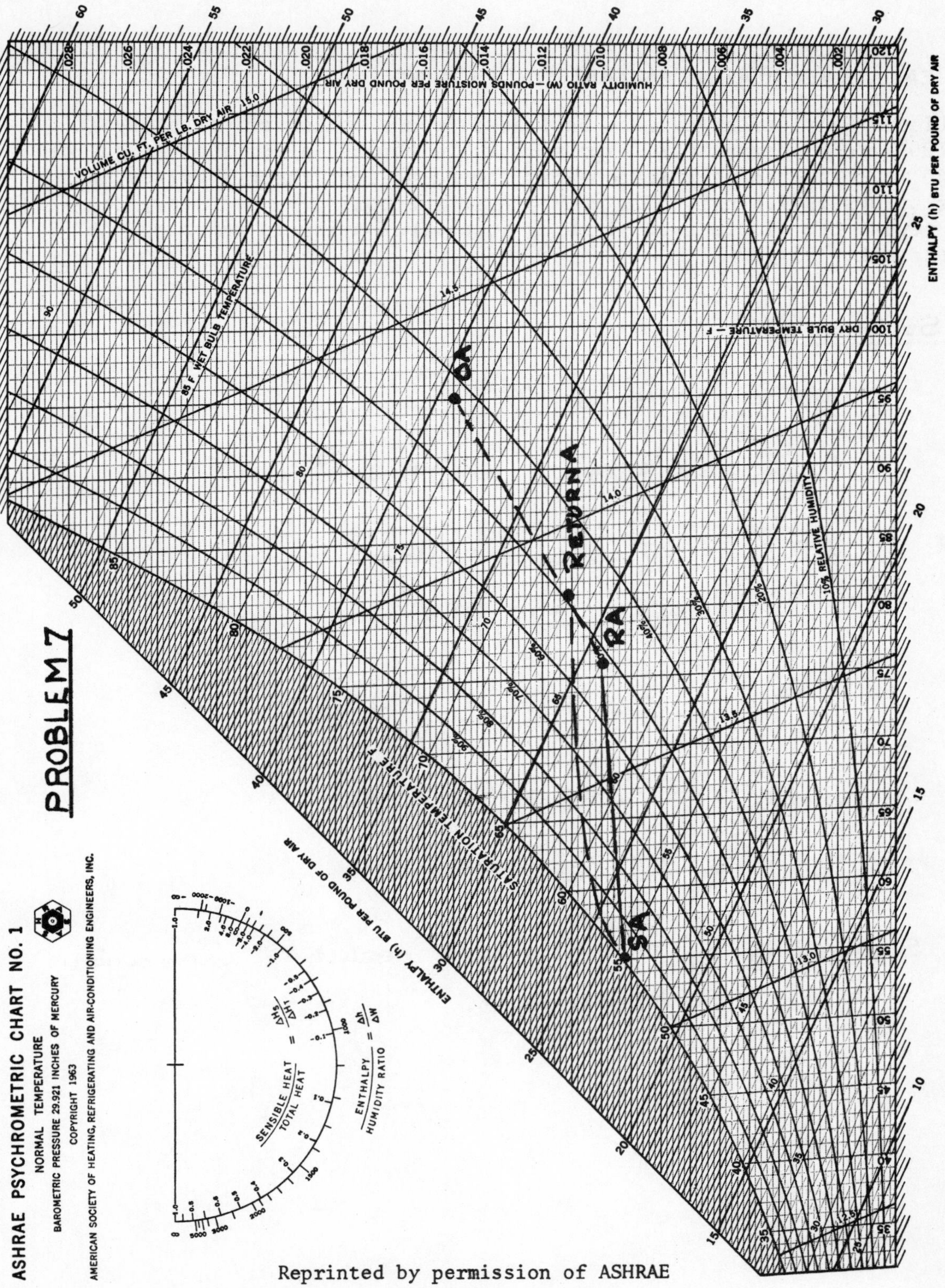

Reprinted by permission of ASHRAE

HEATING, VENTILATING AND AIR CONDITIONING 8

Liquid Refrigerant 12 must be lifted from a condenser operating at 100°F to a thermostatic expansion valve through a vertical height of 40 ft. The piping friction loss including that through valves and fittings is 7.5 psi. How much liquid subcooling is necessary to prevent the liquid from flashing into vapor prior to entering the expansion valve?

Solution

For refrigerant 12 @ 100°F:

$$P_{sat} = 131.9 \text{ psia} \qquad \rho = 78.8 \text{ ft}^3/\text{lbm}$$

$$\Delta P_{Total} = \Delta P_{friction} + \Delta P_{elevation}$$

$$= 7.5 + \frac{40 \times 78.8}{144}$$

$$= 7.5 + 21.9 = 29.4 \text{ psi}$$

$$\therefore P_{valve} = 131.9 - 29.4 = 102.5 \text{ psia}$$

$$T_{sat} = 82.5\ ^\circ\text{F} \ @\ 102.5 \text{ psia}$$

Subcooling by 17.5 F should result in saturation state at throttling valve.

HEATING, VENTILATING AND AIR CONDITIONING 9

A room is to be held at 80°F DB and a maximum of 50% RH when the outside temperature is 96°F DB and 77°F WB. The cubical contents of the room equals 120,000 cubic feet. The internal head load including transmission, occupants, lights, etc., is 286,000 Btu sensible heat and 15,000 Btu latent heat per hour. It is desired to provide one and one half air changes per hour of fresh air; 18,000 cfm will be circulated through the cooling coil under maximum load conditions. The coil will have face and by-pass dampers. The cooling coil will be supplied with chilled water entering at 46°F and leaving at 54°F; assume the characteristics of the coil such that the coil temperature is the average between the entering water and leaving wet bulb temperature.

Determine:

(a) The DB and WB entering the coil.

(b) The required DB and WB leaving the coil.

Solution

Initially assume outside air and return air are premixed in the proportions shown in the figure. Then check to see if resulting air supply state is a possible state from coil. If not, adjust accordingly.

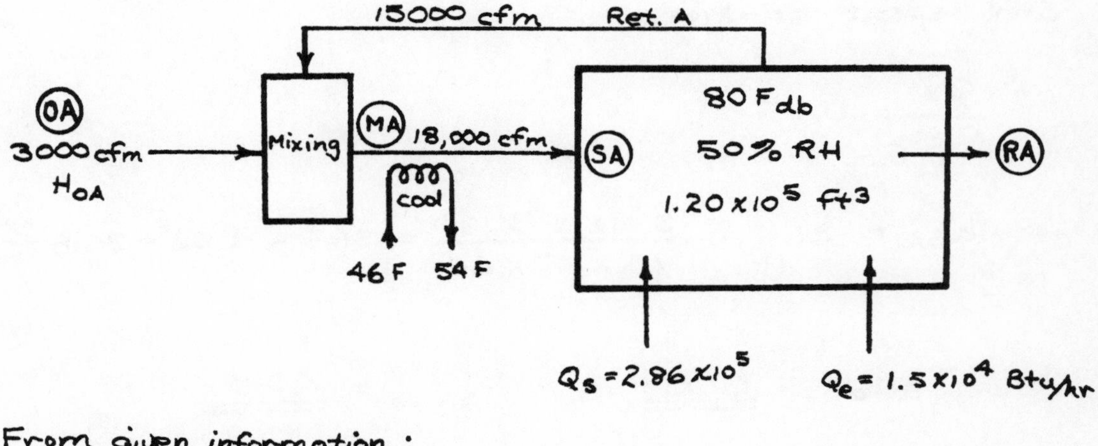

From given information:

$$\frac{q_s}{q_t} = \frac{2.86 \times 10^5}{3.01 \times 10^5} = 0.95$$

–continued–

$$q_t = \frac{cfm}{v}(h_{RA} - h_{SA}) \quad \therefore h_{SA} = h_{RA} - \frac{q_t v}{cfm}$$

$$h_{SA} = 31.4 - \frac{3.01 \times 10^5 \times 13.7}{1.8 \times 10^4 \times 60} = 31.4 - 3.82 = 27.6 \frac{Btu}{lbm}$$

$\therefore$ for $q_s/q_t = 0.95$ & $h_i = 27.6$ Btu,

$T_{SA\,db} = 66.5$ F $\quad T_{SA\,wb} = 61.5$ F

Since $\frac{OA}{Ret.A} = \frac{3000}{1500} = \frac{1}{5}$, use psychrometric chart

to find $T_{MA\,db} = 83.0$ F $\quad T_{MA\,wb} = 68.6$ F

For a chill coil temperature of 50°F, the process line will be as shown on psychrometric chart. For supply air to originate on this line means that air supply rate must be reduced by about 20%. Therefore, reduce return air to about 12,000 cfm and repeat solution.

$$\frac{q_s}{q_t} = 0.95$$

$$h_{SA} = 31.4 - \frac{3.01 \times 10^5 \times 13.7}{1.5 \times 10^4 \times 60} = 31.4 - 4.58 = 26.8 \frac{Btu}{lbm}$$

b) $T'_{SA\,db} = \underline{\underline{61.5 \text{ F}}}$ $\quad T'_{SA\,wb} = \underline{\underline{60.0 \text{ F}}}$

a) $\frac{OA}{Ret.\ A} = \frac{3000}{12000} = \frac{1}{4}$ $\quad T'_{MA\,db} = \underline{\underline{84.5 \text{ F}}}$ $\quad T'_{MA\,wb} = \underline{\underline{69.5°\text{F}}}$

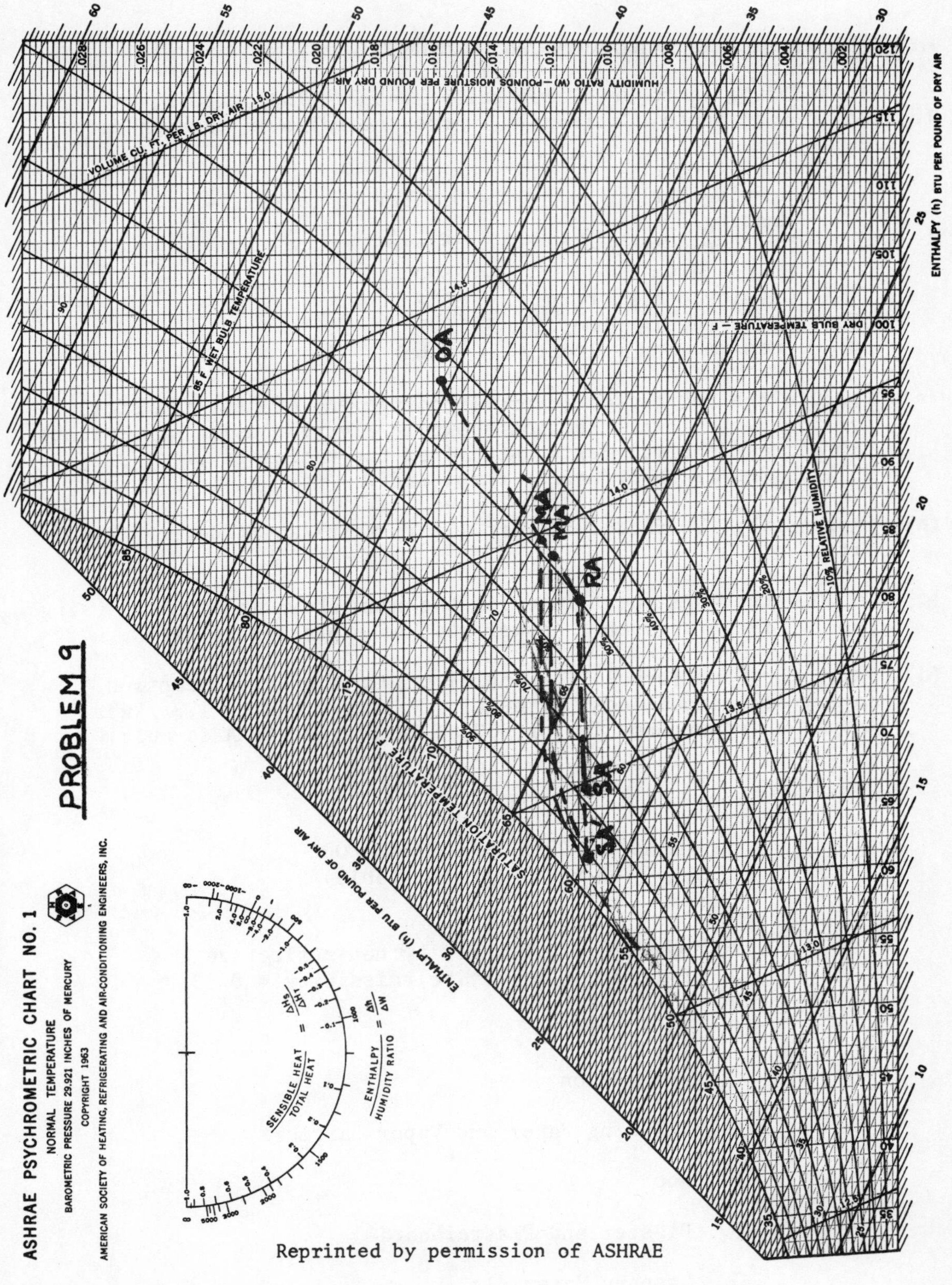

Reprinted by permission of ASHRAE

HEATING, VENTILATING AND AIR CONDITIONING 10

A small building is to be electrically heated. The building is located in Sacramento, California, where the following design data applies:

Heating season	- October 16 to May 14
Heating days	- 211
Avg. mean temp.	- 52.7°F
Degree days	- 2600
Outside design temp.	- 30°F
Indoor design temp.	- 70°F
Outside wind velocity	- 15 mph

(a) Compute the "U" factor (overall coefficient of heat transfer) for the typical wall section shown in Figure 1 using appropriate "R" values (thermal resistance) from the tables given. Units: Btu/hr/sq ft/degree temp. diff.

(b) Express the "U" factor of Part (a) in watts/sq ft/degree temp. diff.

(c) Compute the total heat loss for the room shown in Figure 2 using the "U" factors given. Answer must be expressed in watts.

(d) Using the NEMA formula for annual kilowatt hour consumption, compute the annual cost of heating. Energy cost: 1.5¢/kwhr. Assume an experience factor "C" of 14 for the NEMA formula.

THERMAL RESISTANCE TABLE CODE
(see next page for table)

A. Air Spaces (bounded by non-reflective materials, effective emissivity = 0.82

B. Air Surfaces

C. Insulation

D. Building Paper and Vapor Barriers

E. Wood

F. Plaster and Plasterboard

G. Masonry Materials

A.

Position of air space	Direction of heat flow	Thickness of air space (in.)	Resistance (R)
Horizontal	Up	¾ to 4	0.85
	Down	¾	1.02
		1½	1.15
		4	1.23
		8	1.25
Sloping (45°)	Up	¾ to 4	0.90
Vertical	Horizontal	¾ to 4	0.86

*Bounded by non-reflective materials, effective emissivity = 0.82

B.

Wind speed (mph)	Position of surface	Direction of heat flow	Resistance (R)
No wind; still air	Horizontal	Up	0.61
		Down	0.92
	Sloping (45°)	Up	0.62
	Vertical	Horizontal	0.68
7½	Any position	Any direction	0.25
15	Any position	Any direction	0.17

C.

Type	Material	Resistance (R) per inch thickness
Batts or blankets	Cotton fibre	3.85
	Mineral wool	3.70
	Wood fibre, multilayer	3.70
Loose fill	Macerated paper or pulp products	3.57
	Mineral wool	3.33
	Sawdust or shavings	2.22
	Expanded vermiculite	2.08
	Wood fibre (redwood, hemlock, fir)	3.33
Insulating board or slab	Glass fibre	4.00
	Cellular glass	2.50
	Corkboard (without binder)	3.70
	Hog hair (with asphalt binder)	3.00
	Foamed plastic	3.45
	Shredded wood (cemented)	1.82

D.

Material	Resistance (R)
Vapor-permeable felt	0.06
Vapor seal, 2 layers of mopped 15-lb felt	0.12
Vapor seal, plastic film	Negligible

E.

Material		Thickness of heat path (inches)	Resistance (R)
Wood framing (soft wood)	1 x 2	25/32	0.98
	2 x 2	1 5/8	2.03
	2 x 3	2 5/8	3.28
	2 x 4	3 5/8	4.53
	2 x 6	5 5/8	7.03
	2 x 8	7 5/8	9.53
	2 x 10	9 5/8	12.03
	2 x 12	11 5/8	14.53
Plywood		1/4	0.31
		3/8	0.47
		1/2	0.63
		5/8	0.78
		3/4	0.94
Hardboard (fibreboard)		1/4	0.18
Soft wood (fir, pine, etc.)		25/32	0.98
		1	1.25
Hard wood (maple, oak, etc.)		25/32	0.71
		1	0.91

F.

Material	Thickness (inches)	Resistance (R)
Cement plaster, sand aggregate	1/2	0.10
	3/4	0.15
Gypsum plaster, lightweight aggregate	1/2	0.32
	5/8	0.39
	3/4	0.47
Gypsum board (sheetrock; plasterboard)	3/8	0.32
	1/2	0.45

G.

Material		Density (lb/cu ft)	Resistance (R) per inch thickness
Concrete	Sand and gravel or stone aggregate	140	0.08
	Gypsum fibre	51	0.60
	Lightweight aggregate	100	0.28
		80	0.40
		60	0.59
		40	0.86
		20	1.43
Cement mortar		116	0.20
Stucco		116	0.20
Brick	Common	120	0.20
	Face	130	0.11
Stone, lime, and sand			0.08

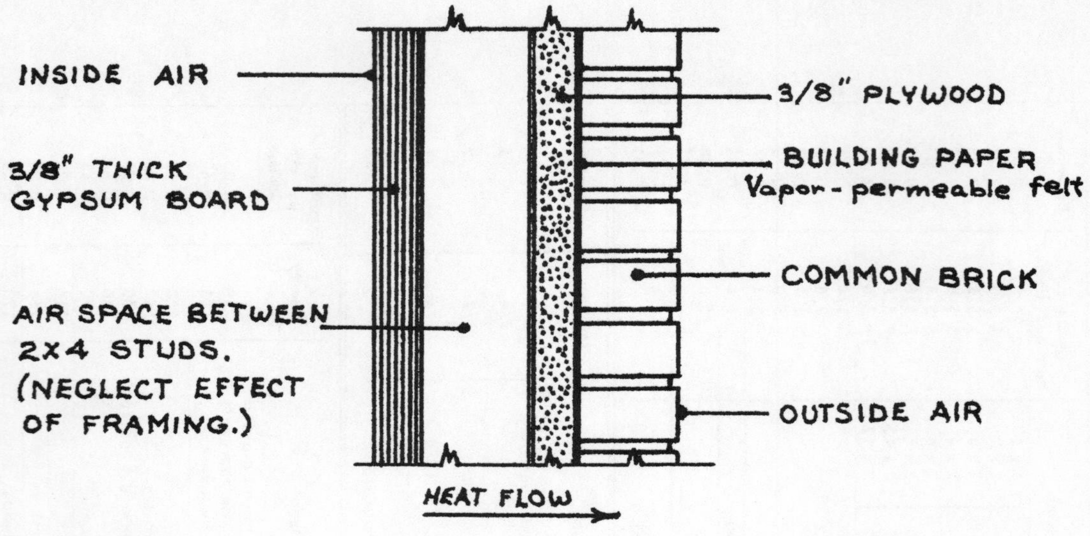

FIGURE 1 TYPICAL WALL SECTION N.T.S.

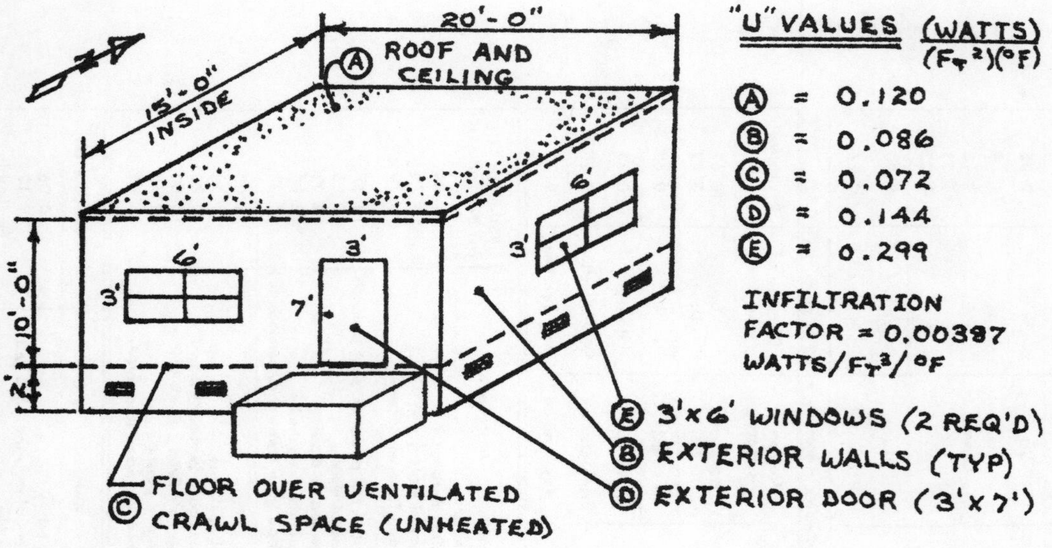

FIGURE 2 TYPICAL ROOM N.T.S.

SPACE HEATING PROBLEM

Solution

a)

$$R_{c,i} = 0.68$$
$$R_{gyp} = 0.32$$
$$R_{air} = 0.86$$
$$R_{ply} = 0.47$$
$$R_{p} = 0.06$$
$$R_{b} = 0.60$$
$$R_{c,o} = 0.17$$

$$\Sigma R = 3.16 \qquad U = \frac{1}{\Sigma R} = 0.32 \ \frac{\text{Btu}}{\text{hr ft}^2\ {}^\circ\text{F}}$$

b) $0.32/3.41 = 0.094 \ \frac{\text{watts}}{\text{ft}^2\ {}^\circ\text{F}}$

c) Total wall area = 700 ft^2 $\quad A_{windows} = 36\ ft^2$

$A_{door} = 21\ ft^2 \quad A_{wall,net} = 643\ ft^2$

Infiltration $= 1.5\ Vol._{room} = 1.5 \times 3000 = 4500\ ft^3/hr$

$$q = \Sigma UA\Delta T$$

$$q_A = 0.120 \times 300\ \Delta T = 36.0\ \Delta T$$
$$q_B = 0.086 \times 643\ \Delta T = 55.3\ \Delta T$$
$$q_C = 0.072 \times 300\ \Delta T = 21.6\ \Delta T$$
$$q_D = 0.144 \times 21\ \Delta T = 3.0\ \Delta T$$
$$q_E = 0.299 \times 36\ \Delta T = 10.8\ \Delta T$$
$$q_{inf.} = 4.5 \times 10^3 \times 3.87 \times 10^{-3} = 17.4\ \Delta T$$
$$144.1\ \Delta T$$

$$q = 144.1(70-30) = 5760 \text{ watts}$$

d) NEMA formula $\quad Kwh = \frac{\text{heat load (KW)} \times \text{degree day} \times C}{\text{Temp. Diff.}}$

$$Kwh = \frac{5.76 \times 2600 \times 14}{40} = 5{,}250$$

$$\text{Cost} = 5{,}250 \times .015 = \$78.70$$

HEATING, VENTILATING AND AIR CONDITIONING 11

The duct system for an office is shown below:

The velocity of the air entering at A is 1800 feet per minute. Total quantity of air entering A is 6000 cubic feet per minute. The static pressure regain is 50%.

(Note: Use the equivalent lengths (includes elbows and fittings) listed in the table for the calculations.)

Static pressure regain is the velocity pressure at the fan outlet minus the velocity pressure at the end of the system.

REQUIRED: (Using the equal friction method)

(a) From the information given in the table below, determine the friction loss in inches water gage per 100 feet of equivalent length.

(b) Determine the following for each section in the table below:

(c) Calculate the static pressure in inches water gage for points A, B, C, D, E. F, G in the duct layout using the equivalent lengths.

(d) What is the total static pressure in inches water gage required by the duct system when the pressure static regain is considered?

Section	Air Qty. cfm	Equivalent length/ft	Duct Size inches/dia.	Velocity ft/min
A-B	6000	78	______	1800
B-C	______	25	______	______
C-D	______	53	______	______
D-E	______	20	______	______
E-F	______	20	______	______
F-G	______	20	______	______
G-H	______	20	______	______

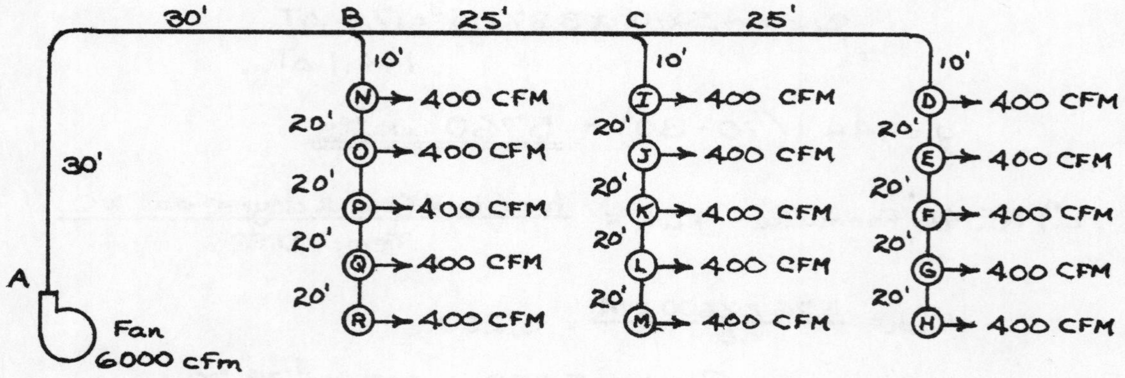

Solution

a) From ASHRAE Handbook "Fundamentals"

$$\Delta p = \underline{\underline{0.16''}} \; H_2O/100 \text{ ft}$$

or $A = Q/v = \frac{6000}{1800}$ $\quad d = 2.05$ ft $\quad$ est. $f = 0.02$

$$\Delta p = f \frac{l}{d} \frac{v^2}{2g} \frac{\rho_{air}}{\rho_{water}} = 0.02 \times \frac{100}{2.05} \times \frac{(30)^2}{64.4} \times \frac{.075}{62.3} \times 12$$

$$= \underline{\underline{0.14'' \; H_2O}}$$

b) Equal friction -- constant $\Delta p_{static}/l$

	Q(cfm)	V(ft/sec.)*	d(in.)*	Eq. l(ft)	Δp ("H_2O)
A-B	6000	1800	25	78	0.124
B-C	4000	1610	23.5	25	0.046
C-D	2000	1360	16.5	53	0.085
D-E	1600	1310	15	20	0.032
E-F	1200	1210	13.5	20	0.032
F-G	800	1100	11.5	20	0.032
G-H	400	910	10	20	0.032

$\Delta p_{total} = 0.377'' H_2O$

* from Handbook

c) $p_A = 0.377 + p_{ref.}$ **

$p_B = 0.253 + p_{ref.}$

$p_C = 0.213 + p_{ref.}$

$p_D = 0.128 + p_{ref.}$

$p_E = 0.096 + p_{ref.}$

$p_F = 0.064 + p_{ref.}$

$p_G = 0.032 + p_{ref.}$

$p_H = 0 + p_{ref.}$

** Static pressure is not given at any point so pref. is not known.

-continued-

$$d)\quad H_{vel,A} = \frac{V_A^2}{2g} \times \frac{\rho_{air}}{\rho_{water}} = \frac{(30)^2}{64.4} \times \frac{.075}{62.3} \times 12 = 0.202''$$

$$H_{vel,H} = 1/4 \times H_{vel,A} = 0.050''$$

$$\Delta P_{static\ recovery} = (0.202 - 0.050)\,0.50 = 0.076''\ H_2O$$

$$\therefore \Delta P_{static,\,net} = 0.377 - .076 = \underline{\underline{0.301''\ H_2O}}$$

12

Engineering Economics

DONALD G. NEWNAN

INTRODUCTION

Engineering economic analysis (often called engineering economy) is a group of techniques for the systematic analysis of alternative courses of action. Here we begin with a detailed review of economic analysis fundamentals, including

Cash Flow	Compound Interest Formulas
Time Value of Money	Nominal and Effective Interest
Equivalence	

An understanding of the fundamentals is prerequisite to the discussion of economic analysis techniques needed for the professional engineer examination. The techniques are grouped as follows.

Present Worth	Benefit-Cost Ratio
Annual Cost	Depreciation
Rate of Return	Income Taxes

Following this, typical professional engineer examination problems are presented together with complete solutions.

CASH FLOW

An initial step in economic analysis problems is to resolve a situation (or each alternative situation) into its favorable and unfavorable consequences. There must be a way of measuring these consequences in some common unit. Typically money is used. A cash flow table shows the money consequences of a situation and its timing.

For example, a simple problem might be to portray the year-by-year consequences of purchasing a used car as shown on the next page.

	Year	Cash Flow	
Beginning of first year	0	-4500	Car purchased "now" for $4500 cash. The minus sign indicates a disbursement.
End of year	1	-350	Maintenance costs are $350 per year.
End of year	2	-350	
End of year	3	-350	
End of year	4	-350 +2000	The car is sold at the end of the 4th year for $2000. The plus sign represents a receipt of money.

This same cash flow may be represented graphically.

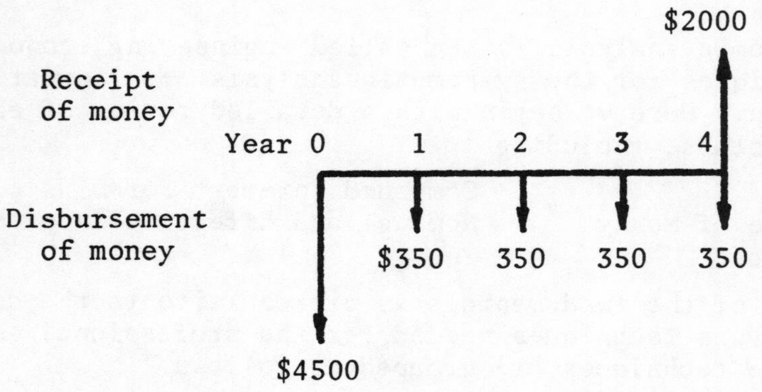

The upward arrow represents a receipt of money, and the downward arrows represent disbursements. The *x*-axis represents the passage of time.

EXAMPLE 1

In January, 19X0 a firm purchased a used typewriter for $500. Repairs cost nothing in 19X0 or 19X1. In 19X2 they are $85; $130 in 19X3; and $140 in 19X4. It is sold in 19X4 for $300.

Compute the cash flow table.

SOLUTION

Unless otherwise stated in problems, the customary assumption is a beginning-of-year purchase, followed by end-of-year receipts or disbursements, and an end-of-year resale or salvage value. Thus the typewriter repairs and the typewriter sale are assumed to occur at the end of the year. Letting a minus sign represent a disbursement of money, and a positive sign a receipt of money, we have

	Year	Cash Flow
Beginning of	19X0	-$500
End of	19X0	0
End of	19X1	0
End of	19X2	-85
End of	19X3	-130
End of	19X4	+160

Notice that at the end of 19X4 the cash flow table shows +160 which is the net sum of -140 and +300. If we define Year 0 as the beginning of 19X0, the cash flow table becomes

Year	Cash Flow
0	-$500
1	0
2	0
3	-85
4	-130
5	+160

From this cash flow table the definitions of Year 0 and Year 1 become clear. Year 0 is defined as the *beginning* of Year 1. Year 1 is the *end* of Year 1. Year 2 is the *end* of Year 2, and so forth.

TIME VALUE OF MONEY

When the money consequences of an alternative occur in a short period of time - say less than one year - we might simply algebraically add up the various sums of money and obtain the net result. But we cannot treat money this same way over longer periods of time. This is because money today is not the same as money at some future time.

Consider this question. Which would you prefer, $100 today or the assurance of receiving $100 a year from now? Clearly you would prefer the $100 today. If you had the money today, rather than a year from now, you could use it for the year. And if you had no use for it, you could lend it to someone who would pay interest for the privilege of using your money for the year.

EQUIVALENCE

In the preceding section we saw that money at different points in time (for example, $100 today or $100 one year hence) may be equal in the sense that they both are $100, but $100 a year hence is not an acceptable substitute for $100 today. When we have acceptable substitutes, we say they are *equivalent* to each other. Thus at 5% interest, $105 a year hence is equivalent to $100 today.

EXAMPLE 2

At a 10% per year interest rate $500 now is *equivalent* to how much three years hence?

SOLUTION

$500 now will increase by 10% in each of the three years.

Now =		$500
End of 1st year =	500 + 10%(500) =	550
End of 2nd year =	550 + 10%(550) =	605
End of 3rd year =	605 + 10%(605) =	665.50

Thus $500 now is *equivalent* to $665.50 at the end of three years.

Equivalence is an essential factor in engineering economic analysis. Suppose we wish to select the better of two alternatives. First, we must compute their cash flows. An example would be

	Alternative	
Year	A	B
0	-$2000	-$2800
1	+800	+1100
2	+800	+1100
3	+800	+1100

The larger investment in Alternative B results in larger subsequent benefits, but we have no direct way of knowing if Alternative B is better than Alternative A. Therefore we do not know which alternative should be selected. To make a decision we must resolve the alternatives into *equivalent* sums so they may be compared accurately and a decision made.

COMPOUND INTEREST FORMULAS

To facilitate equivalence computations a series of compound interest formulas will be derived and their use illustrated.

SYMBOLS

i = interest rate per interest period. In equations the interest rate is stated as a decimal (that is, 5% interest is 0.05).

n = number of interest periods.

P = a present sum of money.

F = a future sum of money. The future sum F is an amount, n interest periods from the present, that is equivalent to P with interest rate i.

A = an end-of-period cash receipt or disbursement in a uniform series continuing for n periods, the entire series equivalent to P or F at interest rate i.

G = uniform period-by-period increase in cash flows; the arithmetic gradient.

FUNCTIONAL NOTATION

	To Find	Given	Functional Notation
Single Payment			
Compound Amount Factor	F	P	(F/P,i%,n)
Present Worth Factor	P	F	(P/F,i%,n)
Uniform Payment Series			
Sinking Fund Factor	A	F	(A/F,i%,n)
Capital Recovery Factor	A	P	(A/P,i%,n)
Compound Amount Factor	F	A	(F/A,i%,n)
Present Worth Factor	P	A	(P/A,i%,n)
Gradient Series			
Gradient Uniform Series	A	G	(A/G,i%,n)
Gradient Present Worth	P	G	(P/G,i%,n)

From the table above we can see that the functional notation scheme is based on writing (To Find/Given,i%,n). Thus if we wished to find the future sum F, given a uniform series of receipts A, the proper compound interest factor to use would be (F/A,i%,n).

SINGLE PAYMENT FORMULAS

Suppose a present sum of money P is invested for one year at interest rate i. At the end of the year we should receive back our initial investment P together with interest equal to Pi or a total amount P+Pi. Factoring P, the sum at the end of one year is P(1+i). If we agree to let our investment remain for subsequent years, the progression is as follows:

	Amount at Beginning of Period	+	Interest for the Period	=	Amount at End of the Period
1st year	P	+	Pi	=	$P(1+i)$
2nd year	$P(1+i)$	+	$Pi(1+i)$	=	$P(1+i)^2$
3rd year	$P(1+i)^2$	+	$Pi(1+i)^2$	=	$P(1+i)^3$
nth year	$P(1+i)^{n-1}$	+	$Pi(1+i)^{n-1}$	=	$P(1+i)^n$

The present sum P increases in n periods to $P(1+i)^n$. This gives us a relationship between a present sum P and its equivalent future sum F.

$$\text{Future Sum} = (\text{Present Sum})(1+i)^n$$
$$F = P(1+i)^n$$

This is the Single Payment Compound Amount Factor.
In functional notation it is written

$$F = P(F/P,i\%,n)$$

The relationship may be rewritten as

$$\text{Present Sum} = (\text{Future Sum})(1+i)^{-n}$$
$$P = F(1+i)^{-n}$$

This is the Single Payment Present Worth Factor.
It is written

$$P = F(P/F,i\%,n)$$

EXAMPLE 3

At a 10% per year interest rate $500 now is *equivalent* to how much three years hence?

SOLUTION

This problem was solved in Example 2. Now it can be solved using a single payment formula.

$$P = \$500$$
$$n = 3 \text{ yr}$$
$$i = 10\%$$
$$F = \text{unknown}$$

$$F = P(1+i)^n = 500(1+0.10)^3 = \$665.50$$

This problem may also be solved using the compound interest tables.

$$F = P(F/P,i\%,n) = 500(F/P,10\%,3yr)$$

From the 10% compound interest table
read (F/P,10%,3yr) = 1.331

$$F = 500(F/P,10\%,3yr) = 500(1.331) = \$665.50$$

EXAMPLE 4

To raise money for a new business, a man asks you to loan him some money. He offers to pay you $3000 at the end of four years. How much should you give him now if you want 12% interest per year on your money?

SOLUTION

$$P = \text{unknown}$$
$$n = 4 \text{ yr}$$
$$i = 12\%$$
$$F = \$3000$$

$$P = F(1+i)^{-n} = 3000(1+0.12)^{-4} = \$1906.55$$

Alternate computation using compound interest tables

$$P = F(P/F,i\%,n) = 3000(P/F,12\%,4yr)$$
$$= 3000(0.6355) = \$1906.50$$

Note that the solution based on the compound interest table is slightly different from the exact solution using a hand calculator. In economic analysis the compound interest tables are always considered to be sufficiently accurate.

UNIFORM SERIES FORMULAS

Consider the following situation.

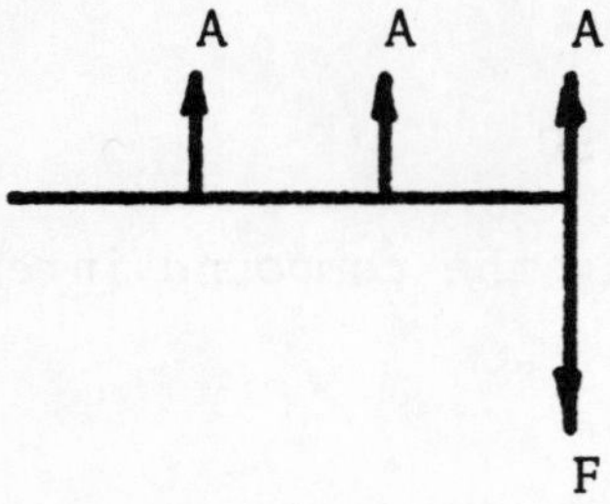

A = end-of-period receipt or disbursement in a uniform series continuing for n periods.
F = a future sum of money.

Using the single payment compound amount factor we can write an equation for F in terms of A.

$$F = A + A(1+i) + A(1+i)^2 \quad (1)$$

In our situation, with n=3, Equation (1) may be written in a more general form:

$$F = A + A(1+i) + A(1+i)^{n-1} \quad (2)$$

Multiply (2) by (1+i)

$$(1+i)F = A(1+i) + A(1+i)^{n-1} + A(1+i)^n \quad (3)$$

Write Eqn (2)

$$F = A + A(1+i) + A(1+i)^{n-1} \quad (2)$$

(3) - (2)

$$iF = -A + A(1+i)^n$$

$$F = A\left[\frac{(1+i)^n - 1}{i}\right]$$ Uniform Series Compound Amount Factor

Solving this equation for A

$$A = F\left[\frac{i}{(1+i)^n - 1}\right]$$ Uniform Series Sinking Fund Factor

Since $F = P(1+i)^n$ we can substitute this expression for F in the equation and obtain

$$A = P\left[\frac{i(1+i)^n}{(1+i)^n - 1}\right]$$ Uniform Series Capital Recovery Factor

Solving the equation for P

$$P = A\left[\frac{(1+i)^n - 1}{i(1+i)^n}\right]$$ Uniform Series Present Worth Factor

In functional notation the uniform series factors are:

Compound Amount	(F/A,i%,n)
Sinking Fund	(A/F,i%,n)
Capital Recovery	(A/P,i%,n)
Present Worth	(P/A,i%,n)

EXAMPLE 5

If $100 is deposited at the end of each year in a savings account that pays 6% interest per year, how much will be in the account at the end of five years?

SOLUTION

$$A = \$100$$
$$F = \text{unknown}$$
$$n = 5 \text{ yr}$$
$$i = 6\%$$

$$F = A(F/A,i\%,n) = 100(F/A,6\%,5\text{yr})$$
$$= 100(5.637) = \$563.70$$

EXAMPLE 6

A man wishes to make a uniform deposit every three months to his savings account so that at the end of 10 years he will have $10,000 in the account. If the account earns 6% annual interest, compounded quarterly, how much should he deposit each three months?

SOLUTION

$$F = \$10{,}000$$
$$A = \text{unknown}$$
$$n = 40 \text{ quarterly deposits}$$
$$i = 1\text{-}1/2\% \text{ per quarter year}$$

Note that i, the interest rate per interest period, is 1-1/2%, and there are 40 deposits.

$$A = F(A/F,i\%,n) = 10{,}000(A/F,1\text{-}1/2\%,40 \text{ interest periods})$$
$$= 10{,}000(0.0184) = \$184$$

EXAMPLE 7

An individual is considering the purchase of an automobile. The total price is $6200 with $1240 as a downpayment and the balance paid in 48 equal monthly payments with interest at 1% per month. The payments are due at the end of each month. Compute the monthly payment.

SOLUTION

The amount to be repaid by the 48 monthly payments is the cost of the automobile minus the $1240 downpayment.

$$
\begin{aligned}
P &= \$4960 \\
A &= \text{unknown} \\
n &= 48 \text{ monthly payments} \\
i &= 1\% \text{ per month} \\
A = P(A/P,i\%,n) &= 4960(A/P,1\%,48) \\
&= 4960(0.0263) = \$130.45
\end{aligned}
$$

EXAMPLE 8

A man sold his home. In addition to cash he took a mortgage on the house. The mortgage will be paid off by monthly payments of $232.50 for 10 years. The man decides to sell the mortgage to a local bank. The bank will buy the mortgage, but requires a 1% per month interest rate on their investment. How much will the bank pay for the mortgage?

SOLUTION

$$
\begin{aligned}
A &= \$232.50 \\
n &= 120 \text{ months} \\
i &= 1\% \text{ per month} \\
P &= \text{unknown} \\
P = A(P/A,i\%,n) &= 232.50(P/A,1\%,120) \\
&= 232.50(69.701) = \$16,205.48
\end{aligned}
$$

GRADIENT SERIES

At times one will encounter a situation where the cash flow series is not a constant amount A. Instead it is an increasing series like:

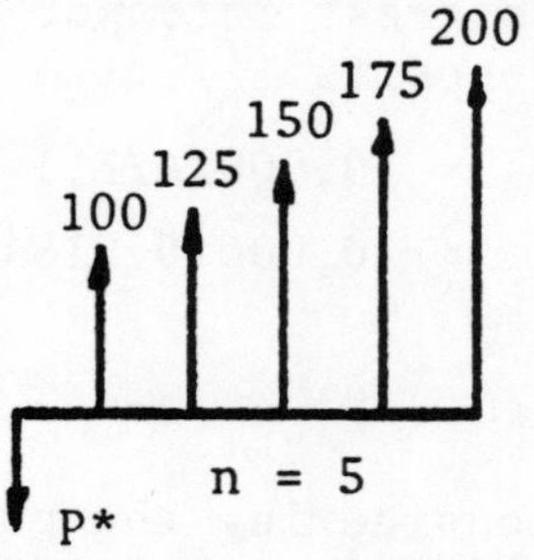

This cash flow may be resolved into two components:

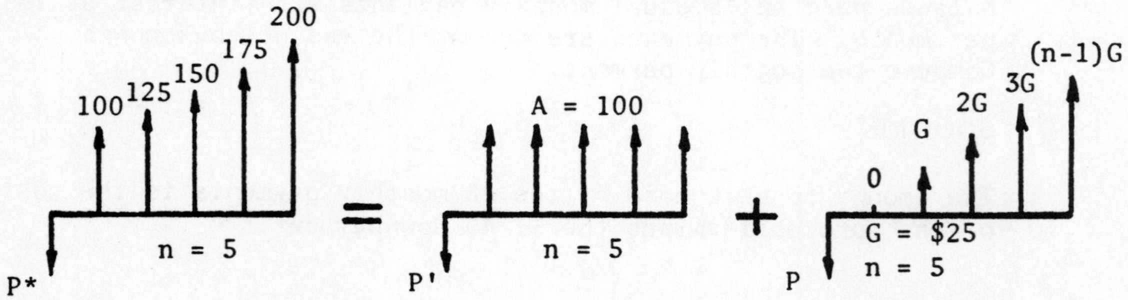

We can compute the value of P* as equal to P' plus P. We already have an equation for P'.

$$P' = A(P/A,i\%,n)$$

From Newnan: *Engineering Economic Analysis* the value for P in the right-hand diagram is:

$$P = \frac{G}{i}\left[\frac{(1+i)^n - 1}{i} - n\right]\left[\frac{1}{(1+i)^n}\right]$$

This is the Gradient Present Worth Factor.
In functional notation the relationship is $P = G(P/G,i\%,n)$

EXAMPLE 9

The maintenance on a machine is expected to be \$155 the end of the first year, and increasing \$35 each year for the following seven years. What present sum of money would need to be set aside now to pay the maintenance for the eight year period? Assume 6% interest.

SOLUTION

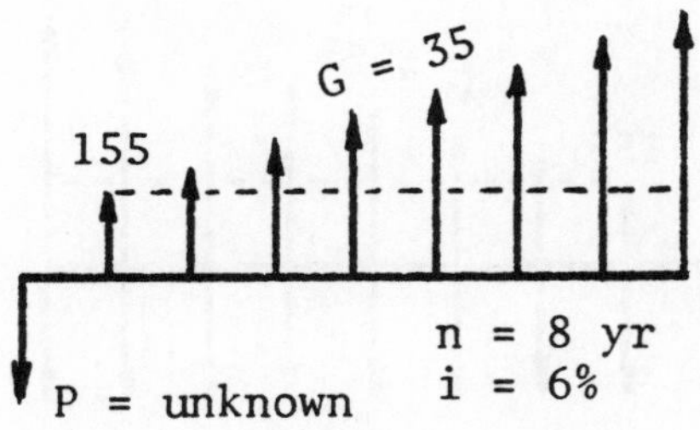

$$P = 155(P/A,6\%,8) + 35(P/G,6\%,8)$$
$$= 155(6.210) + 35(19.842) = \$1657.02$$

In the gradient series if instead of the present sum P, an equivalent uniform series A is desired, the problem becomes:

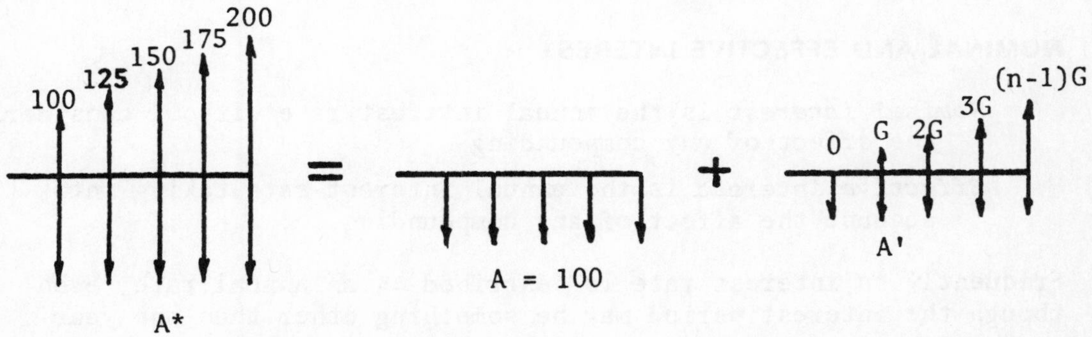

The relationship between A' and G in the right hand diagram is

$$A' = G\left[\frac{1}{i} - \frac{n}{(1+i)^n - 1}\right]$$

In functional notation the Gradient (to) Uniform Series factor is

$$A = G(A/G,i\%,n)$$

It is important to note carefully the diagrams for the two gradient series factors. In both cases the first term in the gradient series is zero and the last term is (n-1)G. But we use n in the equations and functional notation. The derivations (not shown here) were done of this basis and the gradient series compound interest tables are computed this way.

EXAMPLE 10

For the situation in Example 9 we wish now to know the uniform annual maintenance cost. Compute an equivalent A for the maintenance costs.

SOLUTION

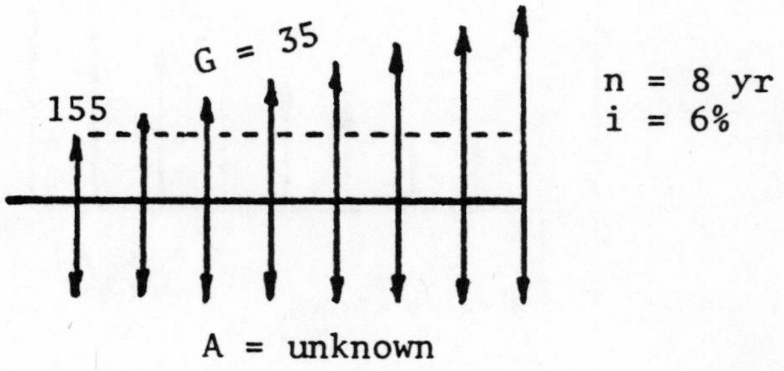

Equivalent uniform annual maintenance cost

$$A = 155 + 35(A/G,6\%,8)$$
$$= 155 + 35(3.195) = \$266.83$$

NOMINAL AND EFFECTIVE INTEREST

Nominal interest is the annual interest rate without considering the affect of any compounding.

Effective interest is the annual interest rate taking into account the affect of any compounding.

Frequently an interest rate is described as an annual rate, even though the interest period may be something other than one year.

A bank may pay 1-1/2% interest on the amount in a savings account every three months. The *nominal* interest rate in this situation is 6% (4 x 1-1/2% = 6%). But if one deposited $1000 in such an account, would you have 106%(1000) = $1060 in the account at the end of one year? The answer is no. You would have more. The amount in the account would increase as follows.

		Amount in Account
	At beginning of year =	$1000.00
End of 3 months:	1000.00 + 1-1/2%(1000.00) =	1015.00
End of 6 months:	1015.00 + 1-1/2%(1015.00) =	1030.23
End of 9 months:	1030.23 + 1-1/2%(1030.23) =	1045.68
End of one year:	1045.68 + 1-1/2%(1045.68) =	1061.37

The actual interest rate on the $1000 would be the interest $61.37 divided by the original $1000 or 6.137%. We call this the *effective* interest rate.

$$\text{Effective interest rate} = (1+i)^m - 1$$

where i = interest rate per interest period.
m = number of compoundings per year.

EXAMPLE 11

A bank charges 1-1/2% per month on the unpaid balance for purchases made on its credit card. What nominal interest rate is it charging? What effective interest rate?

SOLUTION

The nominal interest rate is simply the annual interest ignoring compounding, or 12(1-1/2%) = 18%.

$$\text{Effective interest rate} = (1+0.015)^{12} - 1 = 0.1956 = 19.56\%$$

SOLVING ECONOMIC ANALYSIS PROBLEMS

The techniques presented so far illustrate how to convert single amounts of money, and uniform or gradient series of money, into some equivalent sum at another point in time. These compound interest computations are an essential part of economic analysis problems. The typical situation is that we have a number of alternatives and the question is, which alternative should be selected? The customary method of solution is to resolve each of the alternatives into some common form and then choose the best alternative.

CRITERIA

Economic analysis problems inevitably fall into one of three categories.

1. Fixed Input	The amount of money or other input resources is fixed.	
	Example:	A Project Engineer has a budget of $450,000 to overhaul a plant.
2. Fixed Output	There is a fixed task, or other output to be accomplished.	
	Example:	A mechanical contractor has been awarded a fixed price contract to air condition a building.
3. Neither Input nor Output Fixed	This is the general situation where neither the amount of money or other inputs, nor the amount of benefits or other outputs are fixed.	
	Example:	A consulting engineering firm has more work available than it can handle. It is considering paying the staff for working evenings to increase the amount of design work it can perform.

Four methods of comparing alternatives are present worth, annual cost, rate of return, and benefit-cost ratio. These are presented in the following sections.

PRESENT WORTH

In present worth analysis the approach is to resolve all the money consequences of an alternative into an equivalent present sum. For the three categories given above the criteria are:

Category	Present Worth Criterion
Fixed Input	Maximize the present worth of benefits or other outputs.
Fixed Output	Minimize the present worth of costs or other inputs.
Neither Input nor Output fixed	Maximize [present worth of benefits minus present worth of costs] or stated another way: Maximize Net Present Worth.

APPLICATION OF PRESENT WORTH

Present worth analysis is most frequently used to determine the present value of future money receipts and disbursements. We might want to know, for example, the present worth of an income producing property, like an oil well. This should provide a good estimate of the price at which the property could be bought or sold.

An important restriction in the use of present worth calculations is that there must be a common analysis period when comparing alternatives. It would be incorrect, for example, to compare the present worth (PW) of cost of Pump A, expected to last 6 years, with the PW of cost of Pump B, expected to last 12 years.

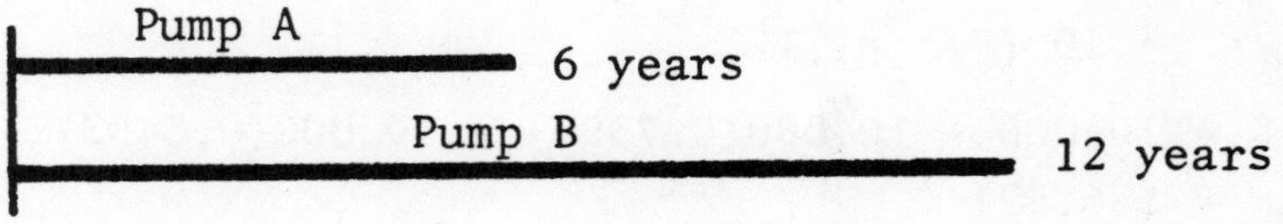

Improper Present Worth Comparison

In situations like this the solution is either to use some other analysis technique* or to restructure the problem so there is a common analysis period. In the example above, a customary assumption would be that there is a need for the pump for 12 years and that Pump A will be replaced by an identical Pump A at the end of 6 years. This gives a 12 year common analysis period.

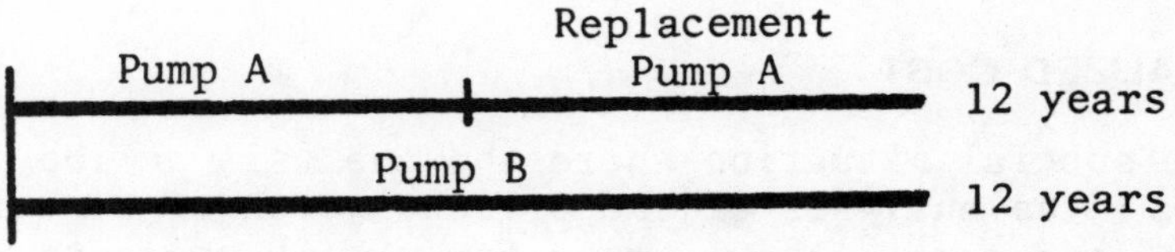

Correct Present Worth Comparison

This approach is easy to use when the different lives of the alternatives have a practical least common multiple life. When this is not true (for example, life of *J* equals 7 years and the life of *K* equals 11 years), some assumptions must be made to select a suitable common analysis period, or the present worth method should not be used.

EXAMPLE 12

Machine X has an initial cost of $10,000, annual maintenance of $500 per year, and no salvage value at the end of its four year useful life.

*Generally the annual cost method is suitable in these situations.

Machine Y costs \$20,000. The first year there is no maintenance cost. The second year maintenance is \$100, and increases \$100 per year in subsequent years. The machine has an anticipated \$5000 salvage value at the end of its 12-year useful life.

If interest is 8%, which machine should be selected?

SOLUTION

The analysis period is not stated in the problem. We therefore select the least common multiple of the lives, or 12 years, as the analysis period.

Present Worth of Cost of 12 years of Machine X

$$= 10{,}000 + 10{,}000(P/F,8\%,4) + 10{,}000(P/F,8\%,8) + 500(P/A,8\%,12)$$

$$= 10{,}000 + 10{,}000(0.7350) + 10{,}000(0.5403) + 500(7.536)$$
$$= \$26{,}521$$

Present Worth of Cost of 12 years of Machine Y

$$= 20{,}000 + 100(P/G,8\%,12) - 5000(P/F,8\%,12)$$

$$= 20{,}000 + 100(34.634) - 5000(0.3971)$$

$$= \$21{,}478$$

Choose Machine Y with its smaller PW of Cost

CAPITALIZED COST

In the special situation where the analysis period is infinite ($n = \infty$), an analysis of the present worth of cost is called *capitalized cost*. There are a few public projects where the analysis period is infinity. Other examples would be permanent endowments and cemetery perpetual care.

When n equals infinity, a present sum P will accrue interest of Pi for every future interest period. For the principal sum P to continue undiminished (an essential requirement for n equal to infinity), the end-of-period sum A that can be disbursed is Pi.

$$P \rightarrow P + Pi \rightarrow P + Pi \rightarrow P + Pi \rightarrow \cdots$$
$$\searrow A \qquad \searrow A \qquad \searrow A$$

When $n = \infty$ the fundamental relationship between P, A, and i is:

$$A = Pi$$

Some form of this equation is used whenever there is a problem with an infinite analysis period.

EXAMPLE 13

In his will a man wishes to establish a perpetual trust to provide for the maintenance of a small local park. If the annual maintenance is \$750 per year and the trust account can earn 5% interest, how much money must be set aside in the trust?

SOLUTION

When $n = \infty$ $A = Pi$ or $P = \frac{A}{i}$

$$\text{Capitalized cost } P = \frac{A}{i} = \frac{\$750}{0.05} = \$15{,}000$$

ANNUAL COST

The annual cost method is more accurately described as the method of Equivalent Uniform Annual Cost (EUAC) or where the computation is of benefits, the method of Equivalent Uniform Annual Benefits (EUAB).

CRITERIA

For each of the three possible categories of problems there is an annual cost criterion for economic efficiency.

Category	Annual Cost Criterion
Fixed Input	Maximize the equivalent uniform annual benefits. That is, maximize EUAB.
Fixed Output	Minimize the equivalent uniform annual cost. That is, minimize EUAC.
Neither Input nor Output fixed	Maximize [EUAB - EUAC]

APPLICATION OF ANNUAL COST ANALYSIS

In the present worth section we pointed out that the present worth method requires that there be a common analysis period for all alternatives. This same restriction does not apply in all annual cost calculations, but it is important to understand the circumstances that justify comparing alternatives with different service lives.

It is frequently the case that an analysis is to provide for a more or less continuing requirement. One might need to pump water from a well, for example, as a continuing requirement. Regardless of whether the pump has a useful service life of 6 years or 12 years, we would select the one whose annual cost is a minimum. And this would still be the case if the pump useful lives were the more troublesome 7 and 11 years, respectively. Thus if we can assume a continuing need for an item, an annual cost comparison among alternatives of differing service lives is valid.

The underlying assumption made in these situations is that when the shorter lived alternative has reached the end of its useful life, it can be replaced with an identical item with identical costs and so forth. This means the EUAC of the initial alternative is equal to the EUAC for the continuing series of replacements.

If, on the other hand, there is a specific requirement in some situation to pump water for 10 years, then each pump must be evaluated to see what costs will be incurred during the analysis period and what salvage value, if any, may be recovered at the end of the analysis period. The annual cost comparison needs to consider the actual circumstances of the situation.

In the professional engineering examination the problems frequently are readily solved by the annual cost method. And the underlying "continuing requirement" is often present, so that an annual cost comparison of unequal lived alternatives is an appropriate method of analysis.

EXAMPLE 14

Consider the following alternatives.

	A	B
First cost	\$5000	\$10,000
Annual maintenance	500	200
End of useful life salvage value	600	1000
Useful life	5 yr	15 yr

Based on an 8% interest rate, which alternative should be selected?

SOLUTION

Assuming both alternatives perform the same task and there is a continuing requirement, the goal is to minimize EUAC.

Alternative A

EUAC = 5000(A/P,8%,5) + 500 - 600(A/F,8%,5)
= 5000(0.2505) + 500 - 600(0.1705) = $1650

Alternative B

EUAC = 10,000(A/P,8%,15) + 200 - 1000(A/F,8%,15)
= 10,000(0.1168) + 200 - 1000(0.0368) = $1331

To minimize EUAC, select Alternative B.

RATE OF RETURN

In the rate of return method the typical situation is a cash flow representing the costs and benefits. The rate of return may be defined as the interest rate where

PW of Cost = PW of Benefit,

EUAC = EUAB,

or PW of Cost - PW of Benefit = 0

EXAMPLE 15

Compute the rate of return for the investment represented by the following cash flow.

Year	Cash Flow
0	-$595
1	+250
2	+200
3	+150
4	+100
5	+50

SOLUTION

This declining gradient series may be separated into two cash flows for which compound interest factors are available.

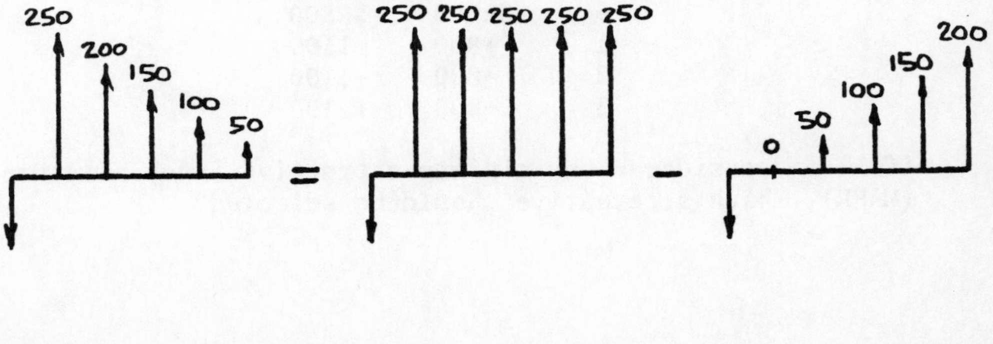

Note that the gradient series factors are based on an *increasing* gradient. When there is a declining cash flow it is solved by substracting an increasing gradient as indicated by the diagram on the previous page.

PW of Cost - PW of Benefits = 0

595 - [250(P/A,i%,5) - 50(P/G,i%,5)] = 0

Try i = 10%

595 - [250(3.791) - 50(6.862)] = -9.65

Try i = 12%

595 - [250(3.605) - 50(6.397)] = +13.60

The rate of return is between 10% and 12%. It may be computed more accurately by linear interpolation.

$$\text{Rate of return} = 10\% + (2\%)\left[\frac{9.65 - 0}{13.60 + 9.65}\right] = 10.83\%$$

RATE OF RETURN ANALYSIS

Rate of Return Criterion for Two Alternatives

Compute the incremental rate of return on the cash flow representing the difference between the higher cost alternative and the lower cost alternative. If this rate of return is greater than or equal to the predetermined minimum attractive rate of return (MARR), choose the higher cost alternative; otherwise choose the lower cost alternative.

EXAMPLE 16

Previously two alternatives were described in terms of cash flows.

	Alternative	
Year	A	B
0	-$2000	-$2800
1	+800	+1100
2	+800	+1100
3	+800	+1100

If 5% is considered the minimum attractive rate of return (MARR), which alternative should be selected?

SOLUTION

First, tabulate the cash flow that represents the difference between the alternatives.

Year	A	B	Difference Between Alternatives B-A
0	-\$2000	-\$2800	-\$800
1	+800	+1100	+300
2	+800	+1100	+300
3	+800	+1100	+300

Compute the rate of return on the difference between the alternatives.

PW of Cost = PW of Benefits

$$800 = 300(P/A,i\%,3)$$

$$(P/A,i\%,3) = \frac{800}{300} = 2.67 \qquad i \approx 6.1\%$$

Since the incremental rate of return exceeds the 5% MARR, the increment of investment is desirable. Choose the higher cost Alternative B.

Before leaving this example problem, one should note something that relates to the rate of return on Alternative A and on Alternative B. These rates of return, if computed, are

	Rate of Return
Alternative A	9.7%
Alternative B	8.7%

The correct answer to this problem has been shown to be Alternative B, and this is true even though Alternative A has a higher rate of return. The higher cost alternative may be thought of as the lower cost alternative, plus the differences between them. Looked at this way, the higher cost alternative B is equal to the desirable lower cost alternative A plus the desirable differences between the alternatives.

The important conclusion is that computing the rate of return for each alternative does not provide the basis for choosing between alternatives. Instead, incremental analysis is required.

Rate of Return Criterion for Three or More Alternatives

When there are three or more mutually exclusive alternatives one must proceed following the same general logic presented for two alternatives. The components of incremental analysis are

1. Compute the rate of return for each alternative. Reject any alternative where the rate of return is less than the given MARR. (This step is not essential, but helps to immediately eliminate unacceptable alternatives.) One must insure that the lowest cost alternative has a rate of return greater than or equal to MARR.
2. Rank the remaining alternatives in their order of increasing cost.
3. Examine the difference between the two lowest cost alternatives as described for the two-alternative problem. Select the best of the two alternatives and reject the other one.
4. Take the preferred alternative from Step 3. Consider the next higher alternative and proceed with another two-alternative comparison.
5. Continue until all alternatives have been examined and the best of the multiple alternatives identified.

EXAMPLE 17

Consider four mutually exclusive alternatives.

	Alternative			
	A	B	C	D
Initial Cost	$400	$100	$200	$500
Uniform Annual Benefit	100.9	27.7	46.2	125.2

Each alternative has a five year useful life and no salvage value. If the minimum attractive rate of return (MARR) is 6%, which alternative should be selected?

SOLUTION

Mutually exclusive is where selecting one alternative precludes selecting any of the other alternatives. This is the typical "textbook" situation.

The solution will follow the several steps in incremental analysis.

1. Rate of return is computed for the four alternatives.

Alternative	A	B	C	D
Computed rate of return	8.3%	12%	5%	8%

Since Alternative C has a rate of return less than MARR, it may be eliminated from further consideration.

2. Rank the remaining alternatives in order of increasing cost and examine the difference between the two lowest cost alternatives.

Alternative	B	A	D
Initial Cost	$100	$400	$500
Uniform Annual Benefit	27.7	100.9	125.2

	A-B
Δ Initial Cost	$300
Δ Uniform Annual Benefit	73.2
Computed Δ rate of return	7%

Since the incremental rate of return exceeds the 6% MARR, the increment is desirable. Alternative A is the better alternative.

3. Take the preferred alternative from the previous step and consider the next higher cost alternative. Do another two-alternative comparison.

	D-A
Δ Initial Cost	$100
Δ Uniform Annual Benefit	24.3
Computed Δ rate of return	6.9%

The incremental rate of return exceeds MARR, hence the increment is desirable. Alternative D is preferred over Alternative A.

Conclusion: Select Alternative D

[Note that once again the alternative with the highest rate of return (Alt. B) is *not* the proper choice.]

BENEFIT—COST RATIO

In public works economics, and in governmental economic analyses generally, the dominant method of analysis is called benefit-cost ratio. It is simply the ratio of benefits divided by costs, taking into account the time value of money.

$$B/C = \frac{\text{PW of Benefits}}{\text{PW of Cost}} = \frac{\text{Equivalent Uniform Annual Benefits}}{\text{Equivalent Uniform Annual Cost}}$$

For a given interest rate, a B/C ratio ≥ 1 reflects an acceptable project. The method of analysis using B/C ratio is parallel to that of rate of return analysis. The same kind of incremental analysis is required.

<u>B/C Ratio Criterion for Two Alternatives</u>

Compute the incremental B/C ratio for the cash flow representing the difference between the higher cost alternative and the lower cost alternative. If this incremental B/C ratio is ≥ 1, choose the higher cost alternative; otherwise choose the lower cost alternative.

<u>B/C Ratio Criterion for Three or More Alternatives</u>

Follow the logic for rate of return, except that the test is whether or not the incremental B/C ratio is ≥ 1.

EXAMPLE 18

Solve Example 17 using Benefit-Cost ratio analysis. Consider four mutually exclusive alternatives.

	Alternative			
	A	B	C	D
Initial Cost	\$400	\$100	\$200	\$500
Uniform Annual Benefit	100.9	27.7	46.2	125.2

Each alternative has a five year useful life and no salvage value. Based on a 6% interest rate, which alternative should be selected?

SOLUTION

1. B/C ratio computed for the alternatives:

Alt. A $B/C = \frac{\text{PW of Benefits}}{\text{PW of Cost}} = \frac{100.9(P/A,6\%,5)}{400} = 1.06$

B $B/C = \frac{27.7(P/A,6\%,5)}{100} = 1.17$

C $B/C = \frac{46.2(P/A,6\%,5)}{200} = 0.97$

D $B/C = \frac{125.2(P/A,6\%,5)}{500} = 1.05$

Alternative C with a B/C ratio less than 1 is eliminated.

2. Rank the remaining alternatives in order of increasing cost and examine the difference between the two lowest cost alternatives.

	Alternative		
	B	A	D
Initial Cost	\$100	\$400	\$500
Uniform Annual Benefit	27.7	100.9	125.2

	A-B
Initial Cost	\$300
Uniform Annual Benefit	72.2

$$\text{Incremental B/C ratio} = \frac{72.2(P/A,6\%,5)}{300} = 1.01$$

The incremental B/C ratio exceeds 1.0 hence the increment is desirable. Alternative A is preferred over B.

3. Do the next two-alternative comparison.

	A	D	Increment D-A
Initial Cost	\$400	\$500	\$100
Uniform Annual Benefit	100.9	125.2	24.3

$$\text{Incremental B/C ratio} = \frac{24.3(P/A,6\%,5)}{100} = 1.02$$

The incremental B/C ratio exceeds 1.0 hence Alternative D is preferred.

Conclusion: Select Alternative D

DEPRECIATION

Depreciation of capital equipment is an important component of many after-tax economic analyses. For this reason one must understand the fundamentals of depreciation accounting.

Depreciation is defined, in its accounting sense, as the systematic allocation of the cost of a capital asset over its useful life.

In computing a schedule of depreciation charges, three items are considered.

1. Cost of the property, P
2. Useful life in years, n
3. Salvage value of the property at the end of its useful life, S

Three principal methods of depreciation are:

Straight Line Depreciation

$$\text{Depreciation charge in any year} = \frac{P - S}{n}$$

Sum-Of-Years-Digits Depreciation

$$\begin{array}{c}\text{Depreciation charge} \\ \text{in any year}\end{array} = \frac{\begin{array}{c}\text{Remaining Useful Life} \\ \text{at Beginning of Year}\end{array}}{\begin{array}{c}\text{Sum of Years Digits} \\ \text{for Total Useful Life}\end{array}} (P - S)$$

$$\text{where Sum-Of-Years-Digits} = 1+2+3+\cdots+n = \frac{n}{2}(n + 1)$$

Double Declining Balance Depreciation

$$\begin{array}{c}\text{Depreciation charge} \\ \text{in any year}\end{array} = \frac{2}{n} (P - \text{Depreciation charges to date})$$

EXAMPLE 19

A piece of machinery costs $5000 and has an anticipated $1000 salvage value at the end of its five year useful life. Compute the depreciation schedule for the machinery by

(*a*) Straight line depreciation
(*b*) Sum-of-years-digits depreciation
(*c*) Double declining balance depreciation

SOLUTION

$$\text{Straight line depreciation} = \frac{P - S}{n} = \frac{5000 - 1000}{5} = \$800$$

Sum-of-years-digits depreciation

$$\text{Sum-of-years-digits} = \frac{n}{2}(n + 1) = \frac{5}{2}(6) = 15$$

$$\text{1st year depreciation} = \frac{5}{15}(5000 - 1000) = \$1333$$

$$\text{2nd year depreciation} = \frac{4}{15}(5000 - 1000) = 1067$$

$$\text{3rd year depreciation} = \frac{3}{15}(5000 - 1000) = 800$$

$$\text{4th year depreciation} = \frac{2}{15}(5000 - 1000) = 533$$

$$\text{5th year depreciation} = \frac{1}{15}(5000 - 1000) = 267$$

$$\$4000$$

Double declining balance depreciation

$$\text{1st year depreciation} = \frac{2}{5}(5000 - 0) = \$2000$$

$$\text{2nd year depreciation} = \frac{2}{5}(5000 - 2000) = 1200$$

$$\text{3rd year depreciation} = \frac{2}{5}(5000 - 3200) = 720$$

$$\text{4th year depreciation} = \frac{2}{5}(5000 - 3920) = 432$$

$$\text{5th year depreciation} = \frac{2}{5}(5000 - 4352) = \underline{259}$$

$$\$4611$$

Since the problem specifies a $1000 salvage value, the total depreciation may not exceed $4000. The double declining balance depreciation must be terminated during the 4th year when its totals $4000.

The depreciation schedules computed by the three methods are as follows.

Year	Straight Line	Sum-Of-Years-Digits	Double Declining Balance
1	$800	$1333	$2000
2	800	1067	1200
3	800	800	720
4	800	533	80
5	800	267	0
	4000	4000	4000

INCOME TAXES

Income taxes represent another of the various kinds of disbursements encountered in an economic analysis. The starting point in an after-tax computation is the before-tax cash flow.

Generally, the before-tax cash flow contains three types of entries.

1. Disbursements of money to purchase capital assets. These expenditures create no direct tax consequence for they are the exchange of one asset (cash) for another (capital equipment).

2. Periodic receipts and/or disbursements representing operating income and/or expenses. These increase or decrease the year-by-year tax liability of the firm.

3. Receipt of money from the sale of capital assets, usually in the form of salvage value when the equipment is removed. The tax consequence depends on the relationship between the book value (cost - depreciation taken) of the asset and its salvage value.

Situation	Tax Consequence
Salvage value > Book value	Capital gain on difference
Salvage value = Book value	No tax consequence
Salvage value < Book value	Capital loss on difference

After the before-tax cash flow, the next step is to compute the depreciation schedule for any capital assets. Taxable income is the taxable component of the before-tax cash flow minus the depreciation. The income tax is the taxable income times the appropriate tax rate. Finally, the after-tax cash flow is the before-tax cash flow adjusted for income taxes.

To organize these data it is customary to arrange them in the form of a cash flow table as follows.

Year	Before Tax Cash Flow	Depreciation	Taxable Income	Income Taxes	After Tax Cash Flow
0	.		.	.	.
1	.	.	.	.	.
.	.	.	.	.	.

EXAMPLE 20

A corporation expects to receive \$32,000 each year for 15 years from the sale of a product. There will be an initial investment of \$150,000. Manufacturing and sales expenses will be \$8067 per year. Assume straight line depreciation, a 15-year useful life and no salvage value. Use a 46% income tax rate.

Determine the projected after-tax rate of return.

SOLUTION

$$\text{Straight line depreciation} = \frac{P - S}{n} = \frac{150{,}000 - 0}{15}$$

$$= \$10{,}000 \text{ per year}$$

Year	Before Tax Cash Flow	Straight Line Depreciation	Taxable Income	46% Income Taxes	After Tax Cash Flow
0	-150,000				-150,000
1	+23,933	10,000	13,933	-6,409	+17,524
2	+23,933	10,000	13,933	-6,409	+17,524
.	.	.	.	.	.
.	.	.	.	.	.
.	.	.	.	.	.
15	+23,933	10,000	13,933	-6,409	+17,524

Take the After Tax Cash Flow and compute the rate of return at which PW of Cost equals PW of Benefits.

$$150{,}000 = 17{,}524(P/A,i\%,15)$$

$$(P/A,i\%,15) = \frac{150{,}000}{17{,}524} = 8.559$$

From interest tables, i = 8%

REFERENCE

Newnan, D. G. *Engineering Economic Analysis*. 2nd ed. Engineering Press, Inc.

ECONOMICS 1

A snow loading machine costing $30,000 requires four operators at $36 per day. The machine can do the work of 50 hand shovelers at $21 per day. Fuel, oil and maintenance for the machine amount to $60 per day. Assume the life of the machine is 8 years with no salvage value. If interest on money is 6%, how many days of snow removal per year are necessary to make purchase of the machine economical?

SOLUTION

Let Annual Cost (EUAC) of hand loading = Annual Cost (EUAC) of machine loading, and solve for X, equal to days of snow removal per year.

$$50(\$21)(X) = 4(\$36)(X) + \$60(X) + \$30{,}000(A/P,6\%,8)$$

$$1050\ X = 144\ X + 60\ X + 30{,}000(0.161)$$

$$\underline{X = 5.7 \text{ days/year}}$$

ECONOMICS 2

A new office building was constructed five years ago by a consulting engineering firm. At that time the firm obtained a bank loan for $100,000 with a 6% annual interest rate, compounded quarterly. The terms of the loan called for equal quarterly payments for a 10 year period with the right of prepayment at any time without penalty.

Due to internal changes in the firm, it is now proposed to refinance the loan through an insurance company. The new loan is planned for a 20 year term with an interest rate of 8% per annum, compounded quarterly. The insurance company has a one time 5% service charge. This new loan also calls for equal quarterly payments.

REQUIRED:

(a) What is the balance due on the original mortgage (principal) if all payments have been made through a full five years?

(b) What will be the difference between the equal quarterly payments in the existing arrangement and the revised proposal?

SOLUTION

(a)

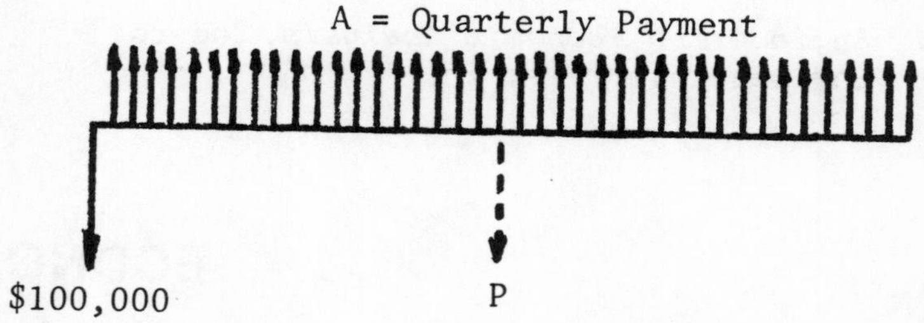

6% per year, compounded quarterly for 10 years

Therefore i = 1-1/2% per interest period

n = 40 interest periods

A = 100,000(A/P,1-1/2%,40) = 100,000(0.0334) = $3340

P = 3340(P/A,1-1/2%,20) = 3340(17.169) = $57,344

The balance due is <u>$57,344</u>

(b)

Service charge = 0.05P

Amount of New Loan = 1.05P = 1.05(57,344) = $60,211

Quarterly payments on new loan = 60,211(A/P,2%,80)

= 60,211(0.0252) = $1517

Difference in quarterly payments between existing loan and new loan is ($3340 - $1517) = <u>$1823</u>

ECONOMICS 3

An investor is considering buying a 20-year corporate bond. The bond has a face value of $1000 and pays 6% interest per year in two semiannual payments. Thus the purchaser of the bond would receive $30 every 6 months and in addition he would receive $1000 at the end of 20 years, along with the last $30 interest payment.

If the investor thought he should receive 8% interest, compounded semiannually, how much would he be willing to pay for the bond?

SOLUTION

The investor would be willing to pay the present worth of the future benefits, computed at 8% compounded semiannually.

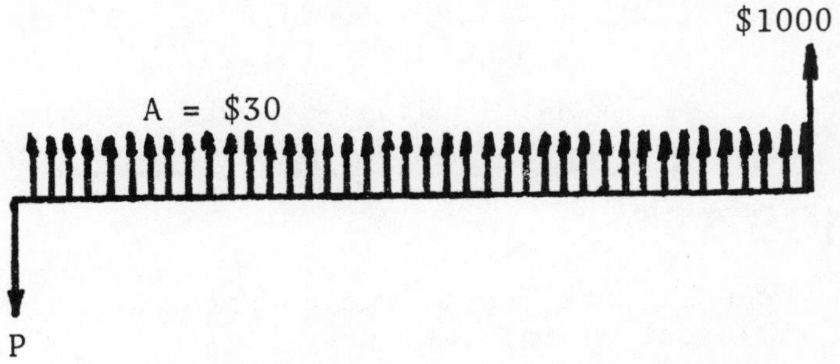

For semiannual interest periods
i = 4% per interest period
n = 40 interest periods

Present Worth = 30(P/A,4%,40) + 1000(P/F,4%,40)
= 30(19.793) + 1000(0.2083)
= 593.79 + 208.30
= <u>$802.09</u>

ECONOMICS 4

Rental equipment is for sale at $110,000. A prospective buyer estimates he would hold the equipment for 12 years, during which time the average annual disbursements for all purposes in connection with its ownership and operation would be $6000.

The prospective buyer believes the equipment could be sold for $80,000 after sales expenses at the end of 12 years. He sets the minimum attractive return on this type of investment to be 7% before income taxes. Estimated average annual receipts from equipment rentals are $14,400.

On the basis of these estimates, what is the maximum price he should pay for the equipment?

SOLUTION

Compute the cash flow.
Let X = maximum purchase price.

Year	Before Tax Cash Flow
0	-X
1	+14,400 - 6,000
2	
3	
.	
.	
.	
12	+80,000

Maximum purchase price = present worth of future benefits

$$X = (14{,}400 - 6000)(P/A,7\%,12) + 80{,}000(P/F,7\%,12)$$
$$= 8400(7.943) + 80{,}000(0.444)$$
$$= 66{,}700 + 35{,}500 = \underline{\$102{,}200}$$

ECONOMICS 5

Imagine that you must take one of two courses of action in the future which will cost as follows:

Year	Alternative A	Alternative B
0	$1300	$ 0
1		100
2		200
3		300
4		400
5		500
Total	$1300	$1500

Using present worth of the two alternatives with interest equal to 6%, which is the more economical?

SOLUTION

Alternative A

$$\text{PW of Cost} = \$1300$$

Alternative B

$$\text{PW of Cost} = 100(P/F,6\%,1) + 200(P/F,6\%,2) + 300(P/F,6\%,3) + 400(P/F,6\%,4) + 500(P/F,6\%,5)$$
$$= 100(0.943) + 200(0.890) + 300(0.840) + 400(0.792) + 500(0.747)$$
$$= 94 + 178 + 252 + 317 + 373$$
$$= \$1214$$

Choose the alternative with the least PW of Cost. <u>Choose B.</u>

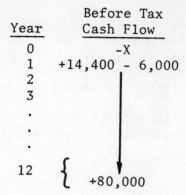

ECONOMICS 6

Two mutually exclusive alternatives are being considered. They have the following cash flows.

Year	Alternative A	Alternative B
0	-$2500	-$6000
1	+746	+1664
2	+746	+1664
3	+746	+1664
4	+746	+1664
5	+746	+1664

The firm has set a minimum attractive rate of return of 8%. Which alternative should be selected?

SOLUTION

The problem may be solved by several methods. Three will be illustrated here.

Rate of Return Analysis

Compute the rate of return on the difference between the alternatives.

Incremental cost = 6000 - 2500 = $3500
Incremental annual benefit = 1664 - 746 = $918

PW of Cost = PW of Benefits

$$3500 = 918(P/A,i\%,5)$$

$$(P/A,i\%,5) = \frac{3500}{918} = 3.81 \qquad i = 9.8\%$$

Since the minimum attractive rate of return is 8%, the increment is desirable. Select the higher cost alternative.

Select Alternative B

Present Worth Analysis

For this situation of neither input nor output fixed,
Maximize (PW of Benefits - PW of Cost)
or Maximize Net Present Worth

Alternative A

$$\text{Net Present Worth} = 746(P/A,8\%,5) - 2500$$
$$= 746(3.993) - 2500$$
$$= +479$$

Alternative B

$$\text{Net Present Worth} = 1664(P/A,8\%,5) - 6000 = +644$$

To maximize Net Present Worth, select Alternative B.

Annual Cost Analysis

The criterion is to maximize (EUAB - EUAC)

Alternative A

EUAB - EUAC = 746 - 2500(A/P,8%,5) = 746 - 2500(0.2505)

= +120

Alternative B

EUAB - EUAC = 1664 - 6000(A/P,8%,5) = 1664 - 6000(0.2505)

= +161

To maximize (EUAB - EUAC), select Alternative B.

ECONOMICS 7

Three mutually exclusive alternatives are being considered.

	Alternative A	Alternative B	Alternative C
Initial Investment	$50,000	$22,000	$15,000
Annual Net Income	5,093	2,077	1,643
Computed Rate of Return	8%	7%	9%

Each alternative has a 20-year useful life with no salvage value. The minimum attractive rate of return is 7%. Which alternative should be selected?

SOLUTION

The rate of return for each alternative is greater than or equal to the minimum attractive rate of return. Proceed with incremental analysis. Order the alternatives by increasing investment.

	Alternative C	Alternative B	Alternative A
Initial Investment	$15,000	$22,000	$50,000
Annual Net Income	1,643	2,077	5,093

Δ Investment	$7000
Δ Annual Income	434

Incremental rate of return on B-C increment:

$$7000 = 434(P/A,i\%,20)$$

$$(P/A,i\%,20) = \frac{7000}{434} = 16.13 \qquad i = 2.1\%$$

The rate of return on the B-C increment is less than the desired 7%. Reject the increment, and select C as preferred over B.

Now we must decide if A is preferred over C. Examine the increment.

	Increment A-C
Investment	\$35,000
Annual Income	3,450

Incremental rate of return on A-C increment:

$$35{,}000 = 3450(P/A,i\%,20)$$

$$(P/A,i\%,20) = \frac{35{,}000}{3{,}450} = 10.14 \qquad i = 7.5\%$$

The A-C increment is desirable, therefore choose the higher cost alternative A.

Conclusion: Select Alternative A.

ECONOMICS 8

A pumping service is required for ten years at a remote location, and gasoline and electric power are being considered on the basis of the following data.

	Gasoline	Electric
First Cost	\$1200	\$3000
Life in Years	5 yrs	10 yrs
Salvage Value	\$150	\$300
Annual Operating Cost	\$600	\$375
Annual Repairs	\$150	\$75
Annual Taxes (% of first cost)	1%	1%

Analyze the two installations on the basis of annual costs using 6 percent interest to determine which is the least expensive.

SOLUTION

Equivalent Uniform Annual Cost (EUAC)

$$= (P-S)(A/P,i\%,n) + Si + \text{Annual Costs}$$

Gasoline

$$\text{EUAC} = (1200 - 150)(A/P,6\%,5) + 150(0.06) + 600 + 150 + 0.01(1200)$$

$$= 1050(0.2374) + 9 + 600 + 150 + 12 = \$1020.27$$

Electric

$$\text{EUAC} = (3000 - 300)(A/P,6\%,10) + 300(0.06) + 375 + 75 + 0.01(3000)$$

$$= 2700(0.1359) + 18 + 375 + 75 + 30 = \$864.93$$

The electric driven pump has the smaller annual cost.

ECONOMICS 9

A nearby city has developed a plan which will provide for future municipal water needs. The plan proposes an aqueduct which passes through 1000 feet of tunnel in a nearby mountain.

Two alternatives are being considered. The first proposes to build a full-capacity tunnel now (10 foot diameter). The second proposes to build a half-capacity (7.5 foot diameter) tunnel now which should be adequate for 20 years, and then to build a second parallel half-capacity tunnel.

The full-capacity tunnel A can be built now for \$556,000 as compared to \$402,000 for one half-capacity tunnel. The costs of repair to the tunnel lining for the full-capacity tunnel are estimated to be \$0.64 per square foot of tunnel every 10 years. For the half-capacity tunnel B the cost is estimated as \$0.70 per square foot every 10 years.

The friction losses in the half-capacity tunnel will be somewhat greater, and it is estimated that the pumping costs in the single half-capacity line will increase by \$2000 per year, and by \$4000 per year after the second half-capacity tunnel has been placed in service.

REQUIRED: Using an annual interest rate of 7%, what is the total capitalized cost for each of the two alternatives?

SOLUTION

Tunnel Lining

Full Capacity Tunnel

Surface Area = 10π(1000 feet) = 31,416 ft^2

Repair Cost = 31,416(\$0.64) = \$20,106

Half Capacity Tunnel

Surface Area = 7.5π(1000 feet) = 23,562 ft^2

Repair Cost = 23,562(\$0.70) = \$16,493 per tunnel

<u>Alternative A - Full Capacity Tunnel</u>

Initial Cost — Lining Repair

$$\text{Capitalized Cost} = \$556{,}000 + \frac{\$20{,}106(A/F,7\%,10)}{0.07}$$

$$= \$556{,}000 + \frac{\$20{,}106(0.0724)}{0.07}$$

$$= \$556{,}000 + \$20{,}800 = \underline{\$576{,}800}$$

Alternative B - Half Capacity Tunnels

First Half Capacity Tunnel

	Initial Cost		Lining Repair		Extra Pumping

$$\text{Capitalized Cost} = \$402{,}000 + \frac{16{,}493(A/F,7\%,10)}{0.07} + \frac{2000}{0.07}$$

$$= \$402{,}000 + \frac{16{,}493(0.0724)}{0.07} + \frac{2000}{0.07}$$

$$= \$447{,}600$$

Second Half Capacity Tunnel

The capitalized cost of the second half-capacity tunnel 20 years hence is equal to the present capitalized cost of the first half-capacity tunnel.

$$\text{The present capitalized cost} = \begin{matrix}\text{capitalized cost}\\ \text{20 years hence}\end{matrix} \times (P/F,7\%,20)$$

$$= \$447{,}600(0.2584)$$

$$= \$115{,}700$$

Capitalized Cost for the half-capacity tunnels = 447,600 + 115,700
= $563,300

ECONOMICS 10

Machine A has been completely overhauled for $9000 and is expected to last another 11 years. The $9000 was treated as an expenditure last year. It can be sold now for $20,000 net after selling expenses, but will have no salvage value 11 years hence. It was bought new 9 years ago for $54,000 and has been depreciated by straight line depreciation using a 12 year depreciable life.

Because capacity requirements have diminished, Machine A can be replaced with smaller Machine B. Machine B costs $42,000, has an anticipated life of 20 years, and would reduce operating costs $2500 per year. It would be depreciated the same as Machine A, that is, by straight line depreciation with a 12 year depreciable life.

For an income tax rate of 40% and a capital gains tax of 25%, compare the after-tax annual cost of the two machines and choose the more economical. Use a 10% after-tax rate of return in the calculations.

SOLUTION

Book value of Machine A now = Cost - Depreciation taken to date
= \$54,000 - 9(54,000 - 0)/12 = \$13,500

Long Term Capital Gain on disposal if sold now
= \$20,000 - 13,500 = \$6,500

Tax on capital gain = 0.25(6,500) = \$1,625

Machine A depreciation

$$\frac{P - S}{n} = \frac{54,000 - 0}{12} = 4,500 \text{ per year}$$

Alternate 1 - Keep Machine A for 11 more years

Year	Before Tax Cash Flow	Straight Line Depreciation	Taxable Income	40% Income Taxes	After Tax Cash Flow
0	-20,000		+6,500*	-1,625**	-18,375
1	0	4,500	-4,500	+1,800	+1,800
2	0	4,500	-4,500	+1,800	+1,800
3	0	4,500	-4,500	+1,800	+1,800
4-11	0	0	0	0	0

*Capital gain **A minus represents a disbursement for taxes.

After tax annual cost = [18,375 - 1800(P/A,10%,3)][(A/P,10%,11)]
= [18,375 - 1800(2.487)][0.1540]
= \$2,140

Alternate 2 - Buy Machine B

Machine B depreciation

$$\frac{P - S}{n} = \frac{42,000 - 0}{12} = \$3,500 \text{ per year}$$

Year	Before Tax Cash Flow	Straight Line Depreciation	Taxable Income	40% Income Taxes	After Tax Cash Flow
0	-42,000				-42,000
1	+2,500	3,500	-1,000	-400	+2,900
2-12	+2,500	3,500	-1,000	-400	+2,900
13-20	+2,500	0	+2,500	+1,000	+1,500

The annual cost equation may be written in several different forms. One way is:

After tax annual cost = [42,000-1400(P/A,10%,12)][A/P,10%,20] -1500
= [42,000-1400(6.814)][0.1175] -1500
= \$2,314

One aspect of the problem solution requires comment. In Alternate 1 the cash flow in Year 0 reflects the loss of income after capital gains tax from not selling Machine A. This is the preferred way to handle the current market value of the "defender" Machine A.

Choose the Alternative with the smaller annual cost - Keep A

ECONOMICS 11

GIVEN: The Highridge Water District needs an additional supply of water from Steep Creek. The engineer has selected two plans for comparison:

(A) GRAVITY PLAN: Divert water at a point 10 miles up Steep Creek and carry it through a pipeline by gravity to the District.

(B) PUMPING PLAN: Divert water at a point near the District and pump it through 2 miles of pipeline to the District. The pumping plant can be built in two stages, with one-half capacity installed initially and the other one-half 10 years later.

All costs are to be repaid within 40 years, with interest at 5%. Salvage values can be ignored. During the first 10 years, the average use of water will be less than during the remaining 30 years. Costs are as follows:

	Gravity	Pumping
Initial Investment	$2,800,000	$1,400,000
Additional Investment in 10th year	none	200,000
Operation, Maintenance and Replacements	10,000/yr	25,000/yr
Power Cost		
Average first 10 years	none	50,000/yr
Average next 30 years	none	100,000/yr

REQUIRED: Select the more economical plan. Show your computations.

SOLUTION

Equivalent Uniform Annual Cost Comparison

Gravity Plan

Initial Investment 2,800,000(A/P,5%,40) = 2,800,000(0.0583)	163,240
Operation, Maintenance & Replacements	10,000
Annual Cost =	$173,240

Pumping Plan

Initial Investment 1,400,000(A/P,5%,40) = 1,400,000(0.0583)	81,620
Additional Investment in 10th year 200,000(P/F,5%,10)(A/P,5%,40) 200,000(0.6139)(0.0583)	7,160
Operation, Maintenance & Replacements	25,000
Power Cost 50,000/year for 40 years	50,000
Additional 50,000/year for last 30 years 50,000(F/A,5%,30)(A/F,5%,40) 50,000(66.439)(0.0083)	27,570
Annual Cost =	$191,350

Choose the alternative with the smaller equivalent uniform annual cost. Choose the Gravity Plan.

Alternate Method Computation:

Present Worth Comparison

Gravity Plan

Initial Investment	2,800,000
Operation, Maintenance & Replacements 10,000(P/A,5%,40) = 10,000(17.159)	171,590
Present Worth =	$2,971,590

Pumping Plan

Initial Investment	1,400,000
Additional Investment in 10th year 200,000(P/F,5%,10) = 200,000(0.6139)	122,780
Operation, Maintenance & Replacements 25,000(P/A,5%,40) = 25,000(17.159)	428,980
Power First 50,000/year 50,000(P/A,5%,40) = 50,000(17.159)	857,950
Additional 50,000/year for last 30 years 50,000(P/A,5%,30)(P/F,5%,10) 50,000(15.372)(0.6139)	471,840
Present Worth =	$3,281,550

Choose the alternative with the smaller present worth of cost. Choose the Gravity Plan.

ECONOMICS 12

Five years ago a dam was constructed on a stream for the purpose of impounding irrigation water. It was also expected to provide flood protection for the area below the dam. Last winter a one-hundred-year flood caused extensive damage both to the dam and to the surrounding area. This may not have been surprising in view of the fact that the dam was designed for a fifty-year flood.

It is estimated that the cost to repair the dam now will be $250,000. Damage in the valley below probably amounts to $750,000. If the spillway is redesigned and certain other improvements are made at a total cost of $500,000 (including repair), the dam may be expected to withstand a one-hundred-year flood without sustaining damage. However, since the storage capacity will not be increased, the probability of damage to the valley will be unchanged. Additional storage would be required to eliminate concern for possible future damage. A second dam can be constructed up the river from the existing dam for $1,000,000. The capacity of the second dam would be more than adequate to provide the desired protection. If the second dam is built, redesign of the existing dam will not be necessary. Repairs to the existing dam will be required regardless.

The valley development is expected to be complete in ten years. A new flood in the meantime may average a $1,000,000 loss, after that time you assume the probable cost will remain at $2,000,000. This is an obvious oversimplification, but is felt to be a reasonable approximation.

In reviewing the decision that must be made now, the investigating board is weighing the recommendation of another consulting engineer who recommends no other action than the repair of the spillway. His argument is that another such flood would not occur for another one hundred years. He concedes that a fifty-year flood is also likely to cause damage, but the spillway would be adequate and the total damage would be about $200,000. Likewise, a twenty-five-year flood would also cause damage amounting to about $50,000.

The consulting engineer also reminds the review board that the community is now taxed to repay a twenty-year serial bond issue of $5,000,000 at 5 percent interest, floated five years ago to finance the original construction. The dam was "sold" to the people then as being good for a fifty-year life, and they are understandably angry over the present situation. The main structure is still good, and its life expectancy now is still fifty years.

REQUIRED:

Give your economic evaluation of the situation. Support your argument with an economic analysis based on an equivalent annual cost comparison of the three alternatives suggested. There may be other alternatives which have not been explored in the above discussion, but you are not expected to extend the problem by considering such, nor are you expected to discuss political or irreducible factors. Point out any errors in the other engineer's reasoning.

SOLUTION

	Flood	Probability of Damage in any year = 1/year	Downstream Damage	Spillway Damage
	25 year	0.04	50,000	-
	50 year	0.02	200,000	-
In next 10 years	100 year	0.01	1,000,000	250,000
Thereafter	100 year	0.01	2,000,000	250,000

Note that we have assumed the $1 million and $2 million in projected damages do not include spillway damage.

Three Alternatives

1. Repair the existing dam ($250,000).
   Make no alterations.
2. Repair the dam and redesign the spillway to carry a 100-year flood ($500,000).
3. Repair the existing dam ($250,000).
   Build a flood control dam upstream ($1,000,000).

Equivalent Uniform Annual Cost

In economic analysis we need be interested only in the differences between the alternatives. Since at least $250,000 worth of work will be done now on the existing dam in all alternatives, this $250,000 may either be included or ignored. Here it will be left out of the analysis.

Alternative 1.

Spillway Damage

The probability that the spillway capacity will be equaled or exceeded in any year is 0.02.

Damage if spillway capacity is exceeded = $250,000.

Expected annual cost of spillway damage
= Damage(Probability of its occurrence)
= 250,000(0.02) = $5000.

Alternative 1 - Continued

Downstream Damage During Next 10 Years

Flood	Probability that flow will be equaled or exceeded*	Damage	Incremental damage over more frequent flood	Annual Cost of flood risk
25 yr	0.04	50,000	50,000	2000
50 yr	0.02	200,000	150,000	3000
100 yr	0.01	1,000,000	800,000	8000

Next 10-year expected annual cost of downstream damage: $13,000

*An N-year flood will be equaled or exceeded at an average interval of N years.

Downstream Damage After 10 Years

Following the same logic as above,

Expected annual cost of downstream damage
= 2000 + 3000 + 0.01(2,000,000 - 200,000) = $23,000

Present Worth of Cost of Expected Spillway Damage plus Expected Downstream Damage

PW = 5000(P/A,7%,50) + 13,000(P/A,7%,10)
+ 23,000(P/A,7%,40)(P/F,7%,10)

= 5000(13.801) + 13,000(7.024) + 23,000(13.332)(0.5083)
= 69,005 + 91,312 + 155,863
= 316,180

Equivalent Annual Cost = 316,180(A/P,7%,50) = 316,180(0.0725)
= $22,920.

Alternative 2.

(Repair the dam and redesign the spillway.)

Additional cost to redesign and reconstruct the spillway = 250,000

Downstream Damage - Same as Alternative 1.

Present Worth of Cost to Reconstruct Spillway plus Expected Downstream Damage

PW = 250,000 + 13,000(P/A,7%,10)
+ 23,000(P/A,7%,40)(P/F,7%,10)

= 250,000 + 13,000(7.024) + 23,000(13.332)(0.5083)
= 250,000 + 91,312 + 155,863
= 497,175

Equivalent Annual Cost = 497,175(A/P,7%,50) = 497,175(0.0725)
= $36,050.

Alternative 3.

(Repair the dam and build flood control dam upstream.)

Cost of flood control dam = 1,000,000

Equivalent Annual Cost = 1,000,000(A/P,7%,50)
= 1,000,000(0.0725)
= $72,500.

Since we are dealing under conditions of risk, it is not possible to make an absolute statement concerning which alternative will result in the least cost to the community. Using a probabilistic approach, however, we can select the alternate which is most likely to result in the least equivalent cost. From the Equivalent Uniform Annual Cost (EUAC) calculations we see that Alternative 1 (Do nothing but repair the damage.) is most likely to result in the least equivalent cost to the community.

One must be careful not to confuse the frequency of a flood and when it might be expected to occur. The occurrence of a 100-year flood this year is no guarantee that it won't happen again next year. In any 50 year period, for example, there are 4 chances in 10 that a 100-year flood (or greater) will occur.

It was suggested by a consulting engineer that no alterations be made now because the project was originally designed to last fifty years. This reasoning is not sound. In the five years since the original construction a number of changes may have taken place that were not anticipated in the earlier planning. More rapid development, for example, might result in substantially higher damage estimates than had been used when the project was planned.

ECONOMICS 13

Disregarding income taxes, calculate the rate of return on the following investment opportunities.

A. Invest $100 now.
Receive two payments of $109.46 - one at the end of year 3 and one at the end of year 6.

B. Invest $100 now.
Receive $30.07 at the end of years 1, 2, 3, 4, 5, & 6.

SOLUTION

A. Set Present Worth of Cost = Present Worth of Benefits

$$100 = 109.46(P/F,i\%,3) + 109.46(P/F,i\%,6)$$

Solve by trial & error

Try i = 15%

$$100 = 109.46(0.6575) + 109.46(0.4323)$$
$$= 109.46(1.0898) = 119.29$$

The PW of Benefits is greater than the PW of Cost, indicating that the interest rate i is too low.

Try i = 20%

$$100 = 109.46(0.5787) + 109.46(0.3349)$$
$$= 109.46(0.9136) = 100.00$$

Here the PW of Cost = PW of Benefits when i = 20%.

Therefore, the rate of return = 20%.

In most situations a trial i will not result in PW of Cost = PW of Benefits. A linear interpolation for i is usually required.

B. Set PW of Cost = PW of Benefits

$$100 = 30.07(P/A,i\%,6)$$

$$(P/A,i\%,6) = \frac{100}{30.07} = 3.326$$

Here we have only one compound interest factor. It can be looked up in interest tables to determine which value of i produces

$$(P/A,i\%,6) = 3.326$$

From interest tables we see that i is exactly 20%

ECONOMICS 14

An engineer is faced with the prospect of a fluctuating future budget for the maintenance of a particular machine. During each of the first five years $1000 per year will be budgeted. During the second five years the annual budget will be $1500 per year. In addition, $3500 will be budgeted for an overhaul of the machine at the end of the fourth year, and another $3500 for an overhaul at the end of the eighth year.

The engineer wonders what uniform annual expenditure would be equivalent to these fluctuating amounts, assuming compound interest at 6% per annum. Compute the equivalent uniform annual expenditure for the 10 year period.

SOLUTION

A diagram of the projected budget amounts is as follows:

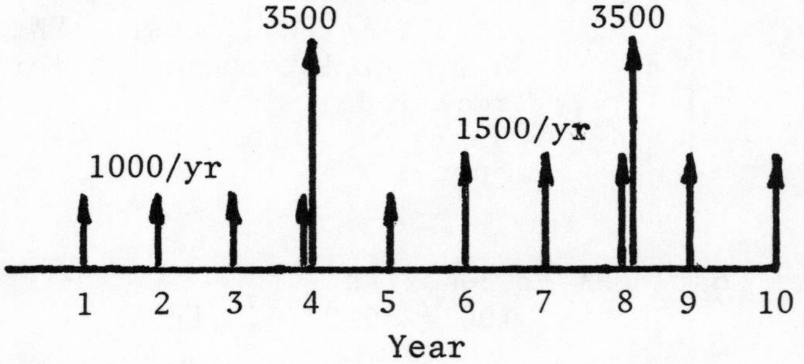

Compute the present worth of the various future amounts. Then compute the equivalent uniform annual amount.

$$
\begin{aligned}
PW &= 1000(P/A,6\%,5) + 3500(P/F,6\%,4) + 1500(P/A,6\%,5)(P/F,6\%,5) \\
&\quad + 3500(P/F,6\%,8) \\
&= 1000(4.212) + 3500(0.7921) + 1500(4.212)(0.7473) \\
&\quad + 3500(0.6274) \\
&= 4212 + 2772 + 4721 + 2196 = \$13{,}901
\end{aligned}
$$

$$
\begin{aligned}
\text{Equivalent uniform annual amount} &= 13{,}901(A/P,6\%,10) \\
&= 13{,}901(0.1359) = \underline{\$1889}
\end{aligned}
$$

ECONOMICS 15

A trust fund is to be established to

(a) Provide \$750,000 for the construction and \$250,000 for the initial equipment of an Electrical Engineering Laboratory.

(b) Pay the laboratory annual operating costs of \$150,000 per year.

(c) Pay for \$100,000 of replacement equipment every four years, beginning four years from now.

At 4% interest, how much money would be required in the trust fund to provide for the laboratory and equipment and its perpetual operation and equipment replacement?

SOLUTION

This problem is composed of three components.

Required money in trust fund = \$1,000,000 initial construction and equipment money
+ money for annual operation
+ money for periodic equipment replacement.

For $n = \infty$

$$A = Pi \quad \text{or } P = \frac{A}{i}$$

For (a) $P = \$1{,}000{,}000$

(b) $P = \frac{A}{i} = \frac{\$150{,}000}{0.04} = \$3{,}750{,}000$

(c) This situation is more complex. We do not know what equivalent amount A will provide for the necessary replacement equipment.

The perpetual series:

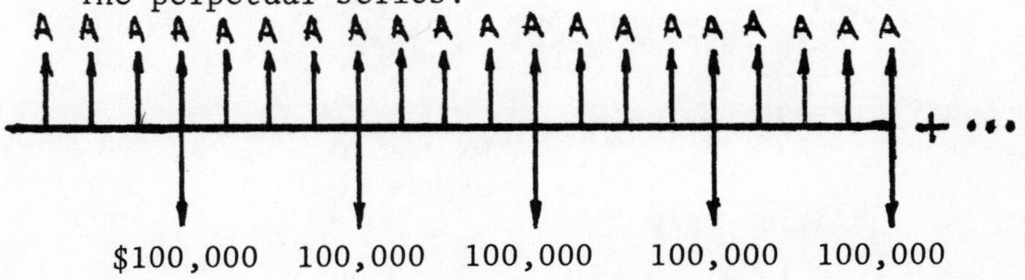

We can solve one portion of the perpetual series for A:

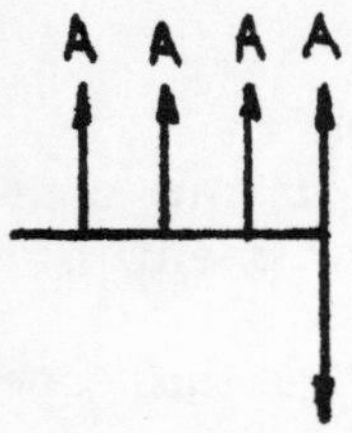

A = \$100,000(A/F,4%,4) = \$100,000(0.2355) = \$23,550

This value of A computed for the four year period is the same as the value of A for the perpetual series.

For the perpetual series:

$$P = \frac{A}{i} = \frac{\$23,550}{0.04} = \$589,000$$

Required money in trust fund = \$1,000,000 + 3,750,000 + 589,000
= \$5,339,000.

ECONOMICS 16

A machine part, operating in a corrosive atmosphere, is made of low-carbon steel, costs \$350 installed and lasts six years. If the part is treated for corrosion resistance it will cost \$700 installed. How long must the treated part last to be a better investment if money is worth 7%?

SOLUTION

The annual cost of the untreated part is
\$350(A/P,7%,6) = 350(0.2098) = \$73.43

The annual cost of the treated part must be at least this low, so
\$73.43 = \$700(A/P,7%,n)

$$(A/P,7\%,n) = \frac{73.43}{700} = 0.1049$$

From compound interest tables:

n	(A/P,7%,n)
16 yrs	0.1059
17 yrs	0.1024

By linear interpolation we get

$$n = 16 + (1)\,\frac{0.1059 - 0.1049}{0.1059 - 0.1024} = 16 + \frac{0.0010}{0.0035} = 16.3 \text{ years}$$

Thus the treated part must last longer than 16.3 years to be a better investment than the untreated part.

ECONOMICS 17

An engineer must select a pump for a particular situation. A supplier offers a pump of all-iron construction with an efficiency of 70% at a price of \$5600. Another supplier offers a pump of all-monel construction with an efficiency of 75% for \$14,700. Iron contamination of the product can be tolerated. The monel pump will last for 20 years and have a \$2000 salvage value.

Installation cost for either pump is \$1000. Both pumps have the same routine maintenance cost. The required input power at 100% efficiency is 42 kilowatts. Operation will be 24 hours per day for 300 days per year. Power costs \$0.005 per million joules.

With interest at 10%, what minimum service life must be obtained from the all-iron pump to justify its selection. Assume it has no salvage value.

SOLUTION

Kilowatt-hours x 3.6×10^6 = joules

Iron Pump

$$\text{Power} = 42 \text{ kW} \times \frac{1}{0.70 \text{ eff}} \times 24 \frac{\text{hrs}}{\text{day}} \times 300 \frac{\text{days}}{\text{year}} = 432{,}000 \text{ kWh}$$

$$\text{Annual Power Cost} = 432{,}000 \times 3.6 \times 10^6 \times \frac{\$0.005}{10^6} = \$7776$$

Monel Pump

$$\text{Annual Power Cost} = 42 \times \frac{1}{0.75} \times 24 \times 300 \times 3.6 \times 10^6 \times \frac{\$0.005}{10^6}$$

$$= \$7256$$

Equivalent Uniform Annual Costs

$$\text{EUAC}_{\text{iron pump}} = (5600 + 1000)(A/P,10\%,n) + 7776$$

$$\text{EUAC}_{\text{monel pump}} = (14{,}700 + 1000)(A/P,10\%,20) - 2000(A/F,10\%,20) + 7256$$

$$= (15{,}700)(0.1175) - 2000(0.0175) + 7256 = \$9066$$

Set $\text{EUAC}_{\text{iron pump}} = \text{EUAC}_{\text{monel pump}}$ and solve for the unknown n.

$$6600(A/P,10\%,n) + 7776 = 9066$$

$$(A/P,10\%,n) = \frac{9066 - 7776}{6600} = 0.1955$$

From the 10% compound interest table, n is about 7-1/2 years.

This is the required minimum service life of the iron pump.

ECONOMICS 18

Given the following data for two machines:

	A	B
Original Cost	\$55,000	\$75,000
Annual Operation Expense	9,500	7,200
Annual Maintenance Expense	5,000	3,000
Property Taxes, Insurance, and Other Annual Costs	1,700	2,250
Total Annual Costs	16,200	12,450

REQUIRED:

(a) With an interest rate of 10%, at what service life do these two pieces of equipment have an equivalent uniform annual cost?

(b) If the interest rate was 6%, what effect would this have on the average service life at which these machines were equivalent?

(c) Give reasons or show computations to support your answer in (b).

SOLUTION

(a) The problem may be most readily solved by setting the Present Worth of Cost of Machine A equal to the Present Worth of Cost of Machine B and solving for the unknown service life, N.

$$PW_A = PW_B$$

$$55{,}000 + 16{,}200(P/A,10\%,N) = 75{,}000 + 12{,}450(P/A,10\%,N)$$

$$(P/A,10\%,N) = \frac{75{,}000 - 55{,}000}{16{,}200 - 12{,}450} = \frac{20{,}000}{3{,}750} = 5.33$$

From the 10% compound interest table:

$$(P/A,10\%,8yrs) = 5.335$$

Therefore, the equivalent uniform annual costs are the same at a service life of 8 years.

(b) At a 0% interest rate the breakeven service life would be $\frac{\$20{,}000}{\$3{,}750} = 5.33$ years.

(At 0% interest the Series Present Worth factor equals N.) Thus Machine A is preferred if the actual service life is expected to be less than 5.33 years.

At a 10% interest rate the effect of future annual costs is reduced, with the result that Machine A is preferred for an expected service life not in excess of 8 years. (For an expected service life of over 8 years, of course, Machine B is preferred.)

(c) Following the logic of (a), we see that the problem has been changed so that now (P/A,6%,N) = 5.33

From compound interest tables: (P/A,6%,6yrs) = 4.92
(P/A,6%,7yrs) = 5.58

By linear interpolation the service life at which the machines have equivalent annual costs is:

$$6 + \frac{5.33 - 4.92}{5.58 - 4.92} = \underline{6.6 \text{ years}}$$

ECONOMICS 19

You have to analyze the economics of four plans available to you for the expansion of your firm's manufacturing facilities. Your market studies indicate that maximum sales under all plans are reasonably assured. Revenues and expenses shown in the tabulation below can be used for analysis with a very high probability of their reflecting the true market and investment potential.

Minimum after-income-tax return that you can accept on your 100% equity investment is 6%. Life of the equipment is estimated at 20 years. This life is also to be used for computing income tax depreciation on a straight line basis. Income taxes are to be computed at an assumed 50% rate. The equipment will have no value at the end of its life.

	Plan I	Plan II	Plan III	Plan IV
Investment: Original cost of equipment	$200,000	$250,000	$350,000	$500,000
Annual Gross Revenue	44,800	60,200	101,300	117,900
Total Operating Expenses, excluding income taxes	20,000	27,500	50,200	52,500
Net Annual Revenue:	$24,800	32,700	51,100	65,400

REQUIRED: Analyze the four plans, giving consideration to both total and incremental investment. Recommend which one, if any, should be adopted. Are there other factors that should be considered?

SOLUTION

Compute the After Tax Rate of Return for each Plan

Where income taxes are to be considered, the normal procedure is as follows:

1. Compute the Net Annual Revenue before income taxes.
2. Compute the schedule of depreciation charges for tax purposes. In this problem straight line depreciation has been specified.
3. Taxable income is the net annual revenue minus the depreciation.
4. Income Tax is the tax rate (50%) times the taxable income.
5. After Tax Cash Flow is the net annual revenue minus income taxes.
6. Knowing the Investment and the After Tax Cash Flow, we can set:

   Present Worth of Cost = Present Worth of Benefits

   In this problem, with a constant After Tax Cash Flow,

   Investment = After Tax Cash Flow x (P/A,i%,20yrs)

   This equation is solved for i, the after tax rate of return.

These steps for a single alternative are normally organized in horizontal tabular form. This multiple alternative problem may be more conveniently presented as a vertical table.

	Plan I	Plan II	Plan III	Plan IV	
Investment	200,000	250,000	350,000	500,000	
Net Annual Revenue	\$24,800	\$32,700	\$51,100	\$65,400	(1)
Annual Depreciation $= \frac{\text{Investment}}{\text{20 yrs}}$	10,000	12,500	17,500	25,000	(2)
Taxable Income = (1) - (2)	14,800	20,200	33,600	40,400	(3)
Income Taxes = 50% of (3)	7,400	10,100	16,800	20,200	(4)
After Tax Cash Flow = (1) - (4)	17,400	22,600	34,300	45,200	(5)
Ratio: $\frac{\text{Investment}}{\text{After Tax Cash Flow}}$	11.49	11.06	10.20	11.06	

Since the ratio $\frac{\text{Investment}}{\text{After Tax Cash Flow}}$ equals (P/A,i%,20yrs) in this problem, interpolate in the compound interest tables to obtain i - the after tax rate of return.

	Plan I	Plan II	Plan III	Plan IV
(P/A,i%,20yrs) =	11.49	11.06	10.20	11.06
After Tax Rate of Return	< 6%	6.5%	7.5%	6.5%

At this point we have computed the overall rate of return for each alternative. We must compare the overall rate of return for each alternative with the minimum acceptable rate of return (6% in this problem) and reject any alternatives that fail to meet this criterion. Since Plan I yields less than 6%, it should be rejected.

Thus we are left with 3 mutually exclusive alternatives. Which one of the three should be selected? At this point in the analysis we do not know. It would be incorrect to conclude that the Plan with the greatest rate of return is the alternative that should be selected.

The proper analysis method is to examine each separable increment of investment (incremental analysis) to see if it provides a satisfactory increment of benefits. Or stated another way: Each separable increment of investment must produce the minimum acceptable rate of return. If it does not yield this rate of return the increment of investment should not be made.

Incremental analysis is as follows:

A. Rank order the remaining alternatives (after discarding any that fail to meet the overall rate of return criterion) in order of increasing investment.

B. Beginning with the lowest acceptable investment (Plan II), compute the incremental investment and incremental benefits between it (Plan II) and the next higher investment (Plan III).

	Plan II	Plan III
Investment	\$250,000	\$350,000
After Tax Cash Flow	22,600	34,300

Incremental Investment = \$350,000 - 250,000 = \$100,000
Incremental Benefits = 34,300 - 22,600 = 11,700

$$\text{Ratio } \frac{\Delta \text{ Investment}}{\Delta \text{ After Tax Cash Flow}} = \frac{100,000}{11,700} = 8.55 = (P/A,i\%,20yrs)$$

By interpolation, Incremental rate of return = 9.9%

Since the incremental rate of return exceeds the minimum acceptable rate of return (6%), the incremental investment in an attractive one. Thus if the choice were limited to either Plan II or Plan III, we would select Plan III.

C. Take the outcome from Step B. In this analysis the increment of investment was justified, but in other analyses this may or may not be the case.

Go to the next higher investment alternative (Plan IV) and compare it to the outcome from Step B.

	Plan III	Plan IV
Investment	\$350,000	\$500,000
After Tax Cash Flow	34,300	45,200

Incremental Investment = \$500,000 - 350,000 = \$150,000
Incremental Benefits = 45,200 - 34,300 = 10,900

$$\text{Ratio } \frac{\Delta \text{Investment}}{\Delta \text{After Tax Cash Flow}} = \frac{150{,}000}{10{,}900} = 13.76 = (P/A, i\%, 20\text{yrs})$$

By interpolation, Incremental rate of return is 3.9%

Here the incremental rate of return is less than the minimum acceptable rate of return, hence the increment of investment is not acceptable. In choosing between Plan III and Plan IV we would select Plan III.

D. The incremental analysis would be continued, always comparing the best one of the alternatives examined to date with the next larger investment alternative. In this problem there is nothing beyond Plan IV, so the analysis has been completed.

Conclusion: Select Plan III.

Other factors to be considered:

Possible constraints that may not have been considered.
- The ability to finance the required investment

Long term marketing considerations.
- Each alternative results in a different share of the total market for this firm.

Intangible considerations
- Safety, etc.

Compound Interest Factors

½%

	SINGLE PAYMENT		UNIFORM PAYMENT SERIES				GRADIENT SERIES	
	Compound Amount Factor	Present Worth Factor	Sinking Fund Factor	Capital Recovery Factor	Compound Amount Factor	Present Worth Factor	Gradient Uniform Series	Gradient Present Worth
n	Find *F* Given *P* *F/P*	Find *P* Given *F* *P/F*	Find *A* Given *F* *A/F*	Find *A* Given *P* *A/P*	Find *F* Given *A* *F/A*	Find *P* Given *A* *P/A*	Find *A* Given *G* *A/G*	Find *P* Given *G* *P/G*
1	1.005	.9950	1.0000	1.0050	1.000	.995	0	0
2	1.010	.9901	.4988	.5038	2.005	1.985	.499	.990
3	1.015	.9851	.3317	.3367	3.015	2.970	.997	2.960
4	1.020	.9802	.2481	.2531	4.030	3.950	1.494	5.901
5	1.025	.9754	.1980	.2030	5.050	4.926	1.990	9.803
6	1.030	.9705	.1646	.1696	6.076	5.896	2.485	14.655
7	1.036	.9657	.1407	.1457	7.106	6.862	2.980	20.449
8	1.041	.9609	.1228	.1278	8.141	7.823	3.474	27.176
9	1.046	.9561	.1089	.1139	9.182	8.779	3.967	34.824
10	1.051	.9513	.0978	.1028	10.228	9.730	4.459	43.386
11	1.056	.9466	.0887	.0937	11.279	10.677	4.950	52.853
12	1.062	.9419	.0811	.0861	12.336	11.619	5.441	63.214
13	1.067	.9372	.0746	.0796	13.397	12.556	5.930	74.460
14	1.072	.9326	.0691	.0741	14.464	13.489	6.419	86.583
15	1.078	.9279	.0644	.0694	15.537	14.417	6.907	99.574
16	1.083	.9233	.0602	.0652	16.614	15.340	7.394	113.424
17	1.088	.9187	.0565	.0615	17.697	16.259	7.880	128.123
18	1.094	.9141	.0532	.0582	18.786	17.173	8.366	143.663
19	1.099	.9096	.0503	.0553	19.880	18.082	8.850	160.036
20	1.105	.9051	.0477	.0527	20.979	18.987	9.334	177.232
21	1.110	.9006	.0453	.0503	22.084	19.888	9.817	195.243
22	1.116	.8961	.0431	.0481	23.194	20.784	10.299	214.061
23	1.122	.8916	.0411	.0461	24.310	21.676	10.781	233.677
24	1.127	.8872	.0393	.0443	25.432	22.563	11.261	254.082
25	1.133	.8828	.0377	.0427	26.559	23.446	11.741	275.269
26	1.138	.8784	.0361	.0411	27.692	24.324	12.220	297.228
27	1.144	.8740	.0347	.0397	28.830	25.198	12.698	319.952
28	1.150	.8697	.0334	.0384	29.975	26.068	13.175	343.433
29	1.156	.8653	.0321	.0371	31.124	26.933	13.651	367.663
30	1.161	.8610	.0310	.0360	32.280	27.794	14.126	392.632
36	1.197	.8356	.0254	.0304	39.336	32.871	16.962	557.560
40	1.221	.8191	.0226	.0276	44.159	36.172	18.836	681.335
48	1.270	.7871	.0185	.0235	54.098	42.580	22.544	959.919
50	1.283	.7793	.0177	.0227	56.645	44.143	23.462	1035.697
52	1.296	.7716	.0169	.0219	59.218	45.690	24.378	1113.816
60	1.349	.7414	.0143	.0193	69.770	51.726	28.006	1448.646
70	1.418	.7053	.0120	.0170	83.566	58.939	32.468	1913.643
72	1.432	.6983	.0116	.0166	86.409	60.340	33.350	2012.348
80	1.490	.6710	.0102	.0152	98.068	65.802	36.847	2424.646
84	1.520	.6577	.0096	.0146	104.074	68.453	38.576	2640.664
90	1.567	.6383	.0088	.0138	113.311	72.331	41.145	2976.077
96	1.614	.6195	.0081	.0131	122.829	76.095	43.685	3324.185
100	1.647	.6073	.0077	.0127	129.334	78.543	45.361	3562.793
104	1.680	.5953	.0074	.0124	135.970	80.942	47.025	3806.286
120	1.819	.5496	.0061	.0111	163.879	90.073	53.551	4823.505
240	3.310	.3021	.0022	.0072	462.041	139.581	96.113	13415.540
360	6.023	.1660	.0010	.0060	1004.515	166.792	128.324	21403.304
480	10.957	.0913	.0005	.0055	1991.491	181.748	151.795	27588.357

1% Compound Interest Factors 1%

	SINGLE PAYMENT		UNIFORM PAYMENT SERIES				GRADIENT SERIES	
	Compound Amount Factor	Present Worth Factor	Sinking Fund Factor	Capital Recovery Factor	Compound Amount Factor	Present Worth Factor	Gradient Uniform Series	Gradient Present Worth
n	Find *F* Given *P* *F/P*	Find *P* Given *F* *P/F*	Find *A* Given *F* *A/F*	Find *A* Given *P* *A/P*	Find *F* Given *A* *F/A*	Find *P* Given *A* *P/A*	Find *A* Given G *A/G*	Find *P* Given G *P/G*
1	1.010	.9901	1.0000	1.0100	1.000	.990	0	0
2	1.020	.9803	.4975	.5075	2.010	1.970	.498	.980
3	1.030	.9706	.3300	.3400	3.030	2.941	.993	2.921
4	1.041	.9610	.2463	.2563	4.060	3.902	1.488	5.804
5	1.051	.9515	.1960	.2060	5.101	4.853	1.980	9.610
6	1.062	.9420	.1625	.1725	6.152	5.795	2.471	14.321
7	1.072	.9327	.1386	.1486	7.214	6.728	2.960	19.917
8	1.083	.9235	.1207	.1307	8.286	7.652	3.448	26.381
9	1.094	.9143	.1067	.1167	9.369	8.566	3.934	33.696
10	1.105	.9053	.0956	.1056	10.462	9.471	4.418	41.843
11	1.116	.8963	.0865	.0965	11.567	10.368	4.901	50.807
12	1.127	.8874	.0788	.0888	12.683	11.255	5.381	60.569
13	1.138	.8787	.0724	.0824	13.809	12.134	5.861	71.113
14	1.149	.8700	.0669	.0769	14.947	13.004	6.338	82.422
15	1.161	.8613	.0621	.0721	16.097	13.865	6.814	94.481
16	1.173	.8528	.0579	.0679	17.258	14.718	7.289	107.273
17	1.184	.8444	.0543	.0643	18.430	15.562	7.761	120.783
18	1.196	.8360	.0510	.0610	19.615	16.398	8.232	134.996
19	1.208	.8277	.0481	.0581	20.811	17.226	8.702	149.895
20	1.220	.8195	.0454	.0554	22.019	18.046	9.169	165.466
21	1.232	.8114	.0430	.0530	23.239	18.857	9.635	181.695
22	1.245	.8034	.0409	.0509	24.472	19.660	10.100	198.566
23	1.257	.7954	.0389	.0489	25.716	20.456	10.563	216.066
24	1.270	.7876	.0371	.0471	26.973	21.243	11.024	234.180
25	1.282	.7798	.0354	.0454	28.243	22.023	11.483	252.894
26	1.295	.7720	.0339	.0439	29.526	22.795	11.941	272.196
27	1.308	.7644	.0324	.0424	30.821	23.560	12.397	292.070
28	1.321	.7568	.0311	.0411	32.129	24.316	12.852	312.505
29	1.335	.7493	.0299	.0399	33.450	25.066	13.304	333.486
30	1.348	.7419	.0287	.0387	34.785	25.808	13.756	355.002
36	1.431	.6989	.0232	.0332	43.077	30.108	16.428	494.621
40	1.489	.6717	.0205	.0305	48.886	32.835	18.178	596.856
48	1.612	.6203	.0163	.0263	61.223	37.974	21.598	820.146
50	1.645	.6080	.0155	.0255	64.463	39.196	22.436	879.418
52	1.678	.5961	.0148	.0248	67.769	40.394	23.269	939.918
60	1.817	.5504	.0122	.0222	81.670	44.955	26.533	1192.806
70	2.007	.4983	.0099	.0199	100.676	50.169	30.470	1528.647
72	2.047	.4885	.0096	.0196	104.710	51.150	31.239	1597.867
80	2.217	.4511	.0082	.0182	121.672	54.888	34.249	1879.877
84	2.307	.4335	.0077	.0177	130.672	56.648	35.717	2023.315
90	2.449	.4084	.0069	.0169	144.863	59.161	37.872	2240.567
96	2.599	.3847	.0063	.0163	159.927	61.528	39.973	2459.430
100	2.705	.3697	.0059	.0159	170.481	63.029	41.343	2605.776
104	2.815	.3553	.0055	.0155	181.464	64.471	42.688	2752.182
120	3.300	.3030	.0043	.0143	230.039	69.701	47.835	3334.115
240	10.893	.0918	.0010	.0110	989.255	90.819	75.739	6878.602
360	35.950	.0278	.0003	.0103	3494.694	97.218	89.699	8720.432
480	118.648	.0084	.0001	.0101	11764.773	99.157	95.920	9511.158

Compound Interest Factors

1½%

	SINGLE PAYMENT		UNIFORM PAYMENT SERIES				GRADIENT SERIES	
	Compound Amount Factor	Present Worth Factor	Sinking Fund Factor	Capital Recovery Factor	Compound Amount Factor	Present Worth Factor	Gradient Uniform Series	Gradient Present Worth
n	Find *F* Given *P* *F/P*	Find *P* Given *F* *P/F*	Find *A* Given *F* *A/F*	Find *A* Given *P* *A/P*	Find *F* Given *A* *F/A*	Find *P* Given *A* *P/A*	Find *A* Given *G* *A/G*	Find *P* Given *G* *P/G*
1	1.015	.9852	1.0000	1.0150	1.000	.985	0	0
2	1.030	.9707	.4963	.5113	2.015	1.956	.496	.971
3	1.046	.9563	.3284	.3434	3.045	2.912	.990	2.883
4	1.061	.9422	.2444	.2594	4.091	3.854	1.481	5.710
5	1.077	.9283	.1941	.2091	5.152	4.783	1.970	9.423
6	1.093	.9145	.1605	.1755	6.230	5.697	2.457	13.996
7	1.110	.9010	.1366	.1516	7.323	6.598	2.940	19.402
8	1.126	.8877	.1186	.1336	8.433	7.486	3.422	25.616
9	1.143	.8746	.1046	.1196	9.559	8.361	3.901	32.612
10	1.161	.8617	.0934	.1084	10.703	9.222	4.377	40.367
11	1.178	.8489	.0843	.0993	11.863	10.071	4.851	48.857
12	1.196	.8364	.0767	.0917	13.041	10.908	5.323	58.057
13	1.214	.8240	.0702	.0852	14.237	11.732	5.792	67.945
14	1.232	.8118	.0647	.0797	15.450	12.543	6.258	78.499
15	1.250	.7999	.0599	.0749	16.682	13.343	6.722	89.697
16	1.269	.7880	.0558	.0708	17.932	14.131	7.184	101.518
17	1.288	.7764	.0521	.0671	19.201	14.908	7.643	113.940
18	1.307	.7649	.0488	.0638	20.489	15.673	8.100	126.943
19	1.327	.7536	.0459	.0609	21.797	16.426	8.554	140.508
20	1.347	.7425	.0432	.0582	23.124	17.169	9.006	154.615
21	1.367	.7315	.0409	.0559	24.471	17.900	9.455	169.245
22	1.388	.7207	.0387	.0537	25.838	18.621	9.902	184.380
23	1.408	.7100	.0367	.0517	27.225	19.331	10.346	200.001
24	1.430	.6995	.0349	.0499	28.634	20.030	10.788	216.090
25	1.451	.6892	.0333	.0483	30.063	20.720	11.228	232.631
26	1.473	.6790	.0317	.0467	31.514	21.399	11.665	249.607
27	1.495	.6690	.0303	.0453	32.987	22.068	12.099	267.000
28	1.517	.6591	.0290	.0440	34.481	22.727	12.531	284.796
29	1.540	.6494	.0278	.0428	35.999	23.376	12.961	302.978
30	1.563	.6398	.0266	.0416	37.539	24.016	13.388	321.531
36	1.709	.5851	.0212	.0362	47.276	27.661	15.901	439.830
40	1.814	.5513	.0184	.0334	54.268	29.916	17.528	524.357
48	2.043	.4894	.0144	.0294	69.565	34.043	20.667	703.546
50	2.105	.4750	.0136	.0286	73.683	35.000	21.428	749.964
52	2.169	.4611	.0128	.0278	77.925	35.929	22.179	796.877
60	2.443	.4093	.0104	.0254	96.215	39.380	25.093	988.167
70	2.835	.3527	.0082	.0232	122.364	43.155	28.529	1231.166
72	2.921	.3423	.0078	.0228	128.077	43.845	29.189	1279.794
80	3.291	.3039	.0065	.0215	152.711	46.407	31.742	1473.074
84	3.493	.2863	.0060	.0210	166.173	47.579	32.967	1568.514
90	3.819	.2619	.0053	.0203	187.930	49.210	34.740	1709.544
96	4.176	.2395	.0047	.0197	211.720	50.702	36.438	1847.473
100	4.432	.2256	.0044	.0194	228.803	51.625	37.530	1937.451
104	4.704	.2126	.0040	.0190	246.934	52.494	38.589	2025.705
120	5.969	.1675	.0030	.0180	331.288	55.498	42.519	2359.711
240	35.633	.0281	.0004	.0154	2308.854	64.796	59.737	3870.691
360	212.704	.0047	.0001	.0151	14113.586	66.353	64.966	4310.716
480	1269.698	.0008		.0150	84579.837	66.614	66.288	4415.741

2% Compound Interest Factors 2%

n	SINGLE PAYMENT: Compound Amount Factor, Find F Given P, F/P	SINGLE PAYMENT: Present Worth Factor, Find P Given F, P/F	UNIFORM PAYMENT SERIES: Sinking Fund Factor, Find A Given F, A/F	UNIFORM PAYMENT SERIES: Capital Recovery Factor, Find A Given P, A/P	UNIFORM PAYMENT SERIES: Compound Amount Factor, Find F Given A, F/A	UNIFORM PAYMENT SERIES: Present Worth Factor, Find P Given A, P/A	GRADIENT SERIES: Gradient Uniform Series, Find A Given G, A/G	GRADIENT SERIES: Gradient Present Worth, Find P Given G, P/G
1	1.020	.9804	1.0000	1.0200	1.000	.980	0	0
2	1.040	.9612	.4950	.5150	2.020	1.942	.495	.961
3	1.061	.9423	.3268	.3468	3.060	2.884	.987	2.846
4	1.082	.9238	.2426	.2626	4.122	3.808	1.475	5.617
5	1.104	.9057	.1922	.2122	5.204	4.713	1.960	9.240
6	1.126	.8880	.1585	.1785	6.308	5.601	2.442	13.680
7	1.149	.8706	.1345	.1545	7.434	6.472	2.921	18.903
8	1.172	.8535	.1165	.1365	8.583	7.325	3.396	24.878
9	1.195	.8368	.1025	.1225	9.755	8.162	3.868	31.572
10	1.219	.8203	.0913	.1113	10.950	8.983	4.337	38.955
11	1.243	.8043	.0822	.1022	12.169	9.787	4.802	46.998
12	1.268	.7885	.0746	.0946	13.412	10.575	5.264	55.671
13	1.294	.7730	.0681	.0881	14.680	11.348	5.723	64.948
14	1.319	.7579	.0626	.0826	15.974	12.106	6.179	74.800
15	1.346	.7430	.0578	.0778	17.293	12.849	6.631	85.202
16	1.373	.7284	.0537	.0737	18.639	13.578	7.080	96.129
17	1.400	.7142	.0500	.0700	20.012	14.292	7.526	107.555
18	1.428	.7002	.0467	.0667	21.412	14.992	7.968	119.458
19	1.457	.6864	.0438	.0638	22.841	15.678	8.407	131.814
20	1.486	.6730	.0412	.0612	24.297	16.351	8.843	144.600
21	1.516	.6598	.0388	.0588	25.783	17.011	9.276	157.796
22	1.546	.6468	.0366	.0566	27.299	17.658	9.705	171.379
23	1.577	.6342	.0347	.0547	28.845	18.292	10.132	185.331
24	1.608	.6217	.0329	.0529	30.422	18.914	10.555	199.630
25	1.641	.6095	.0312	.0512	32.030	19.523	10.974	214.259
26	1.673	.5976	.0297	.0497	33.671	20.121	11.391	229.199
27	1.707	.5859	.0283	.0483	35.344	20.707	11.804	244.431
28	1.741	.5744	.0270	.0470	37.051	21.281	12.214	259.939
29	1.776	.5631	.0258	.0458	38.792	21.844	12.621	275.706
30	1.811	.5521	.0246	.0446	40.568	22.396	13.025	291.716
36	2.040	.4902	.0192	.0392	51.994	25.489	15.381	392.040
40	2.208	.4529	.0166	.0366	60.402	27.355	16.889	461.993
48	2.587	.3865	.0126	.0326	79.354	30.673	19.756	605.966
50	2.692	.3715	.0118	.0318	84.579	31.424	20.442	642.361
52	2.800	.3571	.0111	.0311	90.016	32.145	21.116	678.785
60	3.281	.3048	.0088	.0288	114.052	34.761	23.696	823.698
70	4.000	.2500	.0067	.0267	149.978	37.499	26.663	999.834
72	4.161	.2403	.0063	.0263	158.057	37.984	27.223	1034.056
80	4.875	.2051	.0052	.0252	193.772	39.745	29.357	1166.787
84	5.277	.1895	.0047	.0247	213.867	40.526	30.362	1230.419
90	5.943	.1683	.0040	.0240	247.157	41.587	31.793	1322.170
96	6.693	.1494	.0035	.0235	284.647	42.529	33.137	1409.297
100	7.245	.1380	.0032	.0232	312.232	43.098	33.986	1464.753
104	7.842	.1275	.0029	.0229	342.092	43.624	34.799	1518.087
120	10.765	.0929	.0020	.0220	488.258	45.355	37.711	1710.416
240	115.889	.0086	.0002	.0202	5744.437	49.569	47.911	2374.880
360	1247.561	.0008		.0200	62328.056	49.960	49.711	2483.568
480	13430.199	.0001		.0200	671459.945	49.996	49.964	2498.027

n	SINGLE PAYMENT		UNIFORM PAYMENT SERIES				GRADIENT SERIES	
	Compound Amount Factor	Present Worth Factor	Sinking Fund Factor	Capital Recovery Factor	Compound Amount Factor	Present Worth Factor	Gradient Uniform Series	Gradient Present Worth
	Find F Given P F/P	Find P Given F P/F	Find A Given F A/F	Find A Given P A/P	Find F Given A F/A	Find P Given A P/A	Find A Given G A/G	Find P Given G P/G
1	1.025	.9756	1.0000	1.0250	1.000	.976	0	0
2	1.051	.9518	.4938	.5188	2.025	1.927	.494	.952
3	1.077	.9286	.3251	.3501	3.076	2.856	.984	2.809
4	1.104	.9060	.2408	.2658	4.153	3.762	1.469	5.527
5	1.131	.8839	.1902	.2152	5.256	4.646	1.951	9.062
6	1.160	.8623	.1565	.1815	6.388	5.508	2.428	13.374
7	1.189	.8413	.1325	.1575	7.547	6.349	2.901	18.421
8	1.218	.8207	.1145	.1395	8.736	7.170	3.370	24.167
9	1.249	.8007	.1005	.1255	9.955	7.971	3.836	30.572
10	1.280	.7812	.0893	.1143	11.203	8.752	4.296	37.603
11	1.312	.7621	.0801	.1051	12.483	9.514	4.753	45.225
12	1.345	.7436	.0725	.0975	13.796	10.258	5.206	53.404
13	1.379	.7254	.0660	.0910	15.140	10.983	5.655	62.109
14	1.413	.7077	.0605	.0855	16.519	11.691	6.100	71.309
15	1.448	.6905	.0558	.0808	17.932	12.381	6.540	80.976
16	1.485	.6736	.0516	.0766	19.380	13.055	6.977	91.080
17	1.522	.6572	.0479	.0729	20.865	13.712	7.409	101.595
18	1.560	.6412	.0447	.0697	22.386	14.353	7.838	112.495
19	1.599	.6255	.0418	.0668	23.946	14.979	8.262	123.755
20	1.639	.6103	.0391	.0641	25.545	15.589	8.682	135.350
21	1.680	.5954	.0368	.0618	27.183	16.185	9.099	147.257
22	1.722	.5809	.0346	.0596	28.863	16.765	9.511	159.456
23	1.765	.5667	.0327	.0577	30.584	17.332	9.919	171.923
24	1.809	.5529	.0309	.0559	32.349	17.885	10.324	184.639
25	1.854	.5394	.0293	.0543	34.158	18.424	10.724	197.584
26	1.900	.5262	.0278	.0528	36.012	18.951	11.121	210.740
27	1.948	.5134	.0264	.0514	37.912	19.464	11.513	224.089
28	1.996	.5009	.0251	.0501	39.860	19.965	11.902	237.612
29	2.046	.4887	.0239	.0489	41.856	20.454	12.286	251.295
30	2.098	.4767	.0228	.0478	43.903	20.930	12.667	265.120
31	2.150	.4651	.0217	.0467	46.000	21.395	13.044	279.074
32	2.204	.4538	.0208	.0458	48.150	21.849	13.417	293.141
33	2.259	.4427	.0199	.0449	50.354	22.292	13.786	307.307
34	2.315	.4319	.0190	.0440	52.613	22.724	14.151	321.560
35	2.373	.4214	.0182	.0432	54.928	23.145	14.512	335.887
40	2.685	.3724	.0148	.0398	67.403	25.103	16.262	408.222
45	3.038	.3292	.0123	.0373	81.516	26.833	17.918	480.807
50	3.437	.2909	.0103	.0353	97.484	28.362	19.484	552.608
55	3.889	.2572	.0087	.0337	115.551	29.714	20.961	622.828
60	4.400	.2273	.0074	.0324	135.992	30.909	22.352	690.866
65	4.978	.2009	.0063	.0313	159.118	31.965	23.660	756.281
70	5.632	.1776	.0054	.0304	185.284	32.898	24.888	818.764
75	6.372	.1569	.0047	.0297	214.888	33.723	26.039	878.115
80	7.210	.1387	.0040	.0290	248.383	34.452	27.117	934.218
85	8.157	.1226	.0035	.0285	286.279	35.096	28.123	987.027
90	9.229	.1084	.0030	.0280	329.154	35.666	29.063	1036.550
95	10.442	.0958	.0026	.0276	377.664	36.169	29.938	1082.838
100	11.814	.0846	.0023	.0273	432.549	36.614	30.752	1125.975

3% Compound Interest Factors 3%

	SINGLE PAYMENT		UNIFORM PAYMENT SERIES				GRADIENT SERIES	
	Compound Amount Factor	Present Worth Factor	Sinking Fund Factor	Capital Recovery Factor	Compound Amount Factor	Present Worth Factor	Gradient Uniform Series	Gradient Present Worth
n	Find *F* Given *P* *F/P*	Find *P* Given *F* *P/F*	Find *A* Given *F* *A/F*	Find *A* Given *P* *A/P*	Find *F* Given *A* *F/A*	Find *P* Given *A* *P/A*	Find *A* Given *G* *A/G*	Find *P* Given *G* *P/G*
1	1.030	.9709	1.0000	1.0300	1.000	.971	0	0
2	1.061	.9426	.4926	.5226	2.030	1.913	.493	.943
3	1.093	.9151	.3235	.3535	3.091	2.829	.980	2.773
4	1.126	.8885	.2390	.2690	4.184	3.717	1.463	5.438
5	1.159	.8626	.1884	.2184	5.309	4.580	1.941	8.889
6	1.194	.8375	.1546	.1846	6.468	5.417	2.414	13.076
7	1.230	.8131	.1305	.1605	7.662	6.230	2.882	17.955
8	1.267	.7894	.1125	.1425	8.892	7.020	3.345	23.481
9	1.305	.7664	.0984	.1284	10.159	7.786	3.803	29.612
10	1.344	.7441	.0872	.1172	11.464	8.530	4.256	36.309
11	1.384	.7224	.0781	.1081	12.808	9.253	4.705	43.533
12	1.426	.7014	.0705	.1005	14.192	9.954	5.148	51.248
13	1.469	.6810	.0640	.0940	15.618	10.635	5.587	59.420
14	1.513	.6611	.0585	.0885	17.086	11.296	6.021	68.014
15	1.558	.6419	.0538	.0838	18.599	11.938	6.450	77.000
16	1.605	.6232	.0496	.0796	20.157	12.561	6.874	86.348
17	1.653	.6050	.0460	.0760	21.762	13.166	7.294	96.028
18	1.702	.5874	.0427	.0727	23.414	13.754	7.708	106.014
19	1.754	.5703	.0398	.0698	25.117	14.324	8.118	116.279
20	1.806	.5537	.0372	.0672	26.870	14.877	8.523	126.799
21	1.860	.5375	.0349	.0649	28.676	15.415	8.923	137.550
22	1.916	.5219	.0327	.0627	30.537	15.937	9.319	148.509
23	1.974	.5067	.0308	.0608	32.453	16.444	9.709	159.657
24	2.033	.4919	.0290	.0590	34.426	16.936	10.095	170.971
25	2.094	.4776	.0274	.0574	36.459	17.413	10.477	182.434
26	2.157	.4637	.0259	.0559	38.553	17.877	10.853	194.026
27	2.221	.4502	.0246	.0546	40.710	18.327	11.226	205.731
28	2.288	.4371	.0233	.0533	42.931	18.764	11.593	217.532
29	2.357	.4243	.0221	.0521	45.219	19.188	11.956	229.414
30	2.427	.4120	.0210	.0510	47.575	19.600	12.314	241.361
31	2.500	.4000	.0200	.0500	50.003	20.000	12.668	253.361
32	2.575	.3883	.0190	.0490	52.503	20.389	13.017	265.399
33	2.652	.3770	.0182	.0482	55.078	20.766	13.362	277.464
34	2.732	.3660	.0173	.0473	57.730	21.132	13.702	289.544
35	2.814	.3554	.0165	.0465	60.462	21.487	14.037	301.627
40	3.262	.3066	.0133	.0433	75.401	23.115	15.650	361.750
45	3.782	.2644	.0108	.0408	92.720	24.519	17.156	420.632
50	4.384	.2281	.0089	.0389	112.797	25.730	18.558	477.480
55	5.082	.1968	.0073	.0373	136.072	26.774	19.860	531.741
60	5.892	.1697	.0061	.0361	163.053	27.676	21.067	583.053
65	6.830	.1464	.0051	.0351	194.333	28.453	22.184	631.201
70	7.918	.1263	.0043	.0343	230.594	29.123	23.215	676.087
75	9.179	.1089	.0037	.0337	272.631	29.702	24.163	717.698
80	10.641	.0940	.0031	.0331	321.363	30.201	25.035	756.087
85	12.336	.0811	.0026	.0326	377.857	30.631	25.835	791.353
90	14.300	.0699	.0023	.0323	443.349	31.002	26.567	823.630
95	16.578	.0603	.0019	.0319	519.272	31.323	27.235	853.074
100	19.219	.0520	.0016	.0316	607.288	31.599	27.844	879.854

3½% Compound Interest Factors 3½%

	SINGLE PAYMENT		UNIFORM PAYMENT SERIES				GRADIENT SERIES	
	Compound Amount Factor	Present Worth Factor	Sinking Fund Factor	Capital Recovery Factor	Compound Amount Factor	Present Worth Factor	Gradient Uniform Series	Gradient Present Worth
	Find *F* Given *P* *F/P*	Find *P* Given *F* *P/F*	Find *A* Given *F* *A/F*	Find *A* Given *P* *A/P*	Find *F* Given *A* *F/A*	Find *P* Given *A* *P/A*	Find *A* Given *G* *A/G*	Find *P* Given *G* *P/G*
n								
1	1.035	.9662	1.0000	1.0350	1.000	.966	0	0
2	1.071	.9335	.4914	.5264	2.035	1.900	.491	.934
3	1.109	.9019	.3219	.3569	3.106	2.802	.977	2.737
4	1.148	.8714	.2373	.2723	4.215	3.673	1.457	5.352
5	1.188	.8420	.1865	.2215	5.362	4.515	1.931	8.720
6	1.229	.8135	.1527	.1877	6.550	5.329	2.400	12.787
7	1.272	.7860	.1285	.1635	7.779	6.115	2.863	17.503
8	1.317	.7594	.1105	.1455	9.052	6.874	3.320	22.819
9	1.363	.7337	.0964	.1314	10.368	7.608	3.771	28.689
10	1.411	.7089	.0852	.1202	11.731	8.317	4.217	35.069
11	1.460	.6849	.0761	.1111	13.142	9.002	4.657	41.919
12	1.511	.6618	.0685	.1035	14.602	9.663	5.091	49.198
13	1.564	.6394	.0621	.0971	16.113	10.303	5.520	56.871
14	1.619	.6178	.0566	.0916	17.677	10.921	5.943	64.902
15	1.675	.5969	.0518	.0868	19.296	11.517	6.361	73.259
16	1.734	.5767	.0477	.0827	20.971	12.094	6.773	81.909
17	1.795	.5572	.0440	.0790	22.705	12.651	7.179	90.824
18	1.857	.5384	.0408	.0758	24.500	13.190	7.580	99.977
19	1.923	.5202	.0379	.0729	26.357	13.710	7.975	109.339
20	1.990	.5026	.0354	.0704	28.280	14.212	8.365	118.888
21	2.059	.4856	.0330	.0680	30.269	14.698	8.749	128.600
22	2.132	.4692	.0309	.0659	32.329	15.167	9.128	138.452
23	2.206	.4533	.0290	.0640	34.460	15.620	9.502	148.424
24	2.283	.4380	.0273	.0623	36.667	16.058	9.870	158.497
25	2.363	.4231	.0257	.0607	38.950	16.482	10.233	168.653
26	2.446	.4088	.0242	.0592	41.313	16.890	10.590	178.874
27	2.532	.3950	.0229	.0579	43.759	17.285	10.942	189.144
28	2.620	.3817	.0216	.0566	46.291	17.667	11.289	199.448
29	2.712	.3687	.0204	.0554	48.911	18.036	11.631	209.773
30	2.807	.3563	.0194	.0544	51.623	18.392	11.967	220.106
31	2.905	.3442	.0184	.0534	54.429	18.736	12.299	230.432
32	3.007	.3326	.0174	.0524	57.335	19.069	12.625	240.743
33	3.112	.3213	.0166	.0516	60.341	19.390	12.946	251.026
34	3.221	.3105	.0158	.0508	63.453	19.701	13.262	261.271
35	3.334	.3000	.0150	.0500	66.674	20.001	13.573	271.471
40	3.959	.2526	.0118	.0468	84.550	21.355	15.055	321.491
45	4.702	.2127	.0095	.0445	105.782	22.495	16.417	369.308
50	5.585	.1791	.0076	.0426	130.998	23.456	17.666	414.370
55	6.633	.1508	.0062	.0412	160.947	24.264	18.808	456.353
60	7.878	.1269	.0051	.0401	196.517	24.945	19.848	495.105
65	9.357	.1069	.0042	.0392	238.763	25.518	20.793	530.599
70	11.113	.0900	.0035	.0385	288.938	26.000	21.650	562.896
75	13.199	.0758	.0029	.0379	348.530	26.407	22.423	592.121
80	15.676	.0638	.0024	.0374	419.307	26.749	23.120	618.439
85	18.618	.0537	.0020	.0370	503.367	27.037	23.747	642.037
90	22.112	.0452	.0017	.0367	603.205	27.279	24.308	663.119
95	26.262	.0381	.0014	.0364	721.781	27.484	24.811	681.890
100	31.191	.0321	.0012	.0362	862.612	27.655	25.259	698.555

4% Compound Interest Factors 4%

	SINGLE PAYMENT		UNIFORM PAYMENT SERIES				GRADIENT SERIES	
	Compound Amount Factor	Present Worth Factor	Sinking Fund Factor	Capital Recovery Factor	Compound Amount Factor	Present Worth Factor	Gradient Uniform Series	Gradient Present Worth
n	Find *F* Given *P* *F/P*	Find *P* Given *F* *P/F*	Find *A* Given *F* *A/F*	Find *A* Given *P* *A/P*	Find *F* Given *A* *F/A*	Find *P* Given *A* *P/A*	Find *A* Given *G* *A/G*	Find *P* Given *G* *P/G*
1	1.040	.9615	1.0000	1.0400	1.000	.962	0	0
2	1.082	.9246	.4902	.5302	2.040	1.886	.490	.925
3	1.125	.8890	.3203	.3603	3.122	2.775	.974	2.703
4	1.170	.8548	.2355	.2755	4.246	3.630	1.451	5.267
5	1.217	.8219	.1846	.2246	5.416	4.452	1.922	8.555
6	1.265	.7903	.1508	.1908	6.633	5.242	2.386	12.506
7	1.316	.7599	.1266	.1666	7.898	6.002	2.843	17.066
8	1.369	.7307	.1085	.1485	9.214	6.733	3.294	22.181
9	1.423	.7026	.0945	.1345	10.583	7.435	3.739	27.801
10	1.480	.6756	.0833	.1233	12.006	8.111	4.177	33.881
11	1.539	.6496	.0741	.1141	13.486	8.760	4.609	40.377
12	1.601	.6246	.0666	.1066	15.026	9.385	5.034	47.248
13	1.665	.6006	.0601	.1001	16.627	9.986	5.453	54.455
14	1.732	.5775	.0547	.0947	18.292	10.563	5.866	61.962
15	1.801	.5553	.0499	.0899	20.024	11.118	6.272	69.735
16	1.873	.5339	.0458	.0858	21.825	11.652	6.672	77.744
17	1.948	.5134	.0422	.0822	23.698	12.166	7.066	85.958
18	2.026	.4936	.0390	.0790	25.645	12.659	7.453	94.350
19	2.107	.4746	.0361	.0761	27.671	13.134	7.834	102.893
20	2.191	.4564	.0336	.0736	29.778	13.590	8.209	111.565
21	2.279	.4388	.0313	.0713	31.969	14.029	8.578	120.341
22	2.370	.4220	.0292	.0692	34.248	14.451	8.941	129.202
23	2.465	.4057	.0273	.0673	36.618	14.857	9.297	138.128
24	2.563	.3901	.0256	.0656	39.083	15.247	9.648	147.101
25	2.666	.3751	.0240	.0640	41.646	15.622	9.993	156.104
26	2.772	.3607	.0226	.0626	44.312	15.983	10.331	165.121
27	2.883	.3468	.0212	.0612	47.084	16.330	10.664	174.138
28	2.999	.3335	.0200	.0600	49.968	16.663	10.991	183.142
29	3.119	.3207	.0189	.0589	52.966	16.984	11.312	192.121
30	3.243	.3083	.0178	.0578	56.085	17.292	11.627	201.062
31	3.373	.2965	.0169	.0569	59.328	17.588	11.937	209.956
32	3.508	.2851	.0159	.0559	62.701	17.874	12.241	218.792
33	3.648	.2741	.0151	.0551	66.210	18.148	12.540	227.563
34	3.794	.2636	.0143	.0543	69.858	18.411	12.832	236.261
35	3.946	.2534	.0136	.0536	73.652	18.665	13.120	244.877
40	4.801	.2083	.0105	.0505	95.026	19.793	14.477	286.530
45	5.841	.1712	.0083	.0483	121.029	20.720	15.705	325.403
50	7.107	.1407	.0066	.0466	152.667	21.482	16.812	361.164
55	8.646	.1157	.0052	.0452	191.159	22.109	17.807	393.689
60	10.520	.0951	.0042	.0442	237.991	22.623	18.697	422.997
65	12.799	.0781	.0034	.0434	294.968	23.047	19.491	449.201
70	15.572	.0642	.0027	.0427	364.290	23.395	20.196	472.479
75	18.945	.0528	.0022	.0422	448.631	23.680	20.821	493.041
80	23.050	.0434	.0018	.0418	551.245	23.915	21.372	511.116
85	28.044	.0357	.0015	.0415	676.090	24.109	21.857	526.938
90	34.119	.0293	.0012	.0412	827.983	24.267	22.283	540.737
95	41.511	.0241	.0010	.0410	1012.785	24.398	22.655	552.731
100	50.505	.0198	.0008	.0408	1237.624	24.505	22.980	563.125

Compound Interest Factors

4½%

	SINGLE PAYMENT		UNIFORM PAYMENT SERIES				GRADIENT SERIES	
	Compound Amount Factor	Present Worth Factor	Sinking Fund Factor	Capital Recovery Factor	Compound Amount Factor	Present Worth Factor	Gradient Uniform Series	Gradient Present Worth
n	Find *F* Given *P* *F/P*	Find *P* Given *F* *P/F*	Find *A* Given *F* *A/F*	Find *A* Given *P* *A/P*	Find *F* Given *A* *F/A*	Find *P* Given *A* *P/A*	Find *A* Given *G* *A/G*	Find *P* Given *G* *P/G*
1	1.045	.9569	1.0000	1.0450	1.000	.957	0	0
2	1.092	.9157	.4890	.5340	2.045	1.873	.489	.916
3	1.141	.8763	.3188	.3638	3.137	2.749	.971	2.668
4	1.193	.8386	.2337	.2787	4.278	3.588	1.445	5.184
5	1.246	.8025	.1828	.2278	5.471	4.390	1.912	8.394
6	1.302	.7679	.1489	.1939	6.717	5.158	2.372	12.233
7	1.361	.7348	.1247	.1697	8.019	5.893	2.824	16.642
8	1.422	.7032	.1066	.1516	9.380	6.596	3.269	21.565
9	1.486	.6729	.0926	.1376	10.802	7.269	3.707	26.948
10	1.553	.6439	.0814	.1264	12.288	7.913	4.138	32.743
11	1.623	.6162	.0722	.1172	13.841	8.529	4.562	38.905
12	1.696	.5897	.0647	.1097	15.464	9.119	4.978	45.391
13	1.772	.5643	.0583	.1033	17.160	9.683	5.387	52.163
14	1.852	.5400	.0528	.0978	18.932	10.223	5.789	59.182
15	1.935	.5167	.0481	.0931	20.784	10.740	6.184	66.416
16	2.022	.4945	.0440	.0890	22.719	11.234	6.572	73.833
17	2.113	.4732	.0404	.0854	24.742	11.707	6.953	81.404
18	2.208	.4528	.0372	.0822	26.855	12.160	7.327	89.102
19	2.308	.4333	.0344	.0794	29.064	12.593	7.695	96.901
20	2.412	.4146	.0319	.0769	31.371	13.008	8.055	104.780
21	2.520	.3968	.0296	.0746	33.783	13.405	8.409	112.715
22	2.634	.3797	.0275	.0725	36.303	13.784	8.755	120.689
23	2.752	.3634	.0257	.0707	38.937	14.148	9.096	128.683
24	2.876	.3477	.0240	.0690	41.689	14.495	9.429	136.680
25	3.005	.3327	.0224	.0674	44.565	14.828	9.756	144.665
26	3.141	.3184	.0210	.0660	47.571	15.147	10.077	152.625
27	3.282	.3047	.0197	.0647	50.711	15.451	10.391	160.547
28	3.430	.2916	.0185	.0635	53.993	15.743	10.698	168.420
29	3.584	.2790	.0174	.0624	57.423	16.022	10.999	176.232
30	3.745	.2670	.0164	.0614	61.007	16.289	11.295	183.975
31	3.914	.2555	.0154	.0604	64.752	16.544	11.583	191.640
32	4.090	.2445	.0146	.0596	68.666	16.789	11.866	199.220
33	4.274	.2340	.0137	.0587	72.756	17.023	12.143	206.707
34	4.466	.2239	.0130	.0580	77.030	17.247	12.414	214.096
35	4.667	.2143	.0123	.0573	81.497	17.461	12.679	221.380
40	5.816	.1719	.0093	.0543	107.030	18.402	13.917	256.099
45	7.248	.1380	.0072	.0522	138.850	19.156	15.020	287.732
50	9.033	.1107	.0056	.0506	178.503	19.762	15.998	316.145
55	11.256	.0888	.0044	.0494	227.918	20.248	16.860	341.375
60	14.027	.0713	.0035	.0485	289.498	20.638	17.617	363.571
65	17.481	.0572	.0027	.0477	366.238	20.951	18.278	382.947
70	21.784	.0459	.0022	.0472	461.870	21.202	18.854	399.750
75	27.147	.0368	.0017	.0467	581.044	21.404	19.354	414.242
80	33.830	.0296	.0014	.0464	729.558	21.565	19.785	426.680
85	42.158	.0237	.0011	.0461	914.632	21.695	20.157	437.309
90	52.537	.0190	.0009	.0459	1145.269	21.799	20.476	446.359
95	65.471	.0153	.0007	.0457	1432.684	21.883	20.749	454.039
100	81.589	.0123	.0006	.0456	1790.856	21.950	20.981	460.538

n	SINGLE PAYMENT: Compound Amount Factor, Find F Given P, F/P	SINGLE PAYMENT: Present Worth Factor, Find P Given F, P/F	UNIFORM PAYMENT SERIES: Sinking Fund Factor, Find A Given F, A/F	UNIFORM PAYMENT SERIES: Capital Recovery Factor, Find A Given P, A/P	UNIFORM PAYMENT SERIES: Compound Amount Factor, Find F Given A, F/A	UNIFORM PAYMENT SERIES: Present Worth Factor, Find P Given A, P/A	GRADIENT SERIES: Gradient Uniform Series, Find A Given G, A/G	GRADIENT SERIES: Gradient Present Worth, Find P Given G, P/G
1	1.050	.9524	1.0000	1.0500	1.000	.952	0	0
2	1.102	.9070	.4878	.5378	2.050	1.859	.488	.907
3	1.158	.8638	.3172	.3672	3.152	2.723	.967	2.635
4	1.216	.8227	.2320	.2820	4.310	3.546	1.439	5.103
5	1.276	.7835	.1810	.2310	5.526	4.329	1.903	8.237
6	1.340	.7462	.1470	.1970	6.802	5.076	2.358	11.968
7	1.407	.7107	.1228	.1728	8.142	5.786	2.805	16.232
8	1.477	.6768	.1047	.1547	9.549	6.463	3.245	20.970
9	1.551	.6446	.0907	.1407	11.027	7.108	3.676	26.127
10	1.629	.6139	.0795	.1295	12.578	7.722	4.099	31.652
11	1.710	.5847	.0704	.1204	14.207	8.306	4.514	37.499
12	1.796	.5568	.0628	.1128	15.917	8.863	4.922	43.624
13	1.886	.5303	.0565	.1065	17.713	9.394	5.322	49.988
14	1.980	.5051	.0510	.1010	19.599	9.899	5.713	56.554
15	2.079	.4810	.0463	.0963	21.579	10.380	6.097	63.288
16	2.183	.4581	.0423	.0923	23.657	10.838	6.474	70.160
17	2.292	.4363	.0387	.0887	25.840	11.274	6.842	77.140
18	2.407	.4155	.0355	.0855	28.132	11.690	7.203	84.204
19	2.527	.3957	.0327	.0827	30.539	12.085	7.557	91.328
20	2.653	.3769	.0302	.0802	33.066	12.462	7.903	98.488
21	2.786	.3589	.0280	.0780	35.719	12.821	8.242	105.667
22	2.925	.3418	.0260	.0760	38.505	13.163	8.573	112.846
23	3.072	.3256	.0241	.0741	41.430	13.489	8.897	120.009
24	3.225	.3101	.0225	.0725	44.502	13.799	9.214	127.140
25	3.386	.2953	.0210	.0710	47.727	14.094	9.524	134.228
26	3.556	.2812	.0196	.0696	51.113	14.375	9.827	141.259
27	3.733	.2678	.0183	.0683	54.669	14.643	10.122	148.223
28	3.920	.2551	.0171	.0671	58.403	14.898	10.411	155.110
29	4.116	.2429	.0160	.0660	62.323	15.141	10.694	161.913
30	4.322	.2314	.0151	.0651	66.439	15.372	10.969	168.623
31	4.538	.2204	.0141	.0641	70.761	15.593	11.238	175.233
32	4.765	.2099	.0133	.0633	75.299	15.803	11.501	181.739
33	5.003	.1999	.0125	.0625	80.064	16.003	11.757	188.135
34	5.253	.1904	.0118	.0618	85.067	16.193	12.006	194.417
35	5.516	.1813	.0111	.0611	90.320	16.374	12.250	200.581
40	7.040	.1420	.0083	.0583	120.800	17.159	13.377	229.545
45	8.985	.1113	.0063	.0563	159.700	17.774	14.364	255.315
50	11.467	.0872	.0048	.0548	209.348	18.256	15.223	277.915
55	14.636	.0683	.0037	.0537	272.713	18.633	15.966	297.510
60	18.679	.0535	.0028	.0528	353.584	18.929	16.606	314.343
65	23.840	.0419	.0022	.0522	456.798	19.161	17.154	328.691
70	30.426	.0329	.0017	.0517	588.529	19.343	17.621	340.841
75	38.833	.0258	.0013	.0513	756.654	19.485	18.018	351.072
80	49.561	.0202	.0010	.0510	971.229	19.596	18.353	359.646
85	63.254	.0158	.0008	.0508	1245.087	19.684	18.635	366.801
90	80.730	.0124	.0006	.0506	1594.607	19.752	18.871	372.749
95	103.035	.0097	.0005	.0505	2040.694	19.806	19.069	377.677
100	131.501	.0076	.0004	.0504	2610.025	19.848	19.234	381.749

6% Compound Interest Factors 6%

	SINGLE PAYMENT		UNIFORM PAYMENT SERIES				GRADIENT SERIES	
	Compound Amount Factor	Present Worth Factor	Sinking Fund Factor	Capital Recovery Factor	Compound Amount Factor	Present Worth Factor	Gradient Uniform Series	Gradient Present Worth
n	Find F Given P F/P	Find P Given F P/F	Find A Given F A/F	Find A Given P A/P	Find F Given A F/A	Find P Given A P/A	Find A Given G A/G	Find P Given G P/G
1	1.060	.9434	1.0000	1.0600	1.000	.943	0	0
2	1.124	.8900	.4854	.5454	2.060	1.833	.485	.890
3	1.191	.8396	.3141	.3741	3.184	2.673	.961	2.569
4	1.262	.7921	.2286	.2886	4.375	3.465	1.427	4.946
5	1.338	.7473	.1774	.2374	5.637	4.212	1.884	7.935
6	1.419	.7050	.1434	.2034	6.975	4.917	2.330	11.459
7	1.504	.6651	.1191	.1791	8.394	5.582	2.768	15.450
8	1.594	.6274	.1010	.1610	9.897	6.210	3.195	19.842
9	1.689	.5919	.0870	.1470	11.491	6.802	3.613	24.577
10	1.791	.5584	.0759	.1359	13.181	7.360	4.022	29.602
11	1.898	.5268	.0668	.1268	14.972	7.887	4.421	34.870
12	2.012	.4970	.0593	.1193	16.870	8.384	4.811	40.337
13	2.133	.4688	.0530	.1130	18.882	8.853	5.192	45.963
14	2.261	.4423	.0476	.1076	21.015	9.295	5.564	51.713
15	2.397	.4173	.0430	.1030	23.276	9.712	5.926	57.555
16	2.540	.3936	.0390	.0990	25.673	10.106	6.279	63.459
17	2.693	.3714	.0354	.0954	28.213	10.477	6.624	69.401
18	2.854	.3503	.0324	.0924	30.906	10.828	6.960	75.357
19	3.026	.3305	.0296	.0896	33.760	11.158	7.287	81.306
20	3.207	.3118	.0272	.0872	36.786	11.470	7.605	87.230
21	3.400	.2942	.0250	.0850	39.993	11.764	7.915	93.114
22	3.604	.2775	.0230	.0830	43.392	12.042	8.217	98.941
23	3.820	.2618	.0213	.0813	46.996	12.303	8.510	104.701
24	4.049	.2470	.0197	.0797	50.816	12.550	8.795	110.381
25	4.292	.2330	.0182	.0782	54.865	12.783	9.072	115.973
26	4.549	.2198	.0169	.0769	59.156	13.003	9.341	121.468
27	4.822	.2074	.0157	.0757	63.706	13.211	9.603	126.860
28	5.112	.1956	.0146	.0746	68.528	13.406	9.857	132.142
29	5.418	.1846	.0136	.0736	73.640	13.591	10.103	137.310
30	5.743	.1741	.0126	.0726	79.058	13.765	10.342	142.359
31	6.088	.1643	.0118	.0718	84.802	13.929	10.547	147.286
32	6.453	.1550	.0110	.0710	90.890	14.084	10.799	152.090
33	6.841	.1462	.0103	.0703	97.343	14.230	11.017	156.768
34	7.251	.1379	.0096	.0696	104.184	14.368	11.228	161.319
35	7.686	.1301	.0090	.0690	111.435	14.498	11.432	165.743
40	10.286	.0972	.0065	.0665	154.762	15.046	12.359	185.957
45	13.765	.0727	.0047	.0647	212.744	15.456	13.141	203.110
50	18.420	.0543	.0034	.0634	290.336	15.762	13.796	217.457
55	24.650	.0406	.0025	.0625	394.172	15.991	14.341	229.322
60	32.988	.0303	.0019	.0619	533.128	16.161	14.791	239.043
65	44.145	.0227	.0014	.0614	719.083	16.289	15.160	246.945
70	59.076	.0169	.0010	.0610	967.932	16.385	15.461	253.327
75	79.057	.0126	.0008	.0608	1300.949	16.456	15.706	258.453
80	105.796	.0095	.0006	.0606	1746.600	16.509	15.903	262.549
85	141.579	.0071	.0004	.0604	2342.982	16.549	16.062	265.810
90	189.465	.0053	.0003	.0603	3141.075	16.579	16.189	268.395
95	253.546	.0039	.0002	.0602	4209.104	16.601	16.290	270.437
100	339.302	.0029	.0002	.0602	5638.368	16.618	16.371	272.047

n	SINGLE PAYMENT		UNIFORM PAYMENT SERIES				GRADIENT SERIES	
	Compound Amount Factor	Present Worth Factor	Sinking Fund Factor	Capital Recovery Factor	Compound Amount Factor	Present Worth Factor	Gradient Uniform Series	Gradient Present Worth
	Find *F* Given *P* *F/P*	Find *P* Given *F* *P/F*	Find *A* Given *F* *A/F*	Find *A* Given *P* *A/P*	Find *F* Given *A* *F/A*	Find *P* Given *A* *P/A*	Find *A* Given *G* *A/G*	Find *P* Given *G* *P/G*
1	1.070	.9346	1.0000	1.0700	1.000	.935	0	0
2	1.145	.8734	.4831	.5531	2.070	1.808	.483	.873
3	1.225	.8163	.3111	.3811	3.215	2.624	.955	2.506
4	1.311	.7629	.2252	.2952	4.440	3.387	1.416	4.795
5	1.403	.7130	.1739	.2439	5.751	4.100	1.865	7.647
6	1.501	.6663	.1398	.2098	7.153	4.767	2.303	10.978
7	1.606	.6227	.1156	.1856	8.654	5.389	2.730	14.715
8	1.718	.5820	.0975	.1675	10.260	5.971	3.147	18.789
9	1.838	.5439	.0835	.1535	11.978	6.515	3.552	23.140
10	1.967	.5083	.0724	.1424	13.816	7.024	3.946	27.716
11	2.105	.4751	.0634	.1334	15.784	7.499	4.330	32.466
12	2.252	.4440	.0559	.1259	17.888	7.943	4.703	37.351
13	2.410	.4150	.0497	.1197	20.141	8.358	5.065	42.330
14	2.579	.3878	.0443	.1143	22.550	8.745	5.417	47.372
15	2.759	.3624	.0398	.1098	25.129	9.108	5.758	52.446
16	2.952	.3387	.0359	.1059	27.888	9.447	6.090	57.527
17	3.159	.3166	.0324	.1024	30.840	9.763	6.411	62.592
18	3.380	.2959	.0294	.0994	33.999	10.059	6.722	67.622
19	3.617	.2765	.0268	.0968	37.379	10.336	7.024	72.599
20	3.870	.2584	.0244	.0944	40.995	10.594	7.316	77.509
21	4.141	.2415	.0223	.0923	44.865	10.836	7.599	82.339
22	4.430	.2257	.0204	.0904	49.006	11.061	7.872	87.079
23	4.741	.2109	.0187	.0887	53.436	11.272	8.137	91.720
24	5.072	.1971	.0172	.0872	58.177	11.469	8.392	96.255
25	5.427	.1842	.0158	.0858	63.249	11.654	8.639	100.676
26	5.807	.1722	.0146	.0846	68.676	11.826	8.877	104.981
27	6.214	.1609	.0134	.0834	74.484	11.987	9.107	109.166
28	6.649	.1504	.0124	.0824	80.698	12.137	9.329	113.226
29	7.114	.1406	.0114	.0814	87.347	12.278	9.543	117.162
30	7.612	.1314	.0106	.0806	94.461	12.409	9.749	120.972
31	8.145	.1228	.0098	.0798	102.073	12.532	9.947	124.655
32	8.715	.1147	.0091	.0791	110.218	12.647	10.138	128.212
33	9.325	.1072	.0084	.0784	118.933	12.754	10.322	131.643
34	9.978	.1002	.0078	.0778	128.259	12.854	10.499	134.951
35	10.677	.0937	.0072	.0772	138.237	12.948	10.669	138.135
40	14.974	.0668	.0050	.0750	199.635	13.332	11.423	152.293
45	21.002	.0476	.0035	.0735	285.749	13.606	12.036	163.756
50	29.457	.0339	.0025	.0725	406.529	13.801	12.529	172.905
55	41.315	.0242	.0017	.0717	575.929	13.940	12.921	180.124
60	57.946	.0173	.0012	.0712	813.520	14.039	13.232	185.768
65	81.273	.0123	.0009	.0709	1146.755	14.110	13.476	190.145
70	113.989	.0088	.0006	.0706	1614.134	14.160	13.666	193.519
75	159.876	.0063	.0004	.0704	2269.657	14.196	13.814	196.104
80	224.234	.0045	.0003	.0703	3189.063	14.222	13.927	198.075
85	314.500	.0032	.0002	.0702	4478.576	14.240	14.015	199.572
90	441.103	.0023	.0002	.0702	6287.185	14.253	14.081	200.704
95	618.670	.0016	.0001	.0701	8823.854	14.263	14.132	201.558
100	867.716	.0012	.0001	.0701	12381.662	14.269	14.170	202.200

8% Compound Interest Factors 8%

	SINGLE PAYMENT		UNIFORM PAYMENT SERIES				GRADIENT SERIES	
	Compound Amount Factor	Present Worth Factor	Sinking Fund Factor	Capital Recovery Factor	Compound Amount Factor	Present Worth Factor	Gradient Uniform Series	Gradient Present Worth
n	Find *F* Given *P* *F/P*	Find *P* Given *F* *P/F*	Find *A* Given *F* *A/F*	Find *A* Given *P* *A/P*	Find *F* Given *A* *F/A*	Find *P* Given *A* *P/A*	Find *A* Given *G* *A/G*	Find *P* Given *G* *P/G*
1	1.080	.9259	1.0000	1.0800	1.000	.926	0	0
2	1.166	.8573	.4808	.5608	2.080	1.783	.481	.857
3	1.260	.7938	.3080	.3880	3.246	2.577	.949	2.445
4	1.360	.7350	.2219	.3019	4.506	3.312	1.404	4.650
5	1.469	.6806	.1705	.2505	5.867	3.993	1.846	7.372
6	1.587	.6302	.1363	.2163	7.336	4.623	2.276	10.523
7	1.714	.5835	.1121	.1921	8.923	5.206	2.694	14.024
8	1.851	.5403	.0940	.1740	10.637	5.747	3.099	17.806
9	1.999	.5002	.0801	.1601	12.488	6.247	3.491	21.808
10	2.159	.4632	.0690	.1490	14.487	6.710	3.871	25.977
11	2.332	.4289	.0601	.1401	16.645	7.139	4.240	30.266
12	2.518	.3971	.0527	.1327	18.977	7.536	4.596	34.634
13	2.720	.3677	.0465	.1265	21.495	7.904	4.940	39.046
14	2.937	.3405	.0413	.1213	24.215	8.244	5.273	43.472
15	3.172	.3152	.0368	.1168	27.152	8.559	5.594	47.886
16	3.426	.2919	.0330	.1130	30.324	8.851	5.905	52.264
17	3.700	.2703	.0296	.1096	33.750	9.122	6.204	56.588
18	3.996	.2502	.0267	.1067	37.450	9.372	6.492	60.843
19	4.316	.2317	.0241	.1041	41.446	9.604	6.770	65.013
20	4.661	.2145	.0219	.1019	45.762	9.818	7.037	69.090
21	5.034	.1987	.0198	.0998	50.423	10.017	7.294	73.063
22	5.437	.1839	.0180	.0980	55.457	10.201	7.541	76.926
23	5.871	.1703	.0164	.0964	60.893	10.371	7.779	80.673
24	6.341	.1577	.0150	.0950	66.765	10.529	8.007	84.300
25	6.848	.1460	.0137	.0937	73.106	10.675	8.225	87.804
26	7.396	.1352	.0125	.0925	79.954	10.810	8.435	91.184
27	7.988	.1252	.0114	.0914	87.351	10.935	8.636	94.439
28	8.627	.1159	.0105	.0905	95.339	11.051	8.829	97.569
29	9.317	.1073	.0096	.0896	103.966	11.158	9.013	100.574
30	10.063	.0994	.0088	.0888	113.283	11.258	9.190	103.456
31	10.868	.0920	.0081	.0881	123.346	11.350	9.358	106.216
32	11.737	.0852	.0075	.0875	134.214	11.435	9.520	108.857
33	12.676	.0789	.0069	.0869	145.951	11.514	9.674	111.382
34	13.690	.0730	.0063	.0863	158.627	11.587	9.821	113.792
35	14.785	.0676	.0058	.0858	172.317	11.655	9.961	116.092
40	21.725	.0460	.0039	.0839	259.057	11.925	10.570	126.042
45	31.920	.0313	.0026	.0826	386.506	12.108	11.045	133.733
50	46.902	.0213	.0017	.0817	573.770	12.233	11.411	139.593
55	68.914	.0145	.0012	.0812	848.923	12.319	11.690	144.006
60	101.257	.0099	.0008	.0808	1253.213	12.377	11.902	147.300
65	148.780	.0067	.0005	.0805	1847.248	12.416	12.060	149.739
70	218.606	.0046	.0004	.0804	2720.080	12.443	12.178	151.533
75	321.205	.0031	.0002	.0802	4002.557	12.461	12.266	152.845
80	471.955	.0021	.0002	.0802	5886.935	12.474	12.330	153.800
85	693.456	.0014	.0001	.0801	8655.706	12.482	12.377	154.492
90	1018.915	.0010	.0001	.0801	12723.939	12.488	12.412	154.993
95	1497.121	.0007	.0001	.0801	18701.507	12.492	12.437	155.352
100	2199.761	.0005		.0800	27484.516	12.494	12.455	155.611

9% Compound Interest Factors 9%

	SINGLE PAYMENT		UNIFORM PAYMENT SERIES				GRADIENT SERIES	
	Compound Amount Factor	Present Worth Factor	Sinking Fund Factor	Capital Recovery Factor	Compound Amount Factor	Present Worth Factor	Gradient Uniform Series	Gradient Present Worth
n	Find *F* Given *P* *F/P*	Find *P* Given *F* *P/F*	Find *A* Given *F* *A/F*	Find *A* Given *P* *A/P*	Find *F* Given *A* *F/A*	Find *P* Given *A* *P/A*	Find *A* Given *G* *A/G*	Find *P* Given *G* *P/G*
1	1.090	.9174	1.0000	1.0900	1.000	.917	0	0
2	1.188	.8417	.4785	.5685	2.090	1.759	.478	.842
3	1.295	.7722	.3051	.3951	3.278	2.531	.943	2.386
4	1.412	.7084	.2187	.3087	4.573	3.240	1.393	4.511
5	1.539	.6499	.1671	.2571	5.985	3.890	1.828	7.111
6	1.677	.5963	.1329	.2229	7.523	4.486	2.250	10.092
7	1.828	.5470	.1087	.1987	9.200	5.033	2.657	13.375
8	1.993	.5019	.0907	.1807	11.028	5.535	3.051	16.888
9	2.172	.4604	.0768	.1668	13.021	5.995	3.431	20.571
10	2.367	.4224	.0658	.1558	15.193	6.418	3.798	24.373
11	2.580	.3875	.0569	.1469	17.560	6.805	4.151	28.248
12	2.813	.3555	.0497	.1397	20.141	7.161	4.491	32.159
13	3.066	.3262	.0436	.1336	22.953	7.487	4.818	36.073
14	3.342	.2992	.0384	.1284	26.019	7.786	5.133	39.963
15	3.642	.2745	.0341	.1241	29.361	8.061	5.435	43.807
16	3.970	.2519	.0303	.1203	33.003	8.313	5.724	47.585
17	4.328	.2311	.0270	.1170	36.974	8.544	6.002	51.282
18	4.717	.2120	.0242	.1142	41.301	8.756	6.269	54.886
19	5.142	.1945	.0217	.1117	46.018	8.950	6.524	58.387
20	5.604	.1784	.0195	.1095	51.160	9.129	6.767	61.777
21	6.109	.1637	.0176	.1076	56.765	9.292	7.001	65.051
22	6.659	.1502	.0159	.1059	62.873	9.442	7.223	68.205
23	7.258	.1378	.0144	.1044	69.532	9.580	7.436	71.236
24	7.911	.1264	.0130	.1030	76.790	9.707	7.638	74.143
25	8.623	.1160	.0118	.1018	84.701	9.823	7.832	76.926
26	9.399	.1064	.0107	.1007	93.324	9.929	8.016	79.586
27	10.245	.0976	.0097	.0997	102.723	10.027	8.191	82.124
28	11.167	.0895	.0089	.0989	112.968	10.116	8.357	84.542
29	12.172	.0822	.0081	.0981	124.135	10.198	8.515	86.842
30	13.268	.0754	.0073	.0973	136.308	10.274	8.666	89.028
31	14.462	.0691	.0067	.0967	149.575	10.343	8.808	91.102
32	15.763	.0634	.0061	.0961	164.037	10.406	8.944	93.069
33	17.182	.0582	.0056	.0956	179.800	10.464	9.072	94.931
34	18.728	.0534	.0051	.0951	196.982	10.518	9.193	96.693
35	20.414	.0490	.0046	.0946	215.711	10.567	9.308	98.359
40	31.409	.0318	.0030	.0930	337.882	10.757	9.796	105.376
45	48.327	.0207	.0019	.0919	525.859	10.881	10.160	110.556
50	74.358	.0134	.0012	.0912	815.084	10.962	10.430	114.325
55	114.408	.0087	.0008	.0908	1260.092	11.014	10.626	117.036
60	176.031	.0057	.0005	.0905	1944.792	11.048	10.768	118.968
65	270.846	.0037	.0003	.0903	2998.288	11.070	10.870	120.334
70	416.730	.0024	.0002	.0902	4619.223	11.084	10.943	121.294
75	641.191	.0016	.0001	.0901	7113.232	11.094	10.994	121.965
80	986.552	.0010	.0001	.0901	10950.574	11.100	11.030	122.431
85	1517.932	.0007	.0001	.0901	16854.800	11.104	11.055	122.753
90	2335.527	.0004		.0900	25939.184	11.106	11.073	122.976
95	3593.497	.0003		.0900	39916.635	11.108	11.085	123.129
100	5529.041	.0002		.0900	61422.675	11.109	11.093	123.234

10% Compound Interest Factors 10%

	SINGLE PAYMENT		UNIFORM PAYMENT SERIES				GRADIENT SERIES	
	Compound Amount Factor	Present Worth Factor	Sinking Fund Factor	Capital Recovery Factor	Compound Amount Factor	Present Worth Factor	Gradient Uniform Series	Gradient Present Worth
n	Find *F* Given *P* *F/P*	Find *P* Given *F* *P/F*	Find *A* Given *F* *A/F*	Find *A* Given *P* *A/P*	Find *F* Given *A* *F/A*	Find *P* Given *A* *P/A*	Find *A* Given *G* *A/G*	Find *P* Given *G* *P/G*
1	1.100	.9091	1.0000	1.1000	1.000	.909	0	0
2	1.210	.8264	.4762	.5762	2.100	1.736	.476	.826
3	1.331	.7513	.3021	.4021	3.310	2.487	.937	2.329
4	1.464	.6830	.2155	.3155	4.641	3.170	1.381	4.378
5	1.611	.6209	.1638	.2638	6.105	3.791	1.810	6.862
6	1.772	.5645	.1296	.2296	7.716	4.355	2.224	9.684
7	1.949	.5132	.1054	.2054	9.487	4.868	2.622	12.763
8	2.144	.4665	.0874	.1874	11.436	5.335	3.004	16.029
9	2.358	.4241	.0736	.1736	13.579	5.759	3.372	19.421
10	2.594	.3855	.0627	.1627	15.937	6.145	3.725	22.891
11	2.853	.3505	.0540	.1540	18.531	6.495	4.064	26.396
12	3.138	.3186	.0468	.1468	21.384	6.814	4.388	29.901
13	3.452	.2897	.0408	.1408	24.523	7.103	4.699	33.377
14	3.797	.2633	.0357	.1357	27.975	7.367	4.996	36.800
15	4.177	.2394	.0315	.1315	31.772	7.606	5.279	40.152
16	4.595	.2176	.0278	.1278	35.950	7.824	5.549	43.416
17	5.054	.1978	.0247	.1247	40.545	8.022	5.807	46.582
18	5.560	.1799	.0219	.1219	45.599	8.201	6.053	49.640
19	6.116	.1635	.0195	.1195	51.159	8.365	6.286	52.583
20	6.727	.1486	.0175	.1175	57.275	8.514	6.508	55.407
21	7.400	.1351	.0156	.1156	64.002	8.649	6.719	58.110
22	8.140	.1228	.0140	.1140	71.403	8.772	6.919	60.689
23	8.954	.1117	.0126	.1126	79.543	8.883	7.108	63.146
24	9.850	.1015	.0113	.1113	88.497	8.985	7.288	65.481
25	10.835	.0923	.0102	.1102	98.347	9.077	7.458	67.696
26	11.918	.0839	.0092	.1092	109.182	9.161	7.619	69.794
27	13.110	.0763	.0083	.1083	121.100	9.237	7.770	71.777
28	14.421	.0693	.0075	.1075	134.210	9.307	7.914	73.650
29	15.863	.0630	.0067	.1067	148.631	9.370	8.049	75.415
30	17.449	.0573	.0061	.1061	164.494	9.427	8.176	77.077
31	19.194	.0521	.0055	.1055	181.943	9.479	8.296	78.640
32	21.114	.0474	.0050	.1050	201.138	9.526	8.409	80.108
33	23.225	.0431	.0045	.1045	222.252	9.569	8.515	81.486
34	25.548	.0391	.0041	.1041	245.477	9.609	8.615	82.777
35	28.102	.0356	.0037	.1037	271.024	9.644	8.709	83.987
40	45.259	.0221	.0023	.1023	442.593	9.779	9.096	88.953
45	72.890	.0137	.0014	.1014	718.905	9.863	9.374	92.454
50	117.391	.0085	.0009	.1009	1163.909	9.915	9.570	94.889
55	189.059	.0053	.0005	.1005	1880.591	9.947	9.708	96.562
60	304.482	.0033	.0003	.1003	3034.816	9.967	9.802	97.701
65	490.371	.0020	.0002	.1002	4893.707	9.980	9.867	98.471
70	789.747	.0013	.0001	.1001	7887.470	9.987	9.911	98.987
75	1271.895	.0008	.0001	.1001	12708.954	9.992	9.941	99.332
80	2048.400	.0005		.1000	20474.002	9.995	9.961	99.561
85	3298.969	.0003		.1000	32979.690	9.997	9.974	99.712
90	5313.023	.0002		.1000	53120.226	9.998	9.983	99.812
95	8556.676	.0001		.1000	85556.761	9.999	9.989	99.877
100	13780.612	.0001		.1000	137796.123	9.999	9.993	99.920

12% Compound Interest Factors 12%

	SINGLE PAYMENT		UNIFORM PAYMENT SERIES				GRADIENT SERIES	
	Compound Amount Factor	Present Worth Factor	Sinking Fund Factor	Capital Recovery Factor	Compound Amount Factor	Present Worth Factor	Gradient Uniform Series	Gradient Present Worth
n	Find F Given P F/P	Find P Given F P/F	Find A Given F A/F	Find A Given P A/P	Find F Given A F/A	Find P Given A P/A	Find A Given G A/G	Find P Given G P/G
1	1.120	.8929	1.0000	1.1200	1.000	.893	0	0
2	1.254	.7972	.4717	.5917	2.120	1.690	.472	.797
3	1.405	.7118	.2963	.4163	3.374	2.402	.925	2.221
4	1.574	.6355	.2092	.3292	4.779	3.037	1.359	4.127
5	1.762	.5674	.1574	.2774	6.353	3.605	1.775	6.397
6	1.974	.5066	.1232	.2432	8.115	4.111	2.172	8.930
7	2.211	.4523	.0991	.2191	10.089	4.564	2.551	11.644
8	2.476	.4039	.0813	.2013	12.300	4.968	2.913	14.471
9	2.773	.3606	.0677	.1877	14.776	5.328	3.257	17.356
10	3.106	.3220	.0570	.1770	17.549	5.650	3.585	20.254
11	3.479	.2875	.0484	.1684	20.655	5.938	3.895	23.129
12	3.896	.2567	.0414	.1614	24.133	6.194	4.190	25.952
13	4.363	.2292	.0357	.1557	28.029	6.424	4.468	28.702
14	4.887	.2046	.0309	.1509	32.393	6.628	4.732	31.362
15	5.474	.1827	.0268	.1468	37.280	6.811	4.980	33.920
16	6.130	.1631	.0234	.1434	42.753	6.974	5.215	36.367
17	6.866	.1456	.0205	.1405	48.884	7.120	5.435	38.697
18	7.690	.1300	.0179	.1379	55.750	7.250	5.643	40.908
19	8.613	.1161	.0158	.1358	63.440	7.366	5.838	42.998
20	9.646	.1037	.0139	.1339	72.052	7.469	6.020	44.968
21	10.804	.0926	.0122	.1322	81.699	7.562	6.191	46.819
22	12.100	.0826	.0108	.1308	92.503	7.645	6.351	48.554
23	13.552	.0738	.0096	.1296	104.603	7.718	6.501	50.178
24	15.179	.0659	.0085	.1285	118.155	7.784	6.641	51.693
25	17.000	.0588	.0075	.1275	133.334	7.843	6.771	53.105
26	19.040	.0525	.0067	.1267	150.334	7.896	6.892	54.418
27	21.325	.0469	.0059	.1259	169.374	7.943	7.005	55.637
28	23.884	.0419	.0052	.1252	190.699	7.984	7.110	56.767
29	26.750	.0374	.0047	.1247	214.583	8.022	7.207	57.814
30	29.960	.0334	.0041	.1241	241.333	8.055	7.297	58.782
31	33.555	.0298	.0037	.1237	271.293	8.085	7.381	59.676
32	37.582	.0266	.0033	.1233	304.848	8.112	7.459	60.501
33	42.092	.0238	.0029	.1229	342.429	8.135	7.530	61.261
34	47.143	.0212	.0026	.1226	384.521	8.157	7.596	61.961
35	52.800	.0189	.0023	.1223	431.663	8.176	7.658	62.605
40	93.051	.0107	.0013	.1213	767.091	8.244	7.899	65.116
45	163.988	.0061	.0007	.1207	1358.230	8.283	8.057	66.734
50	289.002	.0035	.0004	.1204	2400.018	8.304	8.160	67.762
55	509.321	.0020	.0002	.1202	4236.005	8.317	8.225	68.408
60	897.597	.0011	.0001	.1201	7471.641	8.324	8.266	68.810
65	1581.872	.0006	.0001	.1201	13173.937	8.328	8.292	69.058
70	2787.800	.0004		.1200	23223.332	8.330	8.308	69.210
75	4913.056	.0002		.1200	40933.799	8.332	8.318	69.303
80	8658.483	.0001		.1200	72145.692	8.332	8.324	69.359
85	15259.206	.0001		.1200	127151.714	8.333	8.328	69.393
90	26891.934			.1200	224091.118	8.333	8.330	69.414
95	47392.777			.1200	394931.471	8.333	8.331	69.426
100	83522.266			.1200	696010.547	8.333	8.332	69.434

15% Compound Interest Factors 15%

	SINGLE PAYMENT		UNIFORM PAYMENT SERIES				GRADIENT SERIES	
	Compound Amount Factor	Present Worth Factor	Sinking Fund Factor	Capital Recovery Factor	Compound Amount Factor	Present Worth Factor	Gradient Uniform Series	Gradient Present Worth
n	Find *F* Given *P* *F/P*	Find *P* Given *F* *P/F*	Find *A* Given *F* *A/F*	Find *A* Given *P* *A/P*	Find *F* Given *A* *F/A*	Find *P* Given *A* *P/A*	Find *A* Given *G* *A/G*	Find *P* Given *G* *P/G*
1	1.150	.8696	1.0000	1.1500	1.000	.870	0	0
2	1.323	.7561	.4651	.6151	2.150	1.626	.465	.756
3	1.521	.6575	.2880	.4380	3.472	2.283	.907	2.071
4	1.749	.5718	.2003	.3503	4.993	2.855	1.326	3.786
5	2.011	.4972	.1483	.2983	6.742	3.352	1.723	5.775
6	2.313	.4323	.1142	.2642	8.754	3.784	2.097	7.937
7	2.660	.3759	.0904	.2404	11.067	4.160	2.450	10.192
8	3.059	.3269	.0729	.2229	13.727	4.487	2.781	12.481
9	3.518	.2843	.0596	.2096	16.786	4.772	3.092	14.755
10	4.046	.2472	.0493	.1993	20.304	5.019	3.383	16.979
11	4.652	.2149	.0411	.1911	24.349	5.234	3.655	19.129
12	5.350	.1869	.0345	.1845	29.002	5.421	3.908	21.185
13	6.153	.1625	.0291	.1791	34.352	5.583	4.144	23.135
14	7.076	.1413	.0247	.1747	40.505	5.724	4.362	24.972
15	8.137	.1229	.0210	.1710	47.580	5.847	4.565	26.693
16	9.358	.1069	.0179	.1679	55.717	5.954	4.752	28.296
17	10.761	.0929	.0154	.1654	65.075	6.047	4.925	29.783
18	12.375	.0808	.0132	.1632	75.836	6.128	5.084	31.156
19	14.232	.0703	.0113	.1613	88.212	6.198	5.231	32.421
20	16.367	.0611	.0098	.1598	102.444	6.259	5.365	33.582
21	18.822	.0531	.0084	.1584	118.810	6.312	5.488	34.645
22	21.645	.0462	.0073	.1573	137.632	6.359	5.601	35.615
23	24.891	.0402	.0063	.1563	159.276	6.399	5.704	36.499
24	28.625	.0349	.0054	.1554	184.168	6.434	5.798	37.302
25	32.919	.0304	.0047	.1547	212.793	6.464	5.883	38.031
26	37.857	.0264	.0041	.1541	245.712	6.491	5.961	38.692
27	43.535	.0230	.0035	.1535	283.569	6.514	6.032	39.289
28	50.066	.0200	.0031	.1531	327.104	6.534	6.096	39.828
29	57.575	.0174	.0027	.1527	377.170	6.551	6.154	40.315
30	66.212	.0151	.0023	.1523	434.745	6.566	6.207	40.753
31	76.144	.0131	.0020	.1520	500.957	6.579	6.254	41.147
32	87.565	.0114	.0017	.1517	577.100	6.591	6.297	41.501
33	100.700	.0099	.0015	.1515	664.666	6.600	6.336	41.818
34	115.805	.0086	.0013	.1513	765.365	6.609	6.371	42.103
35	133.176	.0075	.0011	.1511	881.170	6.617	6.402	42.359
40	267.864	.0037	.0006	.1506	1779.090	6.642	6.517	43.283
45	538.769	.0019	.0003	.1503	3585.128	6.654	6.583	43.805
50	1083.657	.0009	.0001	.1501	7217.716	6.661	6.620	44.096
55	2179.622	.0005	.0001	.1501	14524.148	6.664	6.641	44.256
60	4383.999	.0002		.1500	29219.992	6.665	6.653	44.343
65	8817.787	.0001		.1500	58778.583	6.666	6.659	44.390
70	17735.720	.0001		.1500	118231.467	6.666	6.663	44.416
75	35672.868			.1500	237812.453	6.666	6.665	44.429
80	71750.879			.1500	478332.529	6.667	6.666	44.436
85	144316.647			.1500	962104.313	6.667	6.666	44.440
90	290272.325			.1500	1935142.168	6.667	6.666	44.442
95	583841.328			.1500	3892268.851	6.667	6.667	44.443
100	1174313.451			.1500	7828749.671	6.667	6.667	44.444

	SINGLE PAYMENT		UNIFORM PAYMENT SERIES				GRADIENT SERIES	
	Compound Amount Factor	Present Worth Factor	Sinking Fund Factor	Capital Recovery Factor	Compound Amount Factor	Present Worth Factor	Gradient Uniform Series	Gradient Present Worth
n	Find *F* Given *P* *F/P*	Find *P* Given *F* *P/F*	Find *A* Given *F* *A/F*	Find *A* Given *P* *A/P*	Find *F* Given *A* *F/A*	Find *P* Given *A* *P/A*	Find *A* Given *G* *A/G*	Find *P* Given *G* *P/G*
1	1.180	.8475	1.0000	1.1800	1.000	.847	0	0
2	1.392	.7182	.4587	.6387	2.180	1.566	.459	.718
3	1.643	.6086	.2799	.4599	3.572	2.174	.890	1.935
4	1.939	.5158	.1917	.3717	5.215	2.690	1.295	3.483
5	2.288	.4371	.1398	.3198	7.154	3.127	1.673	5.231
6	2.700	.3704	.1059	.2859	9.442	3.498	2.025	7.083
7	3.185	.3139	.0824	.2624	12.142	3.812	2.353	8.967
8	3.759	.2660	.0652	.2452	15.327	4.078	2.656	10.829
9	4.435	.2255	.0524	.2324	19.086	4.303	2.936	12.633
10	5.234	.1911	.0425	.2225	23.521	4.494	3.194	14.352
11	6.176	.1619	.0348	.2148	28.755	4.656	3.430	15.972
12	7.288	.1372	.0286	.2086	34.931	4.793	3.647	17.481
13	8.599	.1163	.0237	.2037	42.219	4.910	3.845	18.877
14	10.147	.0985	.0197	.1997	50.818	5.008	4.025	20.158
15	11.974	.0835	.0164	.1964	60.965	5.092	4.189	21.327
16	14.129	.0708	.0137	.1937	72.939	5.162	4.337	22.389
17	16.672	.0600	.0115	.1915	87.068	5.222	4.471	23.348
18	19.673	.0508	.0096	.1896	103.740	5.273	4.592	24.212
19	23.214	.0431	.0081	.1881	123.414	5.316	4.700	24.988
20	27.393	.0365	.0068	.1868	146.628	5.353	4.798	25.681
21	32.324	.0309	.0057	.1857	174.021	5.384	4.885	26.300
22	38.142	.0262	.0048	.1848	206.345	5.410	4.963	26.851
23	45.008	.0222	.0041	.1841	244.487	5.432	5.033	27.339
24	53.109	.0188	.0035	.1835	289.494	5.451	5.095	27.772
25	62.669	.0160	.0029	.1829	342.603	5.467	5.150	28.155
26	73.949	.0135	.0025	.1825	405.272	5.480	5.199	28.494
27	87.260	.0115	.0021	.1821	479.221	5.492	5.243	28.791
28	102.967	.0097	.0018	.1818	566.481	5.502	5.281	29.054
29	121.501	.0082	.0015	.1815	669.447	5.510	5.315	29.284
30	143.371	.0070	.0013	.1813	790.948	5.517	5.345	29.486
31	169.177	.0059	.0011	.1811	934.319	5.523	5.371	29.664
32	199.629	.0050	.0009	.1809	1103.496	5.528	5.394	29.819
33	235.563	.0042	.0008	.1808	1303.125	5.532	5.415	29.955
34	277.964	.0036	.0006	.1806	1538.688	5.536	5.433	30.074
35	327.997	.0030	.0006	.1806	1816.652	5.539	5.449	30.177
40	750.378	.0013	.0002	.1802	4163.213	5.548	5.502	30.527
45	1716.684	.0006	.0001	.1801	9531.577	5.552	5.529	30.701
50	3927.357	.0003		.1800	21813.094	5.554	5.543	30.786
55	8984.841	.0001		.1800	49910.228	5.555	5.549	30.827
60	20555.140			.1800	114189.666	5.555	5.553	30.846
65	47025.181			.1800	261245.449	5.555	5.554	30.856
70	107582.222			.1800	597673.458	5.556	5.555	30.860
75	246122.064			.1800	1367339.243	5.556	5.555	30.862
80	563067.660			.1800	3128148.114	5.556	5.555	30.863
85	1288162.408			.1800	7156452.266	5.556	5.555	30.864

20% Compound Interest Factors 20%

	SINGLE PAYMENT		UNIFORM PAYMENT SERIES				GRADIENT SERIES	
	Compound Amount Factor	Present Worth Factor	Sinking Fund Factor	Capital Recovery Factor	Compound Amount Factor	Present Worth Factor	Gradient Uniform Series	Gradient Present Worth
n	Find F Given P F/P	Find P Given F P/F	Find A Given F A/F	Find A Given P A/P	Find F Given A F/A	Find P Given A P/A	Find A Given G A/G	Find P Given G P/G
1	1.200	.8333	1.0000	1.2000	1.000	.833	0	0
2	1.440	.6944	.4545	.6545	2.200	1.528	.455	.694
3	1.728	.5787	.2747	.4747	3.640	2.106	.879	1.852
4	2.074	.4823	.1863	.3863	5.368	2.589	1.274	3.299
5	2.488	.4019	.1344	.3344	7.442	2.991	1.641	4.906
6	2.986	.3349	.1007	.3007	9.930	3.326	1.979	6.581
7	3.583	.2791	.0774	.2774	12.916	3.605	2.290	8.255
8	4.300	.2326	.0606	.2606	16.499	3.837	2.576	9.883
9	5.160	.1938	.0481	.2481	20.799	4.031	2.836	11.434
10	6.192	.1615	.0385	.2385	25.959	4.192	3.074	12.887
11	7.430	.1346	.0311	.2311	32.150	4.327	3.289	14.233
12	8.916	.1122	.0253	.2253	39.581	4.439	3.484	15.467
13	10.699	.0935	.0206	.2206	48.497	4.533	3.660	16.588
14	12.839	.0779	.0169	.2169	59.196	4.611	3.817	17.601
15	15.407	.0649	.0139	.2139	72.035	4.675	3.959	18.509
16	18.488	.0541	.0114	.2114	87.442	4.730	4.085	19.321
17	22.186	.0451	.0094	.2094	105.931	4.775	4.198	20.042
18	26.623	.0376	.0078	.2078	128.117	4.812	4.298	20.680
19	31.948	.0313	.0065	.2065	154.740	4.843	4.386	21.244
20	38.338	.0261	.0054	.2054	186.688	4.870	4.464	21.739
21	46.005	.0217	.0044	.2044	225.026	4.891	4.533	22.174
22	55.206	.0181	.0037	.2037	271.031	4.909	4.594	22.555
23	66.247	.0151	.0031	.2031	326.237	4.925	4.647	22.887
24	79.497	.0126	.0025	.2025	392.484	4.937	4.694	23.176
25	95.396	.0105	.0021	.2021	471.981	4.948	4.735	23.428
26	114.475	.0087	.0018	.2018	567.377	4.956	4.771	23.646
27	137.371	.0073	.0015	.2015	681.853	4.964	4.802	23.835
28	164.845	.0061	.0012	.2012	819.223	4.970	4.829	23.999
29	197.814	.0051	.0010	.2010	984.068	4.975	4.853	24.141
30	237.376	.0042	.0008	.2008	1181.882	4.979	4.873	24.263
31	284.852	.0035	.0007	.2007	1419.258	4.982	4.891	24.368
32	341.822	.0029	.0006	.2006	1704.109	4.985	4.906	24.459
33	410.186	.0024	.0005	.2005	2045.931	4.988	4.919	24.537
34	492.224	.0020	.0004	.2004	2456.118	4.990	4.931	24.604
35	590.668	.0017	.0003	.2003	2948.341	4.992	4.941	24.661
40	1469.772	.0007	.0001	.2001	7343.858	4.997	4.973	24.847
45	3657.262	.0003	.0001	.2001	18281.310	4.999	4.988	24.932
50	9100.438	.0001		.2000	45497.191	4.999	4.995	24.970
55	22644.802			.2000	113219.011	5.000	4.998	24.987
60	56347.514			.2000	281732.572	5.000	4.999	24.994
65	140210.647			.2000	701048.235	5.000	5.000	24.998
70	348888.957			.2000	1744439.785	5.000	5.000	24.999
75	868147.369			.2000	4340731.847	5.000	5.000	25.000

25% Compound Interest Factors 25%

	SINGLE PAYMENT		UNIFORM PAYMENT SERIES				GRADIENT SERIES	
	Compound Amount Factor	Present Worth Factor	Sinking Fund Factor	Capital Recovery Factor	Compound Amount Factor	Present Worth Factor	Gradient Uniform Series	Gradient Present Worth
n	Find *F* Given *P* *F/P*	Find *P* Given *F* *P/F*	Find *A* Given *F* *A/F*	Find *A* Given *P* *A/P*	Find *F* Given *A* *F/A*	Find *P* Given *A* *P/A*	Find *A* Given *G* *A/G*	Find *P* Given *G* *P/G*
1	1.250	.8000	1.0000	1.2500	1.000	.800	0	0
2	1.563	.6400	.4444	.6944	2.250	1.440	.444	.640
3	1.953	.5120	.2623	.5123	3.813	1.952	.852	1.664
4	2.441	.4096	.1734	.4234	5.766	2.362	1.225	2.893
5	3.052	.3277	.1218	.3718	8.207	2.689	1.563	4.204,
6	3.815	.2621	.0888	.3388	11.259	2.951	1.868	5.514
7	4.768	.2097	.0663	.3163	15.073	3.161	2.142	6.773
8	5.960	.1678	.0504	.3004	19.842	3.329	2.387	7.947
9	7.451	.1342	.0388	.2888	25.802	3.463	2.605	9.021
10	9.313	.1074	.0301	.2801	33.253	3.571	2.797	9.987
11	11.642	.0859	.0235	.2735	42.566	3.656	2.966	10.864
12	14.552	.0687	.0184	.2684	54.208	3.725	3.115	11.602
13	18.190	.0550	.0145	.2645	68.760	3.780	3.244	12.262
14	22.737	.0440	.0115	.2615	86.949	3.824	3.356	12.833
15	28.422	.0352	.0091	.2591	109.687	3.859	3.453	13.326
16	35.527	.0281	.0072	.2572	138.109	3.887	3.537	13.748
17	44.409	.0225	.0058	.2558	173.636	3.910	3.608	14.108
18	55.511	.0180	.0046	.2546	218.045	3.928	3.670	14.415
19	69.389	.0144	.0037	.2537	273.556	3.942	3.722	14.674
20	86.736	.0115	.0029	.2529	342.945	3.954	3.767	14.893
21	108.420	.0092	.0023	.2523	429.681	3.963	3.805	15.078
22	135.525	.0074	.0019	.2519	538.101	3.970	3.836	15.233
23	169.407	.0059	.0015	.2515	673.626	3.976	3.863	15.362
24	211.758	.0047	.0012	.2512	843.033	3.981	3.886	15.471
25	264.698	.0038	.0009	.2509	1054.791	3.985	3.905	15.562
26	330.872	.0030	.0008	.2508	1319.489	3.988	3.921	15.637
27	413.590	.0024	.0006	.2506	1650.361	3.990	3.935	15.700
28	516.988	.0019	.0005	.2505	2063.952	3.992	3.946	15.752
29	646.235	.0015	.0004	.2504	2580.939	3.994	3.955	15.796
30	807.794	.0012	.0003	.2503	3227.174	3.995	3.963	15.832
31	1009.742	.0010	.0002	.2502	4034.968	3.996	3.969	15.861
32	1262.177	.0008	.0002	.2502	5044.710	3.997	3.975	15.886
33	1577.722	.0006	.0002	.2502	6306.887	3.997	3.979	15.906
34	1972.152	.0005	.0001	.2501	7884.609	3.998	3.983	15.923
35	2465.190	.0004	.0001	.2501	9856.761	3.998	3.986	15.937
40	7523.164	.0001		.2500	30088.655	3.999	3.995	15.977
45	22958.874			.2500	91831.496	4.000	3.998	15.991
50	70064.923			.2500	280255.693	4.000	3.999	15.997
55	213821.177			.2500	855280.707	4.000	4.000	15.999
60	652530.447			.2500	2610117.787	4.000	4.000	16.000

Compound Interest Factors

	SINGLE PAYMENT		UNIFORM PAYMENT SERIES				GRADIENT SERIES	
	Compound Amount Factor	Present Worth Factor	Sinking Fund Factor	Capital Recovery Factor	Compound Amount Factor	Present Worth Factor	Gradient Uniform Series	Gradient Present Worth
	Find *F* Given *P*	Find *P* Given *F*	Find *A* Given *F*	Find *A* Given *P*	Find *F* Given *A*	Find *P* Given *A*	Find *A* Given *G*	Find *P* Given *G*
n	*F/P*	*P/F*	*A/F*	*A/P*	*F/A*	*P/A*	*A/G*	*P/G*
1	1.300	.7692	1.0000	1.3000	1.000	.769	0	0
2	1.690	.5917	.4348	.7348	2.300	1.361	.435	.592
3	2.197	.4552	.2506	.5506	3.990	1.816	.827	1.502
4	2.856	.3501	.1616	.4616	6.187	2.166	1.178	2.552
5	3.713	.2693	.1106	.4106	9.043	2.436	1.490	3.630
6	4.827	.2072	.0784	.3784	12.756	2.643	1.765	4.666
7	6.275	.1594	.0569	.3569	17.583	2.802	2.006	5.622
8	8.157	.1226	.0419	.3419	23.858	2.925	2.216	6.480
9	10.604	.0943	.0312	.3312	32.015	3.019	2.396	7.234
10	13.786	.0725	.0235	.3235	42.619	3.092	2.551	7.887
11	17.922	.0558	.0177	.3177	56.405	3.147	2.683	8.445
12	23.298	.0429	.0135	.3135	74.327	3.190	2.795	8.917
13	30.288	.0330	.0102	.3102	97.625	3.223	2.889	9.314
14	39.374	.0254	.0078	.3078	127.913	3.249	2.969	9.644
15	51.186	.0195	.0060	.3060	167.286	3.268	3.034	9.917
16	66.542	.0150	.0046	.3046	218.472	3.283	3.089	10.143
17	86.504	.0116	.0035	.3035	285.014	3.295	3.135	10.328
18	112.455	.0089	.0027	.3027	371.518	3.304	3.172	10.479
19	146.192	.0068	.0021	.3021	483.973	3.311	3.202	10.602
20	190.050	.0053	.0016	.3016	630.165	3.316	3.228	10.702
21	247.065	.0040	.0012	.3012	820.215	3.320	3.248	10.783
22	321.184	.0031	.0009	.3009	1067.280	3.323	3.265	10.848
23	417.539	.0024	.0007	.3007	1388.464	3.325	3.278	10.901
24	542.801	.0018	.0006	.3006	1806.003	3.327	3.289	10.943
25	705.641	.0014	.0004	.3004	2348.803	3.329	3.298	10.977
26	917.333	.0011	.0003	.3003	3054.444	3.330	3.305	11.005
27	1192.533	.0008	.0003	.3003	3971.778	3.331	3.311	11.026
28	1550.293	.0006	.0002	.3002	5164.311	3.331	3.315	11.044
29	2015.381	.0005	.0001	.3001	6714.604	3.332	3.319	11.058
30	2619.996	.0004	.0001	.3001	8729.985	3.332	3.322	11.069
31	3405.994	.0003	.0001	.3001	11349.981	3.332	3.324	11.078
32	4427.793	.0002	.0001	.3001	14755.975	3.333	3.326	11.085
33	5756.130	.0002	.0001	.3001	19183.768	3.333	3.328	11.090
34	7482.970	.0001		.3000	24939.899	3.333	3.329	11.094
35	9727.860	.0001		.3000	32422.868	3.333	3.330	11.098
40	36118.865			.3000	120392.883	3.333	3.332	11.107
45	134106.817			.3000	447019.389	3.333	3.333	11.110
50	497929.223			.3000	1659760.745	3.333	3.333	11.111
55	1848776.352			.3000	6162584.505	3.333	3.333	11.111
60	6864377.179			.3000	22881253.930	3.333	3.333	11.111

Compound Interest Factors

35%

	SINGLE PAYMENT		UNIFORM PAYMENT SERIES				GRADIENT SERIES	
	Compound Amount Factor	Present Worth Factor	Sinking Fund Factor	Capital Recovery Factor	Compound Amount Factor	Present Worth Factor	Gradient Uniform Series	Gradient Present Worth
n	Find F Given P F/P	Find P Given F P/F	Find A Given F A/F	Find A Given P A/P	Find F Given A F/A	Find P Given A P/A	Find A Given G A/G	Find P Given G P/G
1	1.350	.7407	1.000	1.3500	1.000	.741	0	0
2	1.823	.5487	.4255	.7755	2.350	1.289	.426	.549
3	2.460	.4064	.2397	.5897	4.172	1.696	.803	1.362
4	3.322	.3011	.1508	.5008	6.633	1.997	1.134	2.265
5	4.484	.2230	.1005	.4505	9.954	2.220	1.422	3.157
6	6.053	.1652	.0693	.4193	14.438	2.385	1.670	3.983
7	8.172	.1224	.0488	.3988	20.492	2.508	1.881	4.717
8	11.032	.0906	.0349	.3849	28.664	2.598	2.060	5.352
9	14.894	.0671	.0252	.3752	39.696	2.665	2.209	5.889
10	20.107	.0497	.0183	.3683	54.590	2.715	2.334	6.336
11	27.144	.0368	.0134	.3634	74.697	2.752	2.436	6.705
12	36.644	.0273	.0098	.3598	101.841	2.779	2.520	7.005
13	49.470	.0202	.0072	.3572	138.485	2.799	2.589	7.247
14	66.784	.0150	.0053	.3553	187.954	2.814	2.644	7.442
15	90.158	.0111	.0039	.3539	254.738	2.825	2.689	7.597
16	121.714	.0082	.0029	.3529	344.897	2.834	2.725	7.721
17	164.314	.0061	.0021	.3521	466.611	2.840	2.753	7.818
18	221.824	.0045	.0016	.3516	630.925	2.844	2.776	7.895
19	299.462	.0033	.0012	.3512	852.748	2.848	2.793	7.955
20	404.274	.0025	.0009	.3509	1152.210	2.850	2.808	8.002
21	545.769	.0018	.0006	.3506	1556.484	2.852	2.819	8.038
22	736.789	.0014	.0005	.3505	2102.253	2.853	2.827	8.067
23	994.665	.0010	.0004	.3504	2839.042	2.854	2.834	8.089
24	1342.797	.0007	.0003	.3503	3833.706	2.855	2.839	8.106
25	1812.776	.0006	.0002	.3502	5176.504	2.856	2.843	8.119
26	2447.248	.0004	.0001	.3501	6989.280	2.856	2.847	8.130
27	3303.785	.0003	.0001	.3501	9436.528	2.856	2.849	8.137
28	4460.109	.0002	.0001	.3501	12740.313	2.857	2.851	8.143
29	6021.148	.0002	.0001	.3501	17200.422	2.857	2.852	8.148
30	8128.550	.0001		.3500	23221.570	2.857	2.853	8.152
31	10973.542	.0001		.3500	31350.120	2.857	2.854	8.154
32	14814.281	.0001		.3500	42323.661	2.857	2.855	8.157
33	19999.280	.0001		.3500	57137.943	2.857	2.855	8.158
34	26999.028			.3500	77137.223	2.857	2.856	8.159
35	36448.688			.3500	104136.251	2.857	2.856	8.160
40	163437.135			.3500	466960.385	2.857	2.857	8.163
45	732857.577			.3500	2093875.934	2.857	2.857	8.163
50	3286157.879			.3500	9389019.655	2.857	2.857	8.163

	SINGLE PAYMENT		UNIFORM PAYMENT SERIES				GRADIENT SERIES	
	Compound Amount Factor	Present Worth Factor	Sinking Fund Factor	Capital Recovery Factor	Compound Amount Factor	Present Worth Factor	Gradient Uniform Series	Gradient Present Worth
n	Find F Given P F/P	Find P Given F P/F	Find A Given F A/F	Find A Given P A/P	Find F Given A F/A	Find P Given A P/A	Find A Given G A/G	Find P Given G P/G
1	1.400	.7143	1.0000	1.4000	1.000	.714	0	0
2	1.960	.5102	.4167	.8167	2.400	1.224	.417	.510
3	2.744	.3644	.2294	.6294	4.360	1.589	.780	1.239
4	3.842	.2603	.1408	.5408	7.104	1.849	1.092	2.020
5	5.378	.1859	.0914	.4914	10.946	2.035	1.358	2.764
6	7.530	.1328	.0613	.4613	16.324	2.168	1.581	3.428
7	10.541	.0949	.0419	.4419	23.853	2.263	1.766	3.997
8	14.758	.0678	.0291	.4291	34.395	2.331	1.919	4.471
9	20.661	.0484	.0203	.4203	49.153	2.379	2.042	4.858
10	28.925	.0346	.0143	.4143	69.814	2.414	2.142	5.170
11	40.496	.0247	.0101	.4101	98.739	2.438	2.221	5.417
12	56.694	.0176	.0072	.4072	139.235	2.456	2.285	5.611
12	79.371	.0126	.0051	.4051	195.929	2.469	2.334	5.762
14	111.120	.0090	.0036	.4036	275.300	2.478	2.373	5.879
15	155.568	.0064	.0026	.4026	386.420	2.484	2.403	5.969
16	217.795	.0046	.0018	.4018	541.988	2.489	2.426	6.038
17	304.913	.0033	.0013	.4013	759.784	2.492	2.444	6.090
18	426.879	.0023	.0009	.4009	1064.697	2.494	2.458	6.130
19	597.630	.0017	.0007	.4007	1491.576	2.496	2.468	6.160
20	836.683	.0012	.0005	.4005	2089.206	2.497	2.476	6.183
21	1171.356	.0009	.0003	.4003	2925.889	2.498	2.482	6.200
22	1639.898	.0006	.0002	.4002	4097.245	2.498	2.487	6.213
23	2295.857	.0004	.0002	.4002	5737.142	2.499	2.490	6.222
24	3214.200	.0003	.0001	.4001	8032.999	2.499	2.493	6.229
25	4499.880	.0002	.0001	.4001	11247.199	2.499	2.494	6.235
26	6299.831	.0002	.0001	.4001	15747.079	2.500	2.496	6.239
27	8819.764	.0001		.4000	22046.910	2.500	2.497	6.242
28	12347.670	.0001		.4000	30866.674	2.500	2.498	6.244
29	17286.737	.0001		.4000	43214.344	2.500	2.498	6.245
30	24201.432			.4000	60501.081	2.500	2.499	6.247
31	33882.005			.4000	84702.513	2.500	2.499	6.248
32	47434.807			.4000	118584.519	2.500	2.499	6.248
33	66408.730			.4000	166019.326	2.500	2.500	6.249
34	92972.223			.4000	232428.057	2.500	2.500	6.249
35	130161.112			.4000	325400.279	2.500	2.500	6.249
40	700037.697			.4000	1750091.743	2.500	2.500	6.250
45	3764970.745			.4000	9412424.362	2.500	2.500	6.250
50	20248916.262			.4000	50622288.153	2.500	2.500	6.250

45% Compound Interest Factors 45%

	SINGLE PAYMENT		UNIFORM PAYMENT SERIES				GRADIENT SERIES	
	Compound Amount Factor	Present Worth Factor	Sinking Fund Factor	Capital Recovery Factor	Compound Amount Factor	Present Worth Factor	Gradient Uniform Series	Gradient Present Worth
n	Find *F* Given *P* *F/P*	Find *P* Given *F* *P/F*	Find *A* Given *F* *A/F*	Find *A* Given *P* *A/P*	Find *F* Given *A* *F/A*	Find *P* Given *A* *P/A*	Find *A* Given *G* *A/G*	Find *P* Given *G* *P/G*
1	1.450	.6897	1.0000	1.4500	1.000	.690	0	0
2	2.103	.4756	.4082	.8582	2.450	1.165	.408	.476
3	3.049	.3280	.2197	.6697	4.553	1.493	.758	1.132
4	4.421	.2262	.1316	.5816	7.601	1.720	1.053	1.810
5	6.410	.1560	.0832	.5332	12.022	1.876	1.298	2.434
6	9.294	.1076	.0543	.5043	18.431	1.983	1.499	2.972
7	13.476	.0742	.0361	.4861	27.725	2.057	1.661	3.418
8	19.541	.0512	.0243	.4743	41.202	2.109	1.791	3.776
9	28.334	.0353	.0165	.4665	60.743	2.144	1.893	4.058
10	41.085	.0243	.0112	.4612	89.077	2.168	1.973	4.277
11	59.573	.0168	.0077	.4577	130.162	2.185	2.034	4.445
12	86.381	.0116	.0053	.4553	189.735	2.196	2.082	4.572
13	125.252	.0080	.0036	.4536	276.115	2.204	2.118	4.668
14	181.615	.0055	.0025	.4525	401.367	2.210	2.145	4.740
15	263.342	.0038	.0017	.4517	582.982	2.214	2.165	4.793
16	381.846	.0026	.0012	.4512	846.324	2.216	2.180	4.832
17	553.676	.0018	.0008	.4508	1228.170	2.218	2.191	4.861
18	802.831	.0012	.0006	.4506	1781.846	2.219	2.200	4.882
19	1164.105	.0009	.0004	.4504	2584.677	2.220	2.206	4.898
20	1687.952	.0006	.0003	.4503	3748.782	2.221	2.210	4.909
21	2447.530	.0004	.0002	.4502	5436.734	2.221	2.214	4.917
22	3548.919	.0003	.0001	.4501	7884.264	2.222	2.216	4.923
23	5145.932	.0002	.0001	.4501	11433.182	2.222	2.218	4.927
24	7461.602	.0001	.0001	.4501	16579.115	2.222	2.219	4.930
25	10819.322	.0001		.4500	24040.716	2.222	2.220	4.933
26	15688.017	.0001		.4500	34860.038	2.222	2.221	4.934
27	22747.625			.4500	50548.056	2.222	2.221	4.935
28	32984.056			.4500	73295.681	2.222	2.221	4.936
29	47826.882			.4500	106279.737	2.222	2.222	4.937
30	69348.978			.4500	154106.618	2.222	2.222	4.937
31	100556.019			.4500	223455.597	2.222	2.222	4.938
32	145806.227			.4500	324011.615	2.222	2.222	4.938
33	211419.029			.4500	469817.842	2.222	2.222	4.938
34	306557.592			.4500	681236.871	2.222	2.222	4.938
35	444508.508			.4500	987794.463	2.222	2.222	4.938

Appendix

State Boards Of Registration

ALABAMA
State Board of Registration for Professional Engineers and Land Surveyors
750 Washington Avenue, Suite 212
Montgomery 36130

ALASKA
State Board of Registration for Architects, Engineers and Land Surveyors
Pouch D (State Office Building, 9th floor)
Juneau 99811

ARIZONA
State Board of Technical Registration
1645 W. Jefferson Street, Suite 140
Phoenix, 85007

ARKANSAS
State Board of Registration for Professional Engineers and Land Surveyors
P.O. Box 2541
700 Twin City Bank Bldg.
Little Rock 72203

CALIFORNIA
Board of Registration for Professional Engineers
1006 Fourth Street, 6th floor
Sacramento 95814

COLORADO
State Board of Registration for Professional Engineers and Land Surveyors
600-B State Services Building
1525 Sherman Street
Denver 80203

CONNECTICUT
Board of Registration for Professional Engineers and Land Surveyors
The State Office Bldg. Room G-3A
165 Capitol Ave.
Hartford 06106

DELAWARE
Delaware Association of Professional Engineers
2005 Concord Pike
Wilmington 19803

DISTRICT OF COLUMBIA
Board of Registration for Professional Engineers
614 H Street, N.W., Room 910
Washington 20001

FLORIDA
State Board of Professional Engineers
130 North Monroe Street
Tallahassee 32301

GEORGIA
State Board of Registration for Professional Engineers and Land Surveyors
166 Pryor St., S.W.
Atlanta 30303

GUAM
Territorial Board of Registration for Professional Engineers, Architects and Land Surveyors
Dept. of Public Works, Government of Guam
P.O. Box 2950
Agana 96910

HAWAII
State Board of Registration for Professional Engineers, Architects, Land Surveyors and Landscape Architects
P.O. Box 3469 (1010 Richards St.)
Honolulu 96801

IDAHO
Board of Professional Engineers and Land Surveyors
842 La Cassia Drive
Boise 83705

ILLINOIS
Department of Registration and Education
Professional Engineers' Examining Committee
320 West Washington, 3rd Floor
Springfield 62786

INDIANA
State Board of Registration for Professional Engineers and Land Surveyors
1021 State Office Building
100 N. Senate Avenue
Indianapolis 46204

IOWA
State Board of Engineering Examiners
State Capitol Complex
1209 East Court Ave.
Des Moines 50319

KANSAS
State Board of Technical Professions
214 West Sixth Street, 2nd Floor
Topeka 66603

KENTUCKY
State Board of Registration for Professional Engineers and Land Surveyors
Rt. 3, Millville Road
Frankfort 40601

LOUSIANA
State Board of Registration for Professional Engineers and Land Surveyors
1055 St. Charles Avenue, Suite 415
New Orleans 70130

MAINE
State Board of Registration for Professional Engineers
State House
Augusta 04333

MARYLAND
State Board fo Registration for Professional Engineers
501 St. Paul Place, Rm 902
Baltimore 21202

MASSACHUSETTS
State Board of Registration of Professional Engineers and of Land Surveyors
Room 1512 Leverett Saltonstall Bldg.
100 Cambridge St.
Boston 02202

MICHIGAN
Board of Registration for Professional Engineers
P.O. Box 30018 (611 W. Ottawa)
Lansing 48909

MINNESOTA
State Board of Registration for Architects, Engineers, Land Surveyors and Landscape Architects
Metro Square Bldg., Rm 162
St. Paul 55101

MISSISSIPPI
State Board of Registration for Professional Engineers and Land Surveyors
P.O. Box 3 (200 S. President St., Suite 516)
Jackson 39205

MISSOURI
Board of Architects, Professional Engineers and Land Surveyors
P.O. Box 184 (3523 North Ten Mile Drive)
Jefferson City 65102

MONTANA
State Board of Professional Engineers and Land Surveyors
Dept. of Commerce
1424 9th Avenue
Helena 59620-0407

NEBRASKA
State Board of Examiners for Professional Engineers and Architects
P.O. Box 94751 (301 Centennial Mall, South)
Lincoln 68509

NEVADA
State Board of Registered Professional Engineers and Land Surveyors
1755 East Plum Lane, Ste. 102
Reno 89502

NEW HAMPSHIRE
State Board of Professional Engineers
Storrs St.
Concord 03301

NEW JERSEY
State Board of Professional Engineers and Land Surveyors
1100 Raymond Boulevard, Rm. 317
Newark 07102

NEW MEXICO
State Board of Registration for Professional Engineers and Land Surveyors
P.O. Box 4847 (440 Cerrillos Road)
Santa Fe 87502

NEW YORK
State Board for Engineering and Land Surveying
The State Education Department
Cultural Education Center, Madison Avenue
Albany 12230

NORTH CAROLINA
Board of Registration for Professional Engineers and Land Surveyors
3620 Six Forks Road
Raleigh 27609

NORTH DAKOTA
State Board of Registration for Professional Engineers and Land Surveyors
P.O. Box 1357 (420 Ave. B East)
Minot 58502

OHIO
State Board of Registration for Professional Engineers and Surveyors
65 South Front Street, Room 302
Columbus 43215

OKLAHOMA
State Board of Registration for Professional Engineers and Land Surveyors
Oklahoma Engineering Center, Room 120
201 N.E. 27th Street
Oklahoma City 73105

OREGON
State Board of Engineering Examiners
Department of Commerce
Labor and Industries Building, 4th Floor
Salem 97310

PENNSYLVANIA
State Registration Board for Professional Engineers
P.O. Box 2649 (Transportation & Safety Bldg.)
Commonwealth Ave. & Forester St., 6th Floor
Harrisburg 17105-2649

PUERTO RICO
Board of Examiners of Engineers, Architects, and Surveyors
Box 3271 (Tanca Street, 261, Comer Tetuan)
San Juan 00904

RHODE ISLAND
State Board of Registration for Professional Engineers and Land Surveyors
308 State Office Building
Providence 02903

SOUTH CAROLINA
State Board of Registration for Professional Engineers
2221 Dvine St., Suite 404
Columbia 29205

SOUTH DAKOTA
State Commission of Engineering and Architectural Examiners
2040 West Main Street, Suite 212
Rapid City 57702-2497

TENNESSEE
State Board of Architectural and Engineering Examiners
546 Doctors' Building
706 Church Street
Nashville 37219

TEXAS
State Board of Registration for Professional Engineers
P.O. Drawer 18329 (1917 1H 35 South)
Austin 78760

UTAH
Representative Committee for Professional Engineers and Land Surveyors
Division of Registration
160 East 300 South
Salt Lake City 84145

VERMONT
State Board of Registration for Professional Engineers
Division of Licensing and Registration
Pavilion Building
Montpelier 05602

VIRGINIA
State Board of Architects, Professional Engineers, Land Surveyors and Certified Landscape Arthitects
3600 W. Broad St., 5th Floor
Richmond 23230-4917

VIRGIN ISLANDS
Board for Architects, Engineers and Land Surveyors
Submarine Base
P.O. Box 476
St. Thomas 00801

WASHINGTON
State Board of Registration for Professional Engineers and Land Surveyors
P.O. Box 9649 (3rd floor, 9th & Columbia)
Olympia 98504

WEST VIRGINIA
State Board of Registration for Professional Engineers
608 Union Building
Charleston 25301

WISCONSIN
State Board of Architects, Professional Engineers, Designers, and Land Surveyors
P.O. Box 8936 (1400 E. Washington Avenue)
Madison 53708

WYOMING
State Board of Examining Engineers
Barrett Building
Cheyenne 82002